Understanding Faults

Understanding Faults

Detecting, Dating, and Modeling

Edited by

David Tanner

Christian Brandes

ELSEVIER

Elsevier
Radarweg 29, PO Box 211, 1000 AE Amsterdam, Netherlands
The Boulevard, Langford Lane, Kidlington, Oxford OX5 1GB, United Kingdom
50 Hampshire Street, 5th Floor, Cambridge, MA 02139, United States

Library of Congress Cataloging-in-Publication Data
A catalog record for this book is available from the Library of Congress

British Library Cataloguing-in-Publication Data
A catalogue record for this book is available from the British Library

ISBN: 978-0-12-815985-9

Publisher: Candice Janco
Acquisition Editor: Amy Shapiro
Editorial Project Manager: Sara Valentino
Production Project Manager: Omer Mukthar
Cover Designer: Greg Harris

Typeset by TNQ Technologies

Contents

List of contributors

Christian Brandes, Institut für Geologie, Leibniz Universität Hannover, Hannover, Germany

Hermann Buness, Leibniz Institute for Applied Geophysics (LIAG), Hannover, Germany

Conrad Childs, Fault Analysis Group, School of Geological Sciences, UCD, Dublin, Ireland

Åke Fagereng, School of Earth & Ocean Sciences, Cardiff University, Cardiff, United Kingdom

Gerald Gabriel, Leibniz Institute for Applied Geophysics (LIAG), Hannover, Germany

Nicolai Gestermann, Federal Institute for Geosciences and Natural Resources, Hannover, Germany

Thomas Günther, Leibniz Institute for Applied Geophysics (LIAG), Hannover, Germany

Andreas Henk, Institute of Applied Geosciences, Technical University Darmstadt, Darmstadt, Germany

Jan Igel, Leibniz Institute for Applied Geophysics (LIAG), Hannover, Germany

Matt Ikari, MARUM—Center for Marine Environmental Sciences and Faculty of Geosciences, University of Bremen, Bremen, Germany

Michael Kettermann, Department of Geodynamics and Sedimentology, University of Vienna, Vienna, Austria

Tom Manzocchi, Fault Analysis Group, School of Geological Sciences, UCD, Dublin, Ireland

Christopher K. Morley, Department of Geological Sciences, Faculty of Science, Chiang Mai University, Chiang Mai, Thailand

Andrew Nicol, Department of Geological Sciences, University of Canterbury, Christchurch, New Zealand

Stefan Nielsen, Department of Earth Sciences, Durham University, Durham, United Kingdom

André Niemeijer, Department of Geoscience, Utrecht University, Utrecht, The Netherlands

Thomas Plenefisch, Federal Institute for Geosciences and Natural Resources, Hannover, Germany

Peter Skiba, Leibniz Institute for Applied Geophysics (LIAG), Hannover, Germany

Luca Smeraglia, Dipartimento di Scienze della Terra, Sapienza University of Rome, Rome, Italy

Takahiro Tagami, Division of Earth and Planetary Sciences, Graduate School of Science, Kyoto University, Kyoto, Japan

David C. Tanner, Leibniz Institute for Applied Geophysics (LIAG), Hannover, Germany

Sumiko Tsukamoto, Leibniz Institute for Applied Geophysics (LIAG), Hannover, Germany

Christoph von Hagke, Institute of Geology & Palaeontology, RWTH Aachen University, Aachen, Germany

John Walsh, Fault Analysis Group, School of Geological Sciences, UCD, Dublin, Ireland

Thomas R. Walter, GFZ German Research Centre for Geosciences, Potsdam, Germany

Ernst Willingshofer, Department of Earth Sciences, Utrecht University, Utrecht, The Netherlands

Horst Zwingmann, Department of Geology and Mineralogy, Faculty of Science, Kyoto University, Kyoto, Japan

Preface

This book is the accumulation of many years of research on faults and represents our own personal views of the state of the art in fault analysis. Some years ago, we recognised that, although different branches of geoscience all regard faults and the processes of faulting as important, each group has its own methods and theories. Consequently, a book that focuses on all aspects of faults was missing on the market.

We had two aims when compiling this book. First, we wanted to provide a holistic view on all facets of faulting, with a focus on fault processes and fault detection. Second, this book presents a transdisciplinary approach that unites the different geoscience sub-disciplines that are concerned with faults, in particular showing the advantages of combining the methods. For instance, it is important to connect the way faults are treated in structural geology with seismological methods of fault analysis. We believe that this holistic treatment is the key to understand faults, and to develop advanced predictive fault models. We have attempted to keep the style of the book so that students from any geo-relevant background can read it. Nevertheless, we also tried reach a level between text-book and research article to make the book interesting for the advanced reader. In addition, for reasons of brevity, some chapters are shorter than we would like; therefore we made an effort to cite the background and advanced reading in these subjects.

Discussion with many colleagues has shaped the book. In addition to the authors in the book, they include; Peter Eichhubl, Bob Holdsworth, Catherine Homberg, Christopher A.-L. Jackson, Rüdiger Killian, Charlotte M. Krawczyk, Katharina Müller, Anne Pluymakers, Janos Urai, Jennifer Ziesch. We humbly claim any mistakes for ourselves.

We are very grateful to the reviewers who have critically read certain chapters and provided constructive reviews, namely: István Dunkl, Ingo Heyde, Inga Moeck, Andy Nicol, Shigeru Sueoka, Martin Schöpfer. Till Schierer and Lotta Hanzelmann are thanked for redrawing some of the figures.

Chapter 1

Introduction

David C. Tanner[a], Christian Brandes[b]
[a]*Leibniz Institute for Applied Geophysics (LIAG), Hannover, Germany;* [b]*Institut für Geologie, Leibniz Universität Hannover, Hannover, Germany*

Chapter outline

Since the advent of plate tectonics, geoscience has rapidly developed. Within the field of geoscience, tectonic research on faults represents a highly diverse sub-discipline. It underwent a transformation over the last few decades in its approach to understanding the Earth, by combining observations that are derived from natural rocks, experiments, and modelling studies. This overcame the previous, simple kinematic and steady-state fault descriptions, and allowed the analysis of dynamic and transient processes (Huntington, K.W., Klepeis, K.A., with 66 community contributors, 2018). We follow this path to present a book that delivers a holistic dynamic treatment of faults.

Faults are structural elements in the lithosphere that compensate for deformation under brittle conditions. At greater depth, faults can pass into shear zones, where plastic deformation occurs, which means that deformation mechanisms vary along a fault. Faults are very widespread in the lithosphere and they generally occur in groups, which means that the subsurface structure is often more heterogeneous than expected. In addition, faults are complex structures that are insufficiently described by simple geometrical models. Such models might work on the first-order scale, but faults (especially faults with displacements of more than several 10s of metres) tend to evolve into complex fault zones that are very heterogeneous in terms of geometry, composition and structure. As such, they have a strong control on the subsurface fluid flow and in the case of active faults, significantly influence rupture behaviour. Consequently, very different geoscience sub-disciplines, such as structural geology, geomechanics, seismology, engineering geology, petroleum geology, and Quaternary geology require profound knowledge of faults. Faults are the source of earthquakes and thus they are the connecting elements between

Understanding Faults. https://doi.org/10.1016/B978-0-12-815985-9.00001-1

structural geology and seismology. For instance, although both disciplines focus on faults, they often treat them from very different perspectives and investigate them on different temporal and spatial scales. Whereas, structural geology treats faults very directly, but often based on outcrop studies, seismology often concentrates on the signals (seismic waves) that are emitted during fault activity. This leads to isolated and thus restricted views on faults.

Analysing fault zone heterogeneity is a key task in characterizing a fault. In this context, there are many questions unsolved and understanding faults is a complex problem. Although there has been great progress in fault analysis over the last two decades, a unified fault model is still lacking that can serve as a predictive tool for fault zone composition, structure and for fault slip behaviour. Especially fault behaviour needs to be understood on different spatial and temporal scales.

When active faults move, they may enter a *seismic* phase, during which *earthquakes* occur. The co-related earth movements that take place during the earthquake, such as landslides, tsunami, and the destruction of infrastructure mean that earthquakes are one of the most important global geological hazards (Fig. 1.1). It is this side of faulting that is most well-known. Earthquakes, at the very best, destroy infrastructure and, at the very worst, cause loss of life. Since 1990, earthquakes have cost 27000 lives on average, each year

FIG. 1.1 300 m high landslip caused by the 2016 magnitude 7.8 Kaikoura Earthquake in the South Island of New Zealand. The photo was taken two years after the earthquake, which occurred 2 minutes after midnight on 14 November 2016.

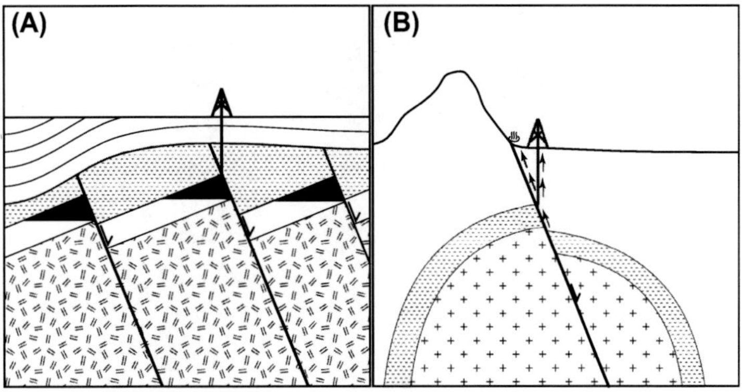

FIG. 1.2 Scenarios in which faults are useful. (A) Hydrocarbons trapped by faulting. (B) fault guiding hydrothermal energy to the surface and a shallow borehole.

(Guha-Sapir et al., 2011). To people who live on plate boundaries, e.g. in New Zealand, Japan, and the west coast of North America, their lives are govern by earthquakes (Fig. 1.1). Even within continental plates, there are less frequent and, for that reason, even more surprising earthquakes.

There is a lesser-known side of faulting, which is clearly beneficial to humankind. Many fault zones are known to act as conduits for the focused migration of fluids and clearly play a central role in determining the location, modes of transport, and emplacement of economically important hydrocarbon and hydrological reservoirs, and hydrothermal mineral deposits (Fig. 1.2A). For instance, water can migrate along a fault damage zone and appear at the surface as hot springs along the fault trace (Fig. 1.2B). The ancient Romans recognised this was the case around Aachen in Germany, and it was for this reason that they settled there (they called it "Aquae Granni" - at the waters, Fig. 1.3). In fact, the springs around Aachen deliver far more thermal power than the SuperC borehole that was drilled in Aachen specifically for geothermal use, showing that the faults are far better at delivering thermal energy than the surrounding rocks (Dijkshoorn et al., 2013). Similar situations, where hot springs are sourced by faults, are found, for instance, in Indonesia (Brehme et al.,. 2014), along the well-named Hot Springs Fault and other faults in California (Onderdonk et al., 2011; Onderdonk, 2012), and along the Alpine Fault in New Zealand (Cox et al., 2015), to name just a few examples. This is an important observation and moves faults into focus for exploration of geothermal energy plays (Moeck, 2014).

Exploration for geothermal energy now often concentrates on finding faults at depth, preferably still active or recently active (Barton et al., 1995; Carewitz and Karson, 1997; Huenges and Ledru, 2011). This is because the faults form both pathways (parallel to the fault surface) and baffles to the flow of hydrothermal water (across the fault; see Chapter 8 - Fault seal). Loveless et al. (2014) suggested that faults could even determine the success or failure of low

FIG. 1.3 The Elisenbrunnen in Aachen, built in neoclassic style in 1827, allows visitors to sample the highly sulphurous, 52°C mineral water that migrates along the many faults in the area (see Chapter 8) photo: Nils Chudalla.

enthalpy geothermal projects. For example, Blackwell et al. (2000) show that 90% or more of major known geothermal systems in the Basin and Range area of America are within 3 km of late Pleistocene or younger faults. Faults can also seal an otherwise open reservoir and trap hydrocarbons or ore minerals that would normally escape and dissipate (Fig. 1.2B). A majority of petroleum traps are due to fault closures and/or fault-rock seals (Sorkhabi and Tsuji, 2005).

The amount of knowledge that is not known about a fault can be shown by the surprises that have resulted from scientific deep-drilling projects that aimed to cross major faults. For instance, the SAFOD project was designed with the purpose of drilling through the San Andreas Fault at a depth of 2.7 km (e.g., Hickman and Zoback, 2004; Zoback et al., 2011). The fault was found to be profoundly weak (coefficient of friction = 0.15; Lockner et al., 2011; Carpenter et al., 2015), which can be attributed to the presence of smectite in shear fractures (Warr et al., 2014).

The German Continental Deep Drilling Program (Abbreviation in German, KTB) drilled through the Franconian Lineament, a major strike-slip fault of western border of the Bohemian Massif in NE Bavaria, Germany (see Section 3.4; Fig. 3.16), The borehole probably crossed the fault at a depth of 6850–7950 m (Emmermann and Lauterjung, 1997). Most surprising was the presence of graphite along the fault plane, making electromagnetics the best geophysical method to determine the position of the fault at depth (Rath et al., 2001). All the basement rocks drilled also contained a surprising amount of free fluids (Emmermann and Lauterjung, 1997), but significant inflows of fluids were noted along the fault zones (Huenges et al., 1997), even at depths down to 9 km.

The Deep Fault Drilling Project (DFDP) has drilled through an inactive portion of the Alpine Fault, New Zealand (Toy et al., 2015). Here, cataclastic processes, in particular, powder lubrication and grain rolling, are considered responsible for deformation on the fault instead of frictional processes (Schuck et al., 2018).

Definition of a fault surface, fault kinematics and displacement

We define a fault surface as:

'A structural discontinuity in a given volume of rock, along which movement has taken place'.

This simple definition implies that the discontinuity is a sharp, three-dimensional surface, i.e., the rock volume underwent brittle deformation. However, the fault surface will acquire, after sufficient movement, a zone around it that will also differ from the original unfaulted medium. During the evolution of a fault, the fault develops into a complex fault zone with a fault core where the slip is concentrated and the surrounding damage zone, which is often characterized by an increase in fracture density towards the fault core. This complex and heterogeneous zone has a strong impact of earthquake behaviour and fluid flow in the lithosphere.

Because a fault, or at least a segment of a fault, is usually sub-planar, it can be described by the strike and dip (or dip and dip azimuth) of the plane (Fig. 1.4). Miners, who worked ore bodies that followed the fault plane, coined the terms *footwall* for the block they were walking on and *hanging-wall* for the beds above them. It is this information, together with the sense of fault block movement, that, in the first instance, classifies the kinematics of the fault (Fig. 1.5).

When the hanging-wall moves downwards, relative to the footwall, the fault is termed a *normal fault*. The opposite sense of movement defines a *reverse fault*. A reverse fault with low angle of dip $< 45°$ is known as a *thrust*. If the fault blocks move laterally, relative to each other, the fault is a *strike-slip fault* (Fig. 1.5). Normal faults place younger beds over older (i.e., maintain the superposition); reverse fault place older beds over younger (i.e., reverse the superposition). These three main fault types represent end members, and real faults may have two components of each kind of movement. Fig. 1.5 also shows that a borehole through a normal fault will 'miss' part of the stratigraphy (this is known as *fault cut*), while the reverse fault replicates the strata at any point on the fault.

The amount of displacement (slip, heave and throw) on a fault can be calculated in two dimensions from the cutoffs of bedding (Fig. 1.6). However, true analysis in two dimensions can only be achieved if the cross-section is parallel to the 3-D fault displacement vector. Otherwise, the displacement vectors will be underestimated (i.e., it is an apparent displacement). This can

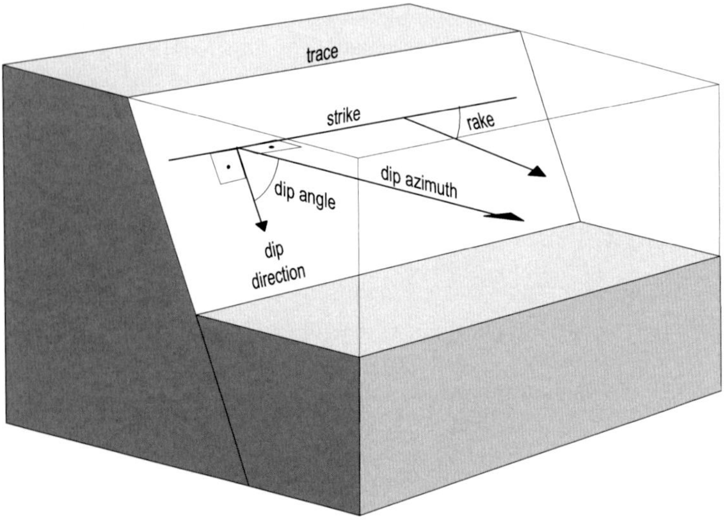

FIG. 1.4 A cutaway faulted block to show the conventions used to measure a fault plane. A fault surface can be represented either by its strike or dip direction and dip, or by the dip and dip azimuth. Note that both dip direction and dip azimuth are both perpendicular to the strike direction. Linear objects on the fault, such as slickensides, can be described the rake angle, i.e., the angle the lineation makes with the strike direction. Rake is measured anticlockwise from the strike direction, thus upward and downward rake angles are positive and negative, respectively. The dark grey side of the block is perpendicular to the strike and therefore shows the true dip angle. Other cross-section directions would give shallower apparent fault dip angles and must be corrected to obtain the true dip angle. The trace of the fault is the intersection of the fault with the topographic surface.

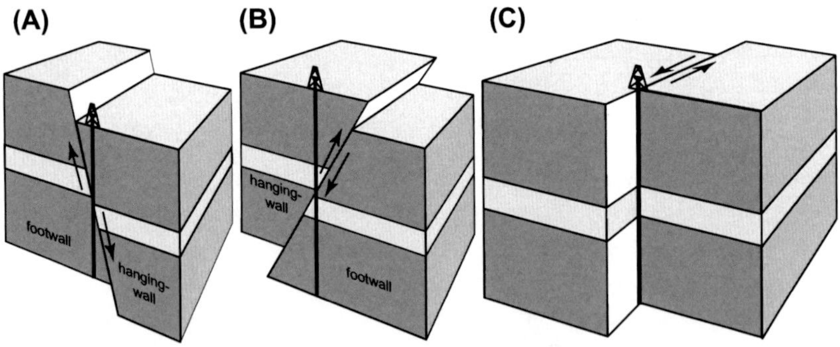

FIG. 1.5 Different classification of faults depending on the sense of movement. (A) a normal fault, (B) a reverse fault, (C) a left-lateral strike-slip fault. Note that in (A) the borehole misses part of the stratigraphy, whereas in (B) the stratum is doubled, and in (C) the borehole sees no change.

be avoided if the fault is observed in three-dimensional space (Fig. 1.7). However, it is not enough to know only the cutoffs of a planar feature, such as a bed. If *piecing points* (i.e. where linear structures on both sides of the fault are known, such as a fold axis, sedimentary feature or a dyke crosscutting bedding) are recognised, then it is possible to calculate the true slip vector

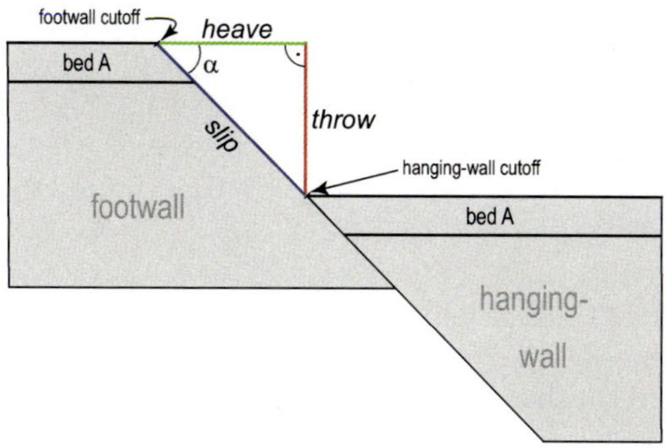

FIG. 1.6 Calculation of the amount of *slip* (displacement on the fault surface), *heave* (horizontal component of displacement) and *throw* (vertical component of displacement) in two dimensions, from the footwall and hanging-wall cutoffs of bed A. α is the fault dip. Since slip, heave and throw, in this case, form a right-angled triangle, it is geometrically easy to derive all the values, if the fault dip and one length are known.

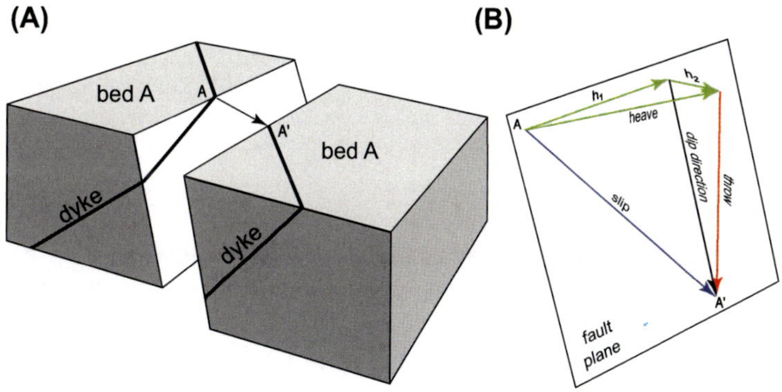

FIG. 1.7 (A) Oblique slip of two fault blocks (a combination of normal and sinistral strike-slip movement). A-A′ represent piecing points caused by the intersection of a dyke with bed A. (B) Decomposition of the slip vector into two heave vectors (h_1 parallel to the strike, h_2 parallel to the dip direction, with respect to the fault orientation) and the throw vector. The heave vector is also known as the *slip azimuth.*

(Fig. 1.7). Of course, such data is rare! In outcrop or 2D seismics, it is usual to determine only the apparent slip.

Fault rarely maintain the same geometry along strike or dip, nor does the amount of fault slip remain constant. The former can be shown as depth or contour maps, whereas the latter is more often shown as heave maps or throw (juxtaposition) diagrams (see Chapters 4–7).

This book follows a systematic approach. In Chapter 2 - "Fault mechanics and earthquakes", the reader will first gain knowledge on fault evolution from an intact rock volume over initial fracture formation to the establishment of a fault that finally separates a rock volume into two individual compartments.

This chapter then focuses on the structure of the fault zone and shows how the view on faults has changed over the last three decades from the simple geometric treatment of faults to the modern view of faults as complex zones that are composed of different structural domains. Based on the mechanics of faulting, earthquake processes are explained. The aim of this chapter is to present the basics of fault mechanics and to connect the theoretical models with field observations. The chapter links the geological outcrop-based fault characterization and the geophysical way of dealing with earthquakes, and thus forms the foundation for the following chapters.

Chapter 3 - "Fault Detection" suggests a number of geophysical methods to detect faults. This includes active methods like reflection seismics, ground penetrating Radar, Electrical Resistivity Tomography surveys and gravimetry, but also passive methods, where the seismic waves that are emitted from a fault during an earthquake are used to detect the presence of a fault and to derive its geometry and kinematics. Remote sensing, thanks to satellites, can be used to detect faulting in any part of the world, but also smaller-scale approaches, such as drone technology, is handled.

In Chapter 4 - "Numerical modelling of faults" different desktop numeral modelling approaches are presented and their application to faults is shown. In the last decades, numerical tools such as the finite difference, finite element and distinct element methods have rapidly spread into geoscience research and application. They allow the simulation of the mechanical development of faults and their effect on the fluid flow in the lithosphere. Finally, the reader is taken through a case study of modelling a fault.

Chapter 5 - "Faulting in the laboratory" describes the growing field of analogue fault (rock) analysis using laboratory experiments. The chapter is subdivided into four sections that focus on frictional experiments and relate the experimental results to natural earthquake processes. In addition, faults in scaled analogue models are discussed in this chapter as well as fault lubrication processes due to frictional melt generation.

In Chapter 6 - "The growth of faults", different models of the spatial evolution of faults are explained. This chapter summarises more than three decades of research in to this subject. It shows how faults grow laterally and how they may interact with each other. Understanding fault growth is important for the characterization of the subsurface structure and for earthquake geology.

Chapter 7 - "Direct dating of fault movement" gives an overview on the different methods that can be utilized to date fault rocks and therefore the fault activity. The focus is set on analytical approaches to derive the age of fault movement based on the minerals that can grow on the fault surface as well as on thermochronological methods.

Finally, Chapter 8 - "Fault seal" deals with fault sealing and permeability. Faults can have a strong impact on the fluid flow in the lithosphere. The basic concepts of fault seal processes and their impact on the permeability are explained. This is flanked by a wide range of case studies that allow the reader to connect the theory with field examples. The reader will learn about the effect that the different host-rock types (clastic rocks, carbonates, evaporites) can have on the structure and composition of a fault that develops within these rocks. Understanding fault seal processes has a strong practical application, especially in hydrocarbon exploration projects and in the growing field of geothermal play assessment.

References

Barton, C.A., Zoback, M.D., Moos, D., 1995. Fluid flow along potentially active faults in crystalline rock. Geology 23, 683–686.

Blackwell, D.D., Wisian, K.W., Richards, M.C., Steele, J.L., 2000. Geothermal resource/reservoir investigations based on heat flow and thermal gradient data for the United States. Unpublished final report for the U.S. Department of Energy, Ref: DOE/ID/13504.

Brehme, M., Moeck, I.S., Kamah, Y., Zimmermann, G., Sauter, M., 2014. A hydrotectonic model of a geothermal reservoir – a study in Lahendong, Indonesia. Geothermics 51, 228–239.

Carpenter, B.M., Saffer, D.M., Marone, C., 2015. Frictional properties of the active San Andreas Fault at SAFOD: implications for fault strength and slip behavior. J. Geophys. Res. Solid Earth 120/7, 5273–5289. https://doi.org/10.1002/2015JB011963.

Cox, S.C., Menzies, C.D., Sutherland, R., Denys, P.H., Chamberlain, C., Teagle, D.A.H., 2015. Changes in hot spring temperature and hydrogeology of the Alpine Fault hanging wall, New Zealand, induced by distal South Island earthquakes. Geofluids 15 (1–2), 216–239. https://doi.org/10.1111/gfl.12093.

Curewitz, D., Karson, J.A., 1997. Structural settings of hydrothermal outflow: Fracture permeability maintained by fault propagation and interaction. J. Volcanol. Geothermal Res. 79, 149–168.

Dijkshoorn, L., Speer, S., Pechnig, R., 2013. Measurements and design calculations for a deep coaxial borehole heat exchanger in Aachen, Germany. Int. J. Geophys. 14. https://doi.org/10.1155/2013/916541. Article ID 916541.

Guha-Sapir, D., Vos, F., Below, R., with Ponserre, S., 2011. Annual Disaster Statistical Review: The Numbers and Trends. CRED, Brussels.

Emmermann, R., Lauterjung, J., 1997. The German continental deep drilling Program KTB: overview and major results. J. Geophys. Res. 102 (B8), 18179–18201. https://doi.org/10.1029/96JB03945.

Hickman, S., Zoback, M.D., 2004. Stress orientations and magnitudes in the SAFOD pilot hole. Geophys. Res. Lett. 31, L15S12. https://doi.org/10.1029/2004GL020043.

Huenges, E.B., Engeser, J., Erzinger, J., Kessels, W., Kück, J., 1997. The permeable crust: geohydraulic properties down to 9100 m depth. J. Geophys. Res. 102 (B8), 18255–18265.

Huenges, E., Ledru, P., 2011. Geothermal Energy Systems: Exploration, Development, and Utilization. John Wiley & Sons, p. 486pp.

Huntington, K.W., Klepeis, K.A., with 66 community contributors, 2018. Challenges and Opportunities for Research in Tectonics: Understanding Deformation and the Processes that Link Earth Systems, from Geologic Time to Human Time. A community vision document submitted to the U.S. National Science Foundation. University of Washington, p. 84. https://doi.org/10.6069/H52R3PQ5.

Lockner, D.A., Morrow, C., Moore, D., Hickman, S., 2011. Low strength of deep San Andreas fault gouge from SAFOD core. Nature 472, 82–85. https://doi.org/10.1038/nature09927.

Loveless, S., Pluymaekers, M., Lagrou, D., De Boever, E., Doornenbal, H., Laenen, B., 2014. Mapping the geothermal potential of fault zones in the Belgium-Netherlands border region. Energy Procedia 59, 351–358. https://doi.org/10.1016/j.egypro.2014.10.388.

Moeck, I.S., 2014. Catalog of geothermal play types based on geologic controls. Renew. Sustain. Energy Rev. 37, 867–882. https://doi.org/10.1016/j.rser.2014.05.032.

Onderdonk, N., 2012. The role of the Hot Springs Fault in the development of the San Jacinto Fault Zone and uplift of the San Jacinto mountains. In: Palms to Pines: Geological and Historical Excursions through the Palm Springs Region, Riverside County, California. San Diego Association of Geologists and South Coast Geological Society Field Trip Guidebook.

Onderdonk, N., Mazzini, A., Shafer, L., Svensen, H., 2011. Controls on the geomorphic expression and evolution of gryphons, pools, and caldera features at hydrothermal seeps in the Salton Sea Geothermal Field, southern California. Geomorphology 130, 327–342. https://doi.org/10.1016/j.geomorph.2011.04.014.

Rath, V., Schwalenberg, K., Brasse, H., 2001. A detailed electromagnetic investigation of the Franconian Lineament/Bavaria. In: Hördt, A., Stoll, J.B. (Eds.), 19. Kolloquium Elektromagnetische Tiefenforschung. Burg Ludwigstein, pp. 335–339, 1.10.-5.10.2001.

Schuck, B., Janssen, C., Schleicher, A.M., Toy, V.G., Dresen, G., 2018. Microstructures imply cataclasis and authigenic mineral formation control geomechanical properties of New Zealand's Alpine Fault. J. Struct. Geol. 110, 172–186. https://doi.org/10.1016/j.jsg.2018.03.001.

Sorkhabi, R., Tsuji, Y., 2005. The place of faults in petroleum traps. In: Sorkhabi, R., Tsuji, Y. (Eds.), Faults, Fluid Flow, and Petroleum Traps, vol. 85. AAPG Memoir, pp. 1–31.

Toy, V.G., Boulton, C.J., Sutherland, R., Townend, J., Norris, R.J., Little, T.A., Prior, D.J., Mariani, E., Faulkner, D., Menzies, C.D., Scott, H., Carpenter, B.M., 2015. Fault rock lithologies and architecture of the central Alpine fault, New Zealand, revealed by DFDP-1 drilling. Lithosphere 7/2, 155–173. https://doi.org/10.1130/L395.1.

Warr, L.N., Wojatschke, J., Carpenter, B.M., Marone, C., Schleicher, A.M., van der Pluijm, B.A., 2014. A "slice-and-view" (FIBeSEM) study of clay gouge from the SAFOD creeping section of the San Andreas Fault at ~2.7 km depth. J. Struct. Geol. 69, 234–244.

Zoback, M., Hickman, S., Ellsworth, W., 2011. Scientific drilling into the San Andreas Fault Zone – an overview of SAFOD's first five years. Sci. Drill. 11, 14–28.

Chapter 2

Fault mechanics and earthquakes

Christian Brandes[a], **David C. Tanner**[b]
[a]*Institut für Geologie, Leibniz Universität Hannover, Hannover, Germany;* [b]*Leibniz Institute for Applied Geophysics (LIAG), Hannover, Germany*

Chapter outline

Understanding Faults. https://doi.org/10.1016/B978-0-12-815985-9.00002-3

2.1 Introduction

Faults can be treated from different perspectives. Faults as static features can be described as, *continuum-Euclidean, fractal or granular* (Ben-Zion and Sammis, 2003) (Fig. 2.1). These three individual views strongly depend on the scale of observation. The standard way faults are described is the continuum-Euclidean view, where faults are smooth and continuous geometric objects in a solid continuum. This view can explain many observations on a first-order scale. The fractal view is largely based on the observation of rough fault surfaces, fault branching and fracture patterns as well as the distribution of earthquakes (Gillespie et al., 1993). The granular view takes observations such as fault block rotation and the development of fault breccias and fault gouge into account (Ben-Zion and Sammis, 2003). The dynamics of faulting on the other hand can be regarded as an energy transformation process, where fault motion extracts strain energy from the surroundings of a fault and transforms it into frictional heat, fracture energy and emitted seismic energy (Husseini, 1977). Movement along a fault is an energy budget that involves work against gravity, internal work within the fault system, work against frictional resistance along the fault surface, energy needed for fault propagation, and energy that is radiated as seismic waves (Cooke and Madden, 2014).

Because the same basic mechanical laws control all fault processes from micro-scale to kilometre-scale, fault mechanics is the key to understanding the formation, development and the long-term behaviour of faults. Treating faults from a mechanical point of view allows to link their static characteristics (fault structure derived from outcrops or reflection seismic) with their dynamic behaviour (fault slip and related seismicity derived from geodetic measurements or the record of seismic signals), and enables us to bridge the gaps in the spatial and temporal scale. This could potentially lead to a unified fault model that integrates the different views of faults. Several grand challenges have been formulated in the last years for the research fields of structural geology and seismology (Lay, 2009; Forsyth et al., 2009; Gudmundsson, 2013; Huntington

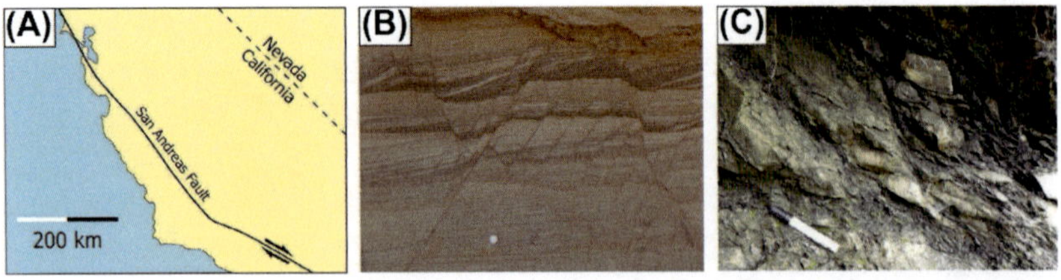

FIG. 2.1 Fault classification, based on the scale of observation. (A) Trace of the San Andreas Fault. In map view and from a large distance, the fault can be regarded as a straight, discrete line/surface (*Euclidean* view). Trace of the San Andreas Fault *based on van der Pluijm and Marshak (2004).* (B) Shear-deformation band network. The deformation bands branch and the branches split into smaller branches (*fractal* view) (a one euro coin for scale). (C) Fault breccia. The core of this fault consists of rock fragments bounded by slip surfaces (*granular* view) (pen for scale).

and Klepeis with 66 community contributors, 2018). Some of these challenges contain questions related to the mechanics of faults:

- How do fault zones behave from the Earth's surface to the base of the lithosphere?
- How are new faults initiated and how are they reactivated throughout Earth history?
- How do fault networks evolve on different spatial and temporal scales?
- How do faults slip?
- What causes large seismogenic fault-slip?
- How do ruptures stop?
- What controls the transition from seismic rupturing to aseismic fault creep?
- How do the mechanical properties of a fault evolve over an earthquake cycle and also over the lifetime of a fault?

All these questions are directly related to fault mechanics and the understanding of the physical processes that control their behaviour. Extensive outcrop analysis of the fault structure and the lithology of fault-rocks, coupled with numerical modelling techniques can help to develop more realistic fault models and to solve these questions.

2.2 Fractures

Fractures are fissures in rocks that include joints and faults. We use the term fracture (as most of the literature does) to describe a mechanical discontinuity in rocks that results from tectonic forces and is characterized by no or very low offset. In contrast, a fault has a distinct offset. The commonly-used term joint describes mode I fractures, i.e., opening of the fracture perpendicular to the fracture wall. The term vein describes the mineral fill between fracture walls, with a composition different from the host rock (Bonnet et al., 2001). The term crack is used to describe an idealized elastic discontinuity (Vermilye and Scholz, 1998).

Fracturing represents the dominant mechanism of rock failure at shallow depths in the upper crust under low confining pressures (Friedman, 1975). A brittle fracture is defined as a macroscopic planar discontinuity in a rock volume (Nelson, 1979, 2001) that has lost most or all of its cohesion due to failure, but can at any time be healed and become completely coherent again (Price and Cosgrove, 1990).

Following the work of Peacock (2001), fractures can pre-date faults, they can develop synchronous to faults or act as their precursors, and they can post-date faults. Fractures that evolve as precursors to faults play a major role in fault development. To fracture an intact rock, the failure strength of the rock volume must be exceeded. This material failure is driven by the stresses acting upon it. Stress is defined as a force per unit area. If the rock undergoes

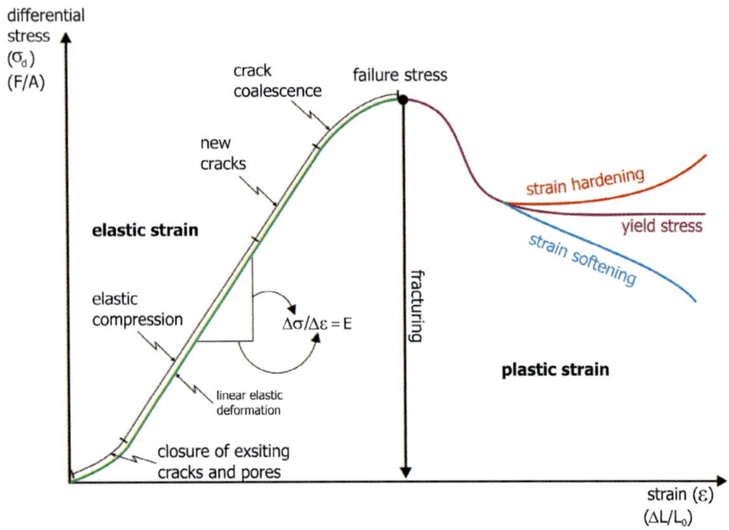

FIG. 2.2 Stress/strain diagram. Stress is shown on the vertical axis and strain on the horizontal axis. The elastic part of the deformation is displayed by the *green curve*. Plastic deformation is shown by the *purple curve* (yield stress). Material that undergoes plastic deformation with strain hardening behaviour is represented by the *red curve*; plastic deformation with strain softening is shown by the *blue curve. Figure is modified after Hajiabdolmajid et al. (2002).*

deformation in response to the applied stress, this is known as strain. Rock deformation can be best explained in a stress/strain diagram (Fig. 2.2).

A summary of deformation features that can be seen in triaxial experiments in which the rock fractures or undergoes plastic strain, after Hajiabdolmajid et al. (2002), are shown in Fig. 2.2. As stress increases, the rock goes through four stages of linear elastic deformation (green line; Fig. 2.2); true elastic behaviour is first reached when existing cracks and pore space are closed. In the elastic domain, the rock follows Hook's Law, for which strain is proportional to stress. The gradient of this part of the curve is:

$$\Delta\sigma/\Delta\varepsilon = E,$$

where E is Young's Modulus. Increasing the stress beyond the failure stress (tensile strength) leads to permanent deformation. This is either manifest as fracturing (black line; Fig. 2.2) or the material continues to deform at a steady, yield stress, lower than the failure stress, which is termed plastic behaviour (purple line). If the material requires increasing or decreasing stress to maintain plastic strain, this is termed strain hardening (red line) and strain softening (blue line), respectively (Fig. 2.2).

In general, three types of fractures can be distinguished by their movement vectors (Lawn, 1993); tensile fractures (mode I; extension or opening) and two types of shear fracture (mode II; in-plane or sliding, and mode III; out-of-plane or tearing mode, Atkinson, 1982; 2001; Fig. 2.3). Hybrid fractures are also

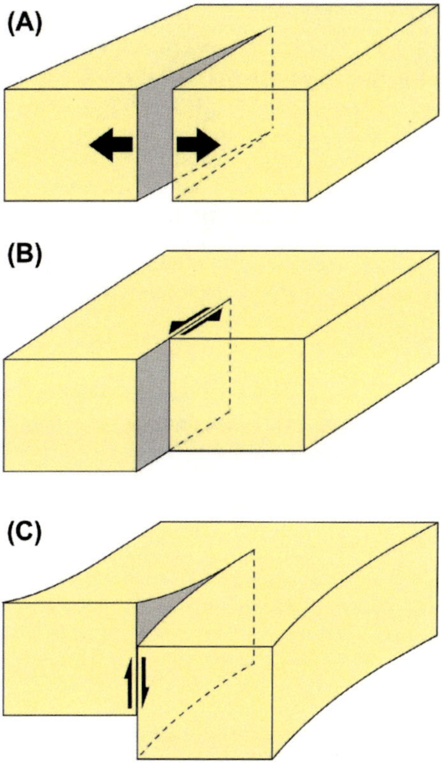

FIG. 2.3 Fracture types. (A) A mode I fracture that opens perpendicular to the direction of extension. (B) A mode II fracture, representing a shear-fracture, shows a lateral displacement parallel to the direction of propagation. (C) A mode III fracture shows a scissor-like opening style. *Modified after Lawn (1993).*

possible, for instance under increasing compressive stress fractures show a continuous transition from extension to shear (Ramsey and Chester, 2004).

There is a systematic relationship between the principle stresses (σ_1, σ_2, and σ_3) and the orientation of the evolving fractures. Under low confining pressure, tensile (mode I fractures) open in the direction of the minimum principle stress (σ_3) and parallel to σ_1, whereas under higher confining pressure, conjugate shear fractures (mode II fractures) often develop, where σ_1 is the bisecting vector between the two conjugate fractures sets (Hancock, 1985). This systematic relationship is often used to derive the paleostress orientation from fracture sets measured in outcrops (e.g., Hancock and Kadhi, 1978; Engelder and Geiser, 1980; Dyer, 1988; Angelier et al., 1990; Brandes et al., 2013; Maerten et al., 2016).

Fracture propagation is controlled by the interaction of the fracture with microscopic structures within the rock (Dyskin and Germanovich, 1993). In the 1980s, first-order models were developed, where fractures are represented by a simple slot in a homogeneous and isotropic material that shows linear elastic behaviour (Lin and Parmentier, 1988; Schultz, 1999). This approach is called linear elastic fracture mechanics (LEFMs) and led to significant progress in the understanding of fractures. During the 1990s, more complex models that are based on an elastic-plastic approach were developed.

2.3 From intact rocks to opening-mode fractures to faults

In general, the evolution of faults begins with the development of small opening-mode fractures in an intact rock volume, which grow in length and later connect to form a continuous shear fracture (Fig. 2.4). Fracture growth is concentrated at the fracture tips. Stress concentrations occur that can lead to the evolution of so-called wing-cracks. However, the term 'intact rock' is not precise, because every rock or even every material, natural or artificial, contains very small discontinuities that can be the nuclei for the development of fractures.

2.3.1 Griffith cracks

At the beginning of the 20th Century, A. A. Griffith analysed the paradox that the tensile strength of a material is lower than its theoretical cohesive strength. In addition, most materials have a lower strength in tension than in compression. Griffith was motivated by the fact that bulk glass fractures under 100 MPa stress, whereas the theoretical stress needed to fracture the atomic bonds of glass is approximately 10 GPa (Lawn, 1993). He discovered that the low tensile strength was caused by micro discontinuities that were already present in the material (Griffith, 1921). These micro discontinuities are now called 'Griffith cracks' and represent microscopic imperfections in crystal lattices. The Griffith approach indicates that the stress amplification is dependent on the aspect ratio of the discontinuity (Fig. 2.5). If we consider the Griffith cracks to be circular holes, then at the edges of the hole, the stress is amplified by a factor three: the stress amplification is even higher at the edges of an elliptic hole. From these stress concentrations, local tensile fractures initiate. This stress magnification at the crack tip is the key to crack growth and depends on the orientation, length and curvature at the crack tip, thus leading to preferential growth of certain cracks (Brace, 1960). Prior to failure, the density of the tensile fractures increases and subsequently the fractures coalesce. This leads to strain localization and the development of a macro-scale shear failure of the material (Hoek and Martin, 2014). Following the work of Griffith, the critical tensional stress that is necessary to cause a fracture depends on the length of the pre-existing micro-cracks and the energy that is needed to create new surfaces during fracture propagation; this is now known as the Griffith criterion (Fig. 2.6). As an approximation, the tensile strength of a rock is about a fifth of its theoretical value.

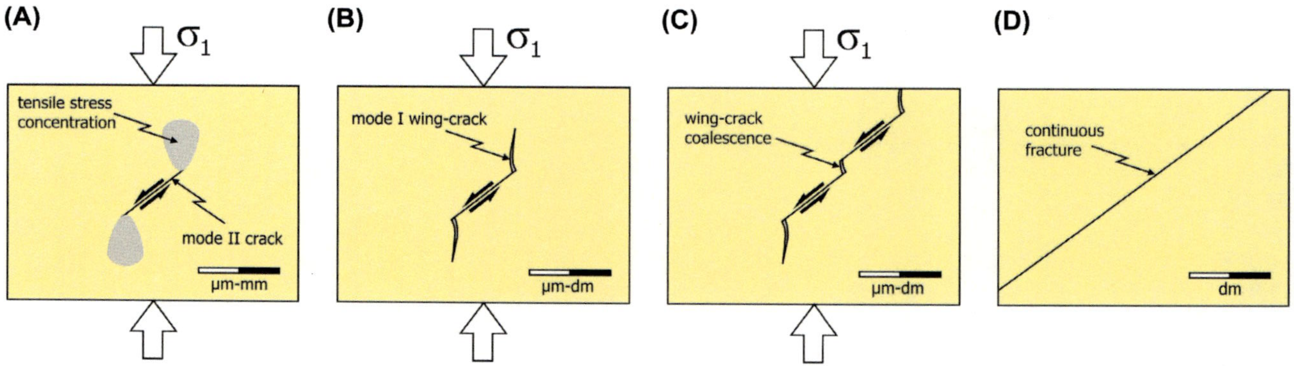

FIG. 2.4 Fracture coalescence. (A) At the tips of the individual fractures tensile stress concentrations are located. (B) In these locations, mode I cracks grow that are called wing-cracks. (C) These cracks can join. (D) After the cracks have joined, a continuous fault surface has developed. *Figure is based on van der Pluijm and Marshak (2004).*

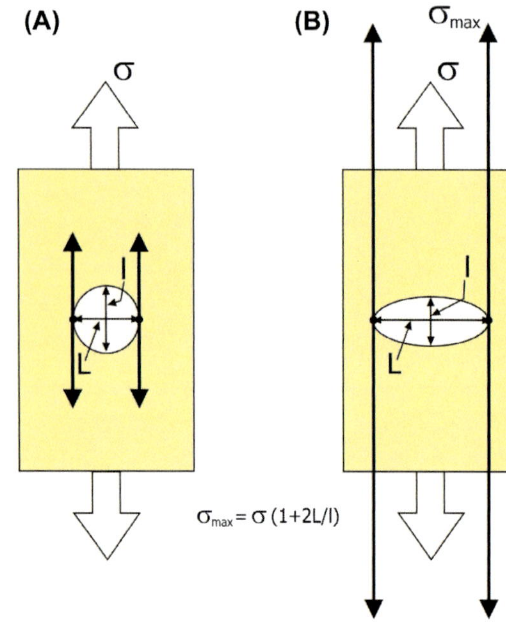

FIG. 2.5 Griffith crack. At the tips of a discontinuity, the applied stress σ is amplified to σ_{max} and can drive fracture opening and lateral crack propagation (A) Circular discontinuity. (B) Elliptic discontinuity. Following the equation $\sigma_{max} = \sigma(1 + 2L/l)$, the stress amplification is higher around an elliptic discontinuity. *Modified after van der Pluijm and Marshak (2004).*

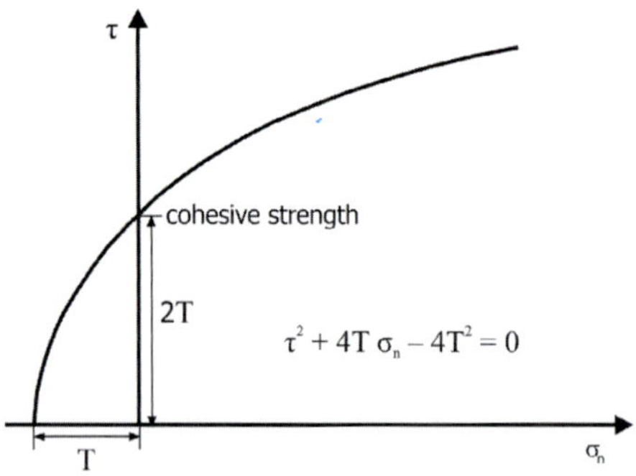

FIG. 2.6 The Griffith criterion. The Griffith criterion is a non-linear function and represents a parabola in the Mohr diagram. T is the tensile strength of the material and the intersection of the envelope and the ordinate gives the cohesive strength (2T). *Modified after Brace (1960) and Fossen (2010).*

2.3.2 The Coulomb failure criterion and the Mohr circle

Charles-Augustin de Coulomb first presented a fundamental concept behind the evolution of shear fractures in the 18th Century (Coulomb, 1776). In his model, a shear fracture will develop in an intact rock, if the applied force exceeds the cohesion of the material, c, as well as overcoming the friction along the developing fracture surface. This is called the 'Coulomb failure criterion' and is expressed as;

$$\tau = \sigma_n \tan(\phi) + c,$$

where c is the cohesion, ϕ is the angle of internal friction (dependent on the deforming material) and τ and σ_n are the shear and normal stresses acting on the fracture surface. The Coulomb failure criterion can be used to derive the stresses necessary to produce a shear fracture, if the angle ϕ is known. New shear fractures develop at the angle θ to σ_3, where θ is typically 60 degrees for most rocks, and at 30° to σ_1 (Ramsey and Chester, 2004). Examples of angles of friction and tensile strength of rocks are given in Tables 2.1 and 2.2.

A common way to visualize the Coulomb failure criterion is the so-called Mohr circle diagram, as developed by Mohr (1900) (Fig. 2.7). The Mohr circle is a graphical representation of the transformation law for the Cauchy stress tensor. As such, the perimeter of the circle represents the shear and normal stress that is applied to any oriented surface within a solid (Fig. 2.7). In a Mohr diagram, the shear stress, τ, is displayed on the ordinate and the normal stress σ_n on the abscissa. A circle that passes through σ_1 and σ_3 is the Mohr circle and thus describes the state of stress in the rock. The failure envelope separates the diagram into stable (yellow) and unstable (grey) fields (Fig. 2.7). The gradient of the failure envelope is the angle of the internal friction, ϕ, and it crosses the ordinate at c, the tensile strength of the rock. The greatest maximum principal stress (σ_1) and the minimum principal stress (σ_3) are shown on the horizontal axis. If the Mohr circle touches the failure envelope (Fig. 2.8), a fracture will occur, the orientation of which is twice the angle θ, with respect to σ_3.

Terzaghi (1923) proposed that the Coulomb criterion should be modified for saturated soils, because the pore pressure plays an important role in lowering the effective stress:

$$\tau = (\sigma_n - p)\tan(\phi) + c,$$

Where p is the pore pressure. Effective stress, σ'_n, is defined as $\sigma_n - p$. Hubbert and Rubey (1959) showed that pore pressure should be also

TABLE 2.1 Typical values of the angle of friction for sands and silts.

Material	Angle of internal friction, ϕ (degrees)
Sand, uniform, round grains	27–34
Sand, well graded, angular	33–45
Sandy gravels	35–50
Silty sand	27–34
Inorganic silt	27–35

After Terzaghi and Peck (1967).

TABLE 2.2 Typical tensile strength (cohesion), and angles of internal friction within ranges of confining pressure.

Rock	Cohesion (MPa)	Angle of internal friction (degrees)	Range of confining pressure (MPa)
Berea sandstone	27.2	27.8	0–200
muddy shale	38.4	14.4	0–200
Sioux quartzite	70.6	48.0	0–203
Georgia marble	21.2	25.3	6–69
chalk	0.0	31.5	10–90
Stone Mt. granite	55.1	51.0	0–69
Indiana limestone	6.7	42.0	0–10

After Goodman (1980).

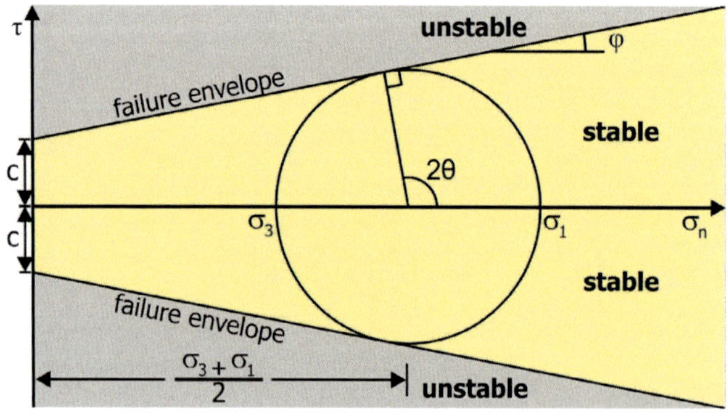

FIG. 2.7 A generalised Mohr circle to demonstrate the shear failure envelope and the state of stress in rock.

considered for rocks. The effect of pore pressure in Mohr space can be seen in Fig. 2.9. Normal stress is reduced by the amount of pore pressure. Graphically, the Mohr circle is shifted to the left by the amount of pore pressure (Fig. 2.9). If the pore pressure is high enough, the Mohr circle will be shifted to the left until it touches the failure envelope. At this point, failure occurs. Mathematically, the ratio of shear stress to normal stress is increased, because the pore fluid pressure reduces the normal stress.

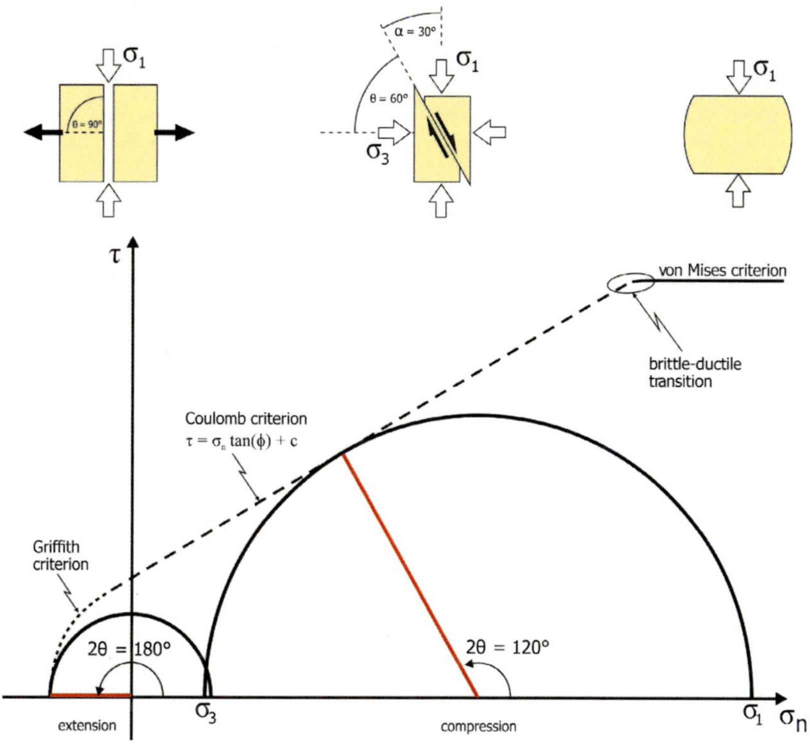

FIG. 2.8 Mohr-Coulomb failure envelope (*long dashes*), including the Griffith failure criterion under tensile stress (*short dashes*) and the von Mises criterion (*solid line*). The ordinate subdivides the diagram into an extensional and a compressional domain. For the large Mohr circle, both principle stress axes are compressive, and shear fractures will be produced with the angle of 60 degrees to σ_3. The smaller Mohr circle lies in both the tensile and compressive fields and touches the envelope so that fractures (red lines) are produced at an angle of 90 degrees to σ_3, i.e., parallel to σ_1 and thus a tensile fracture.

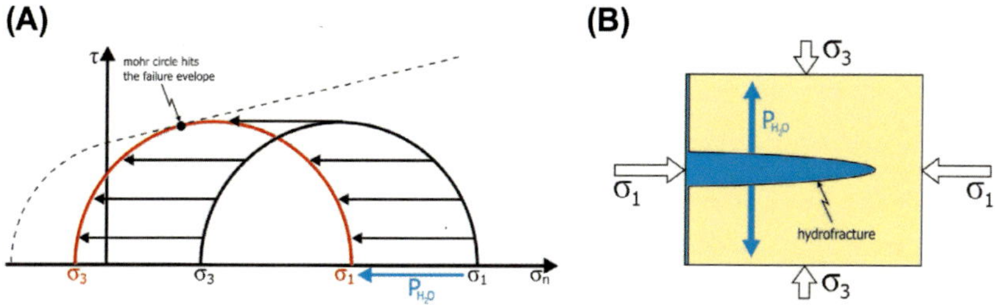

FIG. 2.9 Hydrofractures. (A) The effect of pore pressure is to shift the whole Mohr circle to the left. Because the failure envelope slopes to the left, rocks with a stress state that is close to fracturing can be made to fracture by increasing the pore pressure, while maintaining the same stresses. (B) A hydrofracture that opens due to high fluid pressure. The water pressure (P_{H2O}) exceeds σ_3 and the fracture can open. *(A) Modified after Hobbs et al. (1976). (B) Based on van der Pluijm and Marshak (2004).*

Pore pressure is an important part of the mechanics of faulting, as will be seen later in this chapter. The previous examples of Mohr circles consider homogenous, unfractured rock brought to failure. To allow movement along a pre-existing shear fracture, the amount of shear stress required to overcome the static friction is described by Amonton's Law and is expressed as

$$\tau = \mu_s \cdot \sigma_N,$$

where μ_s is the static friction and σ_N is the force normal to the fracture surface (Blau, 2013). In other words, the force that must exceeded to move on the fracture (i.e., the static friction) is directly proportional to the normal stress applied. In cases where a fracture is already established, cohesion does not have to be taken into account. The Coulomb failure envelope passes into the von Mises criterion (solid line; Fig. 2.8) that describes the behaviour under ductile conditions.

2.3.3 Hydrofractures

Hydrofractures are (pre-existing) fractures that open and propagate due to high fluid pressure (Secor, 1965). They are very common in the upper crust (Fyfe et al., 1978) and almost all are extensional mode I fractures that form major pathways for fluids (Gudmundsson and Brenner, 2001). Hydrofractures can also occur as shear or hybrid shear fractures. The mode of opening depends on the magnitude of principal stresses and the crustal depth. For instance, tensile fractures are generated in Fig. 2.9 because σ_3 is moved in the tensile field.

In the subsurface, fractures are often closed due to the confining pressure, which is a result of the lithostatic pressure and the ambient stress field. Hydrofractures form if the pore fluid pressure in a fracture is high enough to compensate for these surrounding forces (when the fluid pressure exceeds σ_3, Ghani et al., 2013), and the tensile strength of the material (Flekkøy et al., 2002), leading to opening and potential propagation of the fracture. A comparable mechanism acts during the movement of thrust sheets (see Section 2.5.5 "Large overthrusts and the effect of fluid pressure"). This can be visualized in the Mohr diagram, where the increase in fluid pressure reduces the principal stresses, which lead to a shift of the Mohr circle towards the failure envelope (Fig. 2.9). Sources of elevated pore pressure include; compaction due to burial, clay dehydration, organic matter decomposition and aqua-thermal expansion as well as effects related to impermeable rock units that act as barriers to fluid flow (Ghani et al., 2013). Similar to the case of the Griffith cracks, high stress concentrations can occur at the tips of the fracture that support the propagation of the fractures associated with elevated pore pressures.

Pioneering work on hydrofractures was conducted by Hubbert and Willis (1957). They showed that a hydrofracture, in general, opens perpendicular to the minimum principal stress. Consequently, hydrofractures in a normal fault regime, where the minimum principal stress is horizontal, will be vertical.

In reverse faulting stress regimes, where the maximum principal stress is horizontal, and the minimum principle stress is vertical, the newly formed hydrofractures will be horizontal. The fluid pressure lowers the normal stresses, but does not affect the shear stress, thus the Mohr circle moves towards the Mohr-Coulomb envelope, finally leading to rock failure. Hydrofractures have been studied in outcrops (e.g., Gudmundsson and Brenner, 2001), with experiments (e.g., Lockner and Byerlee, 1977) and numerical simulations (e.g., Flekkøy et al., 2002; Al-Busaidi et al., 2005). Hydrofractures may be natural or artificial. Artificial hydrofractures are generated, for instance, to enhance the permeability and thus the production rate of hydrocarbon reservoirs (e.g., Yew, 1997) or geothermal reservoirs (e.g., Legarth et al., 2005).

2.3.4 Stress state and dynamic fault classification of Anderson

Dynamic classification of faults was proposed by E.M. Anderson in the early 20th Century. In this seminal work, he related the orientation and the sense of slip of a fault to the lithospheric stress field. Anderson (1905, 1951) showed that, for newly-formed faults, the bisecting vector of the smallest angle between two conjugate faults defines the position of the maximum principle stress, σ_1. The intersection of the two fault surfaces gives the orientation of σ_2. The vector that bisects the largest angle between two conjugate faults represents the position of σ_3. In the case σ_1 is perpendicular to the Earth's surface, normal faults will develop, whereas reverse and strike-slip faults evolved under a horizontal compressional stress field with σ_1 parallel to the Earth's surface (Anderson, 1905) (Fig. 2.10). In a Coulomb material, a new fault will form at the angle θ to σ_1, which in most cases is approximately 30 degrees. Consequently, the acute angle between two the conjugate faults that enclose σ_1 is approximately 60 degrees and the obtuse angle between the two conjugate faults that enclose σ_3 is consequently 120 degrees. In Anderson's model, the

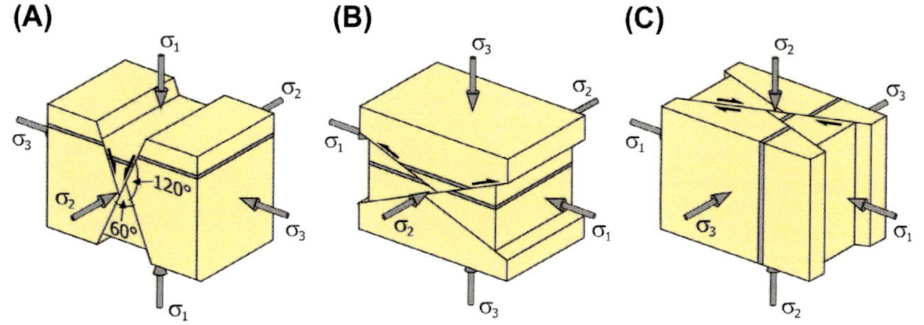

FIG. 2.10 Dynamic fault classification after Anderson that relates a fault to the ambient lithospheric stress field. The orientation and displacement of a fault is controlled by the orientation of the three principle main stresses. (A) If σ_1 is vertical to the Earth's surface, normal faults form. In case of σ_1 parallel to the Earth's surface, (B) reverse or (C) strike-slip fault are developed. With a vertical σ_3, reverse faults form, whereas a horizontal σ_3 leads to the formation of strike-slip faults. *Modified after van der Pluijm and Marshak (2004).*

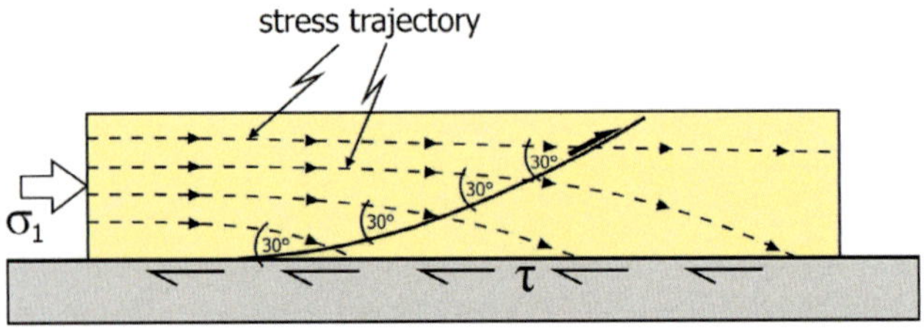

FIG. 2.11 Development of listric reverse faults. The listric geometry is caused by a deflection of the main principle stress due to friction on the basal detachment. Following the Anderson model, a fault forms at a 30 degrees angle to the main principle stress. In case of deflected and curved σ_1 stress trajectories, the angle rotates towards the basal detachment, which leads to a listric geometry of the resulting fault. *Figure is modified after van der Pluijm and Marshak (2004) and Nemčok et al. (2005).*

surface of the Earth is a free boundary. This means no shear or normal stresses occur at it and one of the principle stress axes of the stress tensor is near vertical, and consequently the other two are nearly horizontal, except in cases of stress disturbance due to heterogeneity, structural weakness or topography (Simpson, 1997). This fundamental classification scheme opened the door for an inverse approach, in which the orientation of the paleostress field can be derived from the fault kinematics based on the orientation of the fault surface and the slip direction on this surface, usually derived from slickenside data. Several approaches have been made over the years (e.g., Marrett and Allmendinger, 1990; Delvaux et al., 1995; Žalohar and Vrabec, 2007; Sato, 2012), see Célérier et al. (2012) and Lacombe (2012) for a summary and review of the different techniques. This was initially done with graphical techniques (Angelier and Mechler, 1977), and later carried out with computer-based methods to derive the best-fitting stress tensor (Yamaji, 2000).

The geometry of the stress field can also have a direct control on the fault geometry. In Fig. 2.10, faults are shown as planar features, whereas in nature they also often have a listric, curved geometry in cross-section. This curvature is an effect of the rotation of the local stress field (Fig. 2.11). In a compressional regime, trajectories of σ_1 are parallel to the Earth's surface. Due to friction, e.g., at a basal detachment of a fold-and-thrust belt, the σ_1 trajectories are curved downwards (Nemčok et al., 2005). Following the model of Anderson, the angle between a fault and σ_1 is approximately 30 degrees (see angle α in Fig. 2.8). To keep this angle between the stress trajectories and the fault surface constant, the fault surface has to flatten towards the detachment, which results in a listric geometry of, for instance, a thrust fault.

2.3.5 Wallace-Bott hypothesis

As described above, the orientation and the sense of fault slip are controlled by the lithospheric stress field. This concept was refined by Wallace (1951) and Bott (1959), and is known as the Wallace-Bott hypothesis in the literature.

These authors assume that the slip along a fault occurs parallel to the maximum resolved shear stress on the fault surface. This concept forms the basis for modern paleostress field analysis. As summarized by Pascal (2002), the Wallace-Bott hypothesis also requires that faults are planar, the fault blocks behave rigidly, the applied stress is uniform and no stress perturbations and no block rotations occur along the fault surface. Over the last decades, the validity of the Wallace-Bott hypothesis has been tested and discussed (Pollard et al., 1993; Pascal, 2002; Lisle, 2013). Pollard et al. (1993) showed that stress inversion techniques are valid, if the slip direction is controlled by the orientation of the fault surface and a homogeneous regional stress field is present.

2.4 Fault zone processes and structure

As described above, the development of a fault is a process that starts from small discontinuities in the rock volume (i.e., Griffith cracks) that coalesce to form a continuous fault surface that accommodates slip (Fig. 2.12). There is a linear relationship between the displacement and the length of a fault (Cowie and Scholz, 1992a). Under brittle conditions, faults typically comprise primary slip surfaces, secondary fractures (e.g., joints, veins and small-scale faults) and ductile folding (Rotevatn and Fossen, 2011; Fig. 2.12). Collectively, this fault-related deformation has been described as a fault zone or damage zone, which may range in width or thickness from millimetre- to kilometre-scale. With increasing slip the dimensions of damage zones generally increase. Wider fault zones are associated with longer faults. Discrete faults with offsets in a range of metres are in total tens of metres to hundreds of metres long. With increasing fault length, the fault zone widens, up to a finite point. Below the brittle-ductile transition, faults tend to pass into shear zones, where the deformation is distributed over a wider zone (Fig. 2.13).

2.4.1 The fault zone

Faults are often described and treated as discrete surfaces along which slip occurs (continuum-Euclidean view; Knipe et al., 1998; Chen and Bai, 2006). Over the last three decades, this view has changed and faults are increasingly considered to be 3-D volumes (e.g., Wibberley et al., 2008; Riley et al., 2010), known as a 'fault zone' (Caine et al., 1996; Childs et al., 2009). Hobbs et al. (1976) and Schutz et al. (2010) use the term 'fault zone' to describe a series of closely-spaced faults (which is now often referred to as a fault array; see van der Pluijim and Marshak, 2004), whereas other authors define a fault zone as the sum of all structural-mechanical units of a fault (Chester and Logan, 1986). With a focus on the hydraulic properties, Bense et al. (2013) define a fault zone as a rock volume in which permeability has been altered by fault-related deformation.

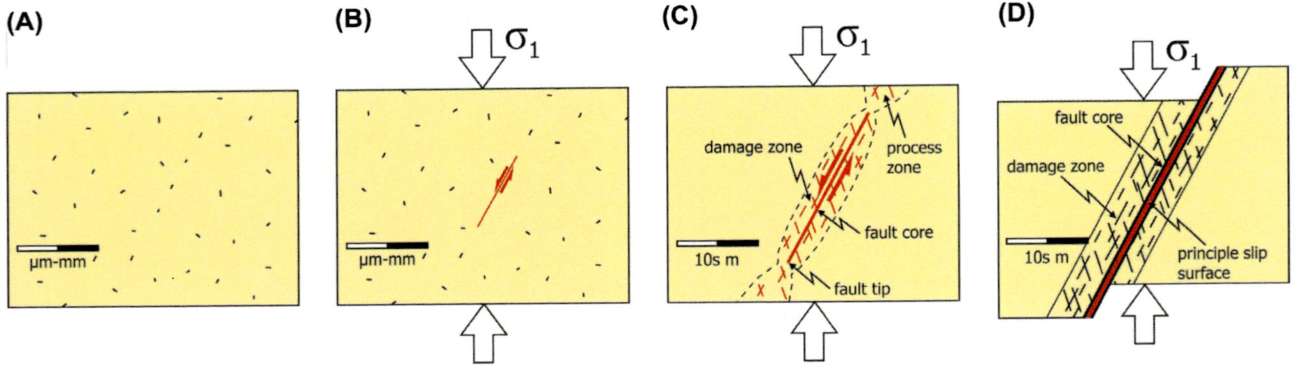

FIG. 2.12 Fault evolution. (A) A rock volume with microscopic discontinuities (Griffith cracks). (B) Under an applied stress, optimally-oriented microfractures connect and form a fracture that (C) develops into a fault by accumulating displacement. The fault is flanked by a damage zone and a process zone develops ahead of the fault tips. (D) Ongoing displacement leads to the formation of a principle slip surface that accommodates most of the strain.

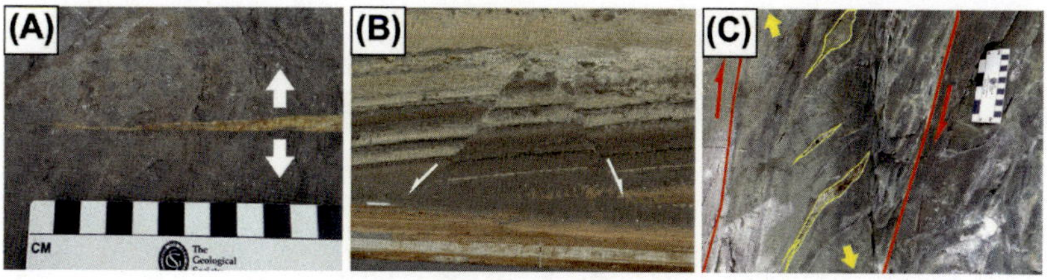

FIG. 2.13 Field examples that show the characteristics of fractures, faults and shear-zones. (A) Mode I fracture healed by calcite (scale is in centimetres), northern Alps. (B) Planar normal faults. In case of these low-offset faults, the faults are discrete surfaces, central eastern Spain. Limiting pole for scale. (C) Ductile shear-zone with tension gashes. Within the shear-zone, deformation is distributed over a wide zone; hosted in metamorphic rocks in northern Germany.

Fault zones are 3-D volumes that consist of different structural features. These zones consist of at least one high strain region that may be centrally located, accommodates most of the displacement, and is characterized by cataclastic material, gouge (Chester and Logan, 1986; Faulkner et al., 2011) or clay smear (e.g., Clausen et al., 2003). These high strain fault rocks are also described as 'fault core' and are the focus of the principal slip, which may be seismic or aseismic (Shipton et al., 2006; Fig. 2.12). Within the fault core, the original rock material and fabric disintegrates and is transformed into so-called 'fault rocks'. This covers a large range of materials that have been sheared, crushed and/or brecciated, which deforms and destroys the original rock fabric (Sibson, 1977).

There is an approximately linear relationship between the total slip along the fault and the thickness of the fault core (Robertson, 1983; Scholz, 1987). A classification scheme for such rocks, based on their textural characteristics, was developed by Higgins (1971) and Sibson (1977) and revised by Killick (2003) and Woodcock and Mort (2008). In the classification of Woodcock and Mort (2008), the definition of a fault breccia is based on the clast-size. Fault breccias contain more than 30% clasts with a size greater than 2 mm, whereas a fault gouge, in contrast, is incohesive and contains less than 30% clasts with a size greater 2 mm (Woodcock and Mort, 2008). Fault breccias and fault gouge form above 10−15 km depth (Sibson, 1986a,b). In the Woodcock and Mort (2008) classification scheme, cataclasites and mylonites are distinguished by the presence of a foliation, which is only formed in mylonites. Cataclasites and mylonites are further subdivided according to their matrix (Fig. 2.14). A protocataclasite and a protomylonite both have a matrix volume between 0% and 50% and a clast size smaller than 0.1 mm. Increasing matrix content defines (meso)catalasite/(meso)mylonite (50%−90% matrix) and ultracataclasite/ultramylonite (90%−100% matrix) (Woodcock and Mort, 2008).

A 'damage zone' flanks the fault core and is defined as the near-field domain that contains all fault-associated deformation (Vermilye and Scholz, 1998), often expressed in the form of fractures (e.g., Childs et al., 2009;

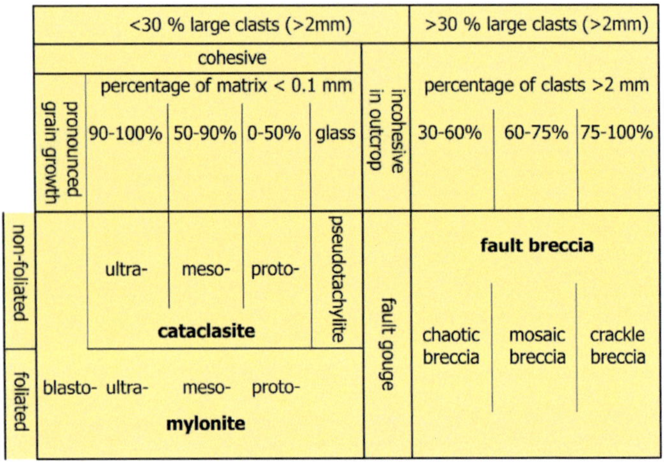

FIG. 2.14 Classification scheme of fault-rocks. *Modified after Woodcock and Mort (2008).*

Faulkner et al., 2003) or deformation bands (Fossen et al., 2007) (Figs 2.12 and 2.15A). A brittle fault zone forms by the progressive incorporation of pre-existing or newly-developed subsidiary shear or extension fractures. The damage zone represents a finite stage with all deformations accumulated during fault evolution and consequently has to be separated from the process zone that temporarily forms at the fault tip during fault propagation. During fault growth, the process zone becomes the damage zone (Fossen, 2010). The damage zone may contain altered material (Schulz and Evans, 1998) or be marked by an alteration halo, as shown in Faulkner et al. (2011).

Based on an analysis of strike-slip faults, Kim et al. (2004) subdivide the damage zone into the three domains; along-fault damage zone (wall-damage zone), tip-damage zone and linking-damage zone (Fig. 2.15B). The deformation pattern within a damage zone is often controlled by the fault tip modes (either mode II or mode III) (Kim and Sanderson, 2010). As illustrated in Fig. 2.16, fracture intensity also decreases with distance from the fault. The example in Fig. 2.16 shows that the fractures in the damage zone often possess the same strike as the fault surface, indicating the close connection of fault movement and fracture development.

Fault damage zones can have an important impact on the subsurface fluid-flow. Fracture density and apertures are controlling factors for the subsurface fluid-flow, whereas the aperture has the greatest impact and opening mode fractures that formed at dilational fault jogs in carbonate rocks are the most effective fluid conduits (Kim and Sanderson, 2010). Consequently, several studies have included damage zones into fluid flow models (e.g., Cacace et al., 2013; Romano et al., 2017). This is important as, for instance, the presence of deformation bands can reduce the effective permeability of a simulated reservoir by a factor of 15–25 (Rotevatn et al., 2013). The hydraulic properties

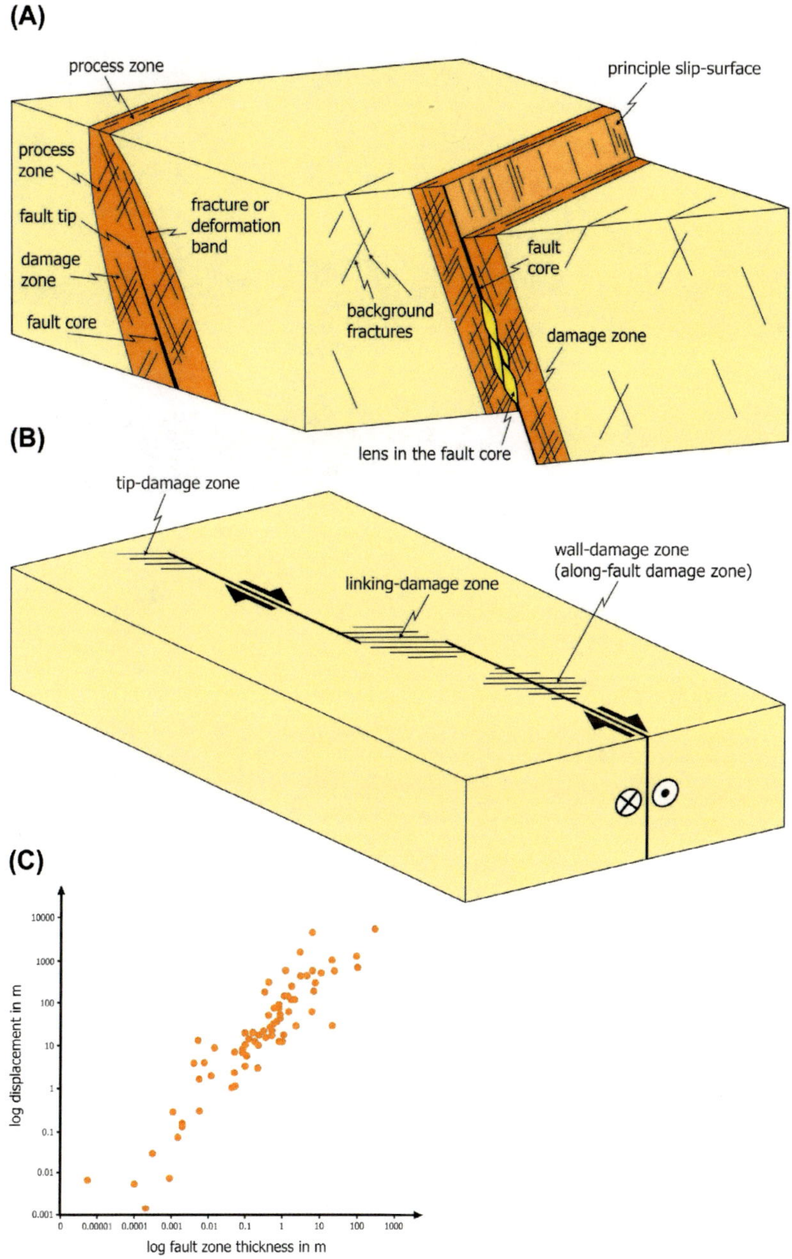

FIG. 2.15 Damage zones. (A) Damage zones developed along dip-slip faults. The fault core with the principle slip-surface is surrounded by a damage zone that is characterized by fractures or deformation bands. Ahead from the tip of the fault, the process zone develops. (B) Damage zone along a strike-slip fault. The damage zone that flanks the fault is named wall-damage zone. Between two overlapping fault segments, a linking-damage zone forms. At the tip of the fault, a tip-damage zone is developed. (C) Plot of fault displacement vs. fault zone thickness. Fault zone width increases with fault displacement. *(B) Modified after Kim et al. (2004). (C) Data from 73 faults, modified after Evans (1990).*

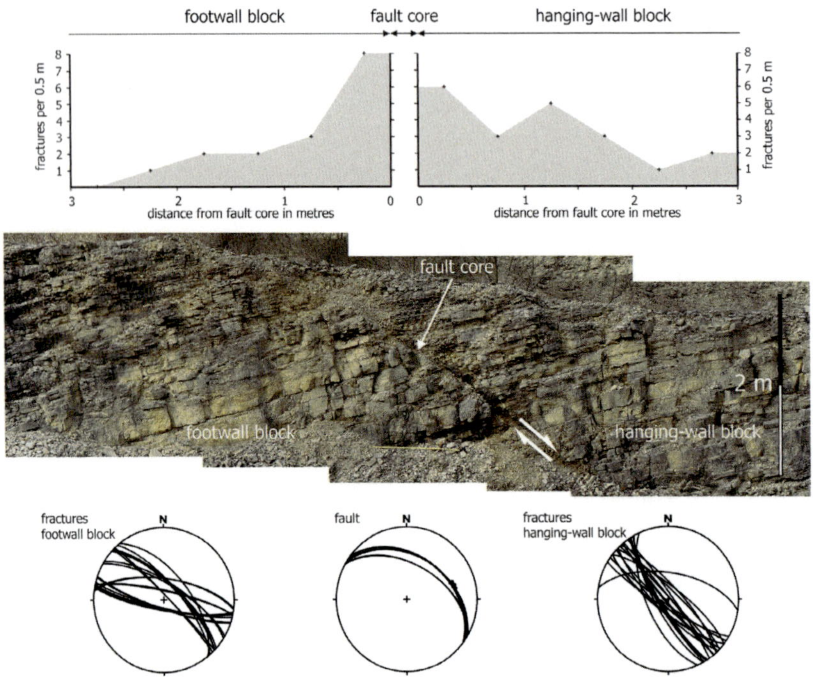

FIG. 2.16 Damage zones developed along a normal fault in Triassic limestone in northern Germany. Fracture density decreases away from the fault core. The orientation of fractures in the hanging wall and footwall block follows the trend of the fault core.

of all fault zones strongly depend on the lithology, the presence of deformation bands or fractures, the fault zone evolution, and depth. Therefore, structural geological characterization of fault zones is important to determine hydraulic properties. Such a structural characterisation is the basis for flow simulation models on all scales.

The width of a damage zone is commonly asymmetrical, with respect to the hanging-wall and footwall of a fault (Fig. 2.12; Choi et al., 2016). This asymmetry is attributed to the different stress conditions that developed in the hanging wall and footwall of the fault (Knott et al., 1996). Asymmetry of the damage zone has been also observed for strike-slip faults. Based on data from faults in southern California, Dor et al. (2006) suggest that the asymmetry is a long-term result from a preferred dynamic direction and propose that studies of damage zones can be utilized to derive the preferred propagation directions of seismic ruptures.

2.4.2 Principal slip surface

The fault surface or principal slip surface is the surface (or narrow high strain zone) on which most of the fault movement is concentrated. This surface is often characterized by slickensides that result from the growth of mineral

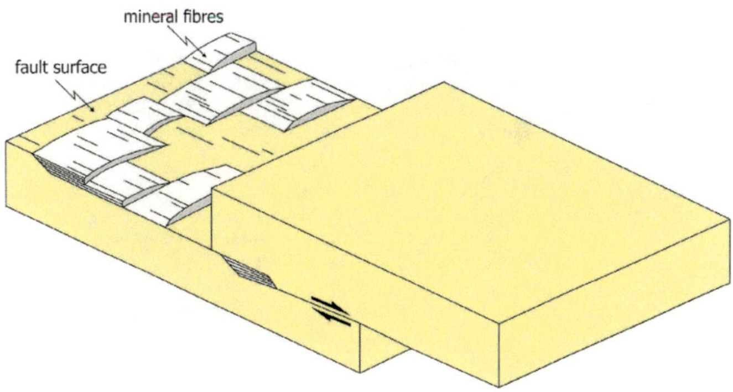

FIG. 2.17 Slickenside formation. During fault movements, mineral fibres grow that form a slickenside.

fibres (Fig. 2.17). Slickensides or slickenlines (slicken = glossy) are striated mineral growth or mechanical abrasions on the surface of faults and in fact, any surface that accommodates shear displacement. Thus, while not an essential component of a fault, they are included here because of their connection with faults. In the case of the mineral slickensides, the type of mineral depends on the host rock and/or fluids that flow along the fault. Typical minerals include quartz, calcite, serpentinite, pyrite, chlorite and tourmaline. Often slickensides stair-step (Fig. 2.18), so that the fractures on which the slickensides grow, step-down in one direction, making it possible to determine the sense of shear. Continued movement on the fault can often lead to the polishing of the mineralised surface. The beauty of this fabric is that the minerals grow parallel to the slip vector and therefore slickensides are parallel to the maximum shear stress and therefore the paleoslip-vector of the fault surface. Slickensides are therefore of great value, for instance, in paleostress analysis.

Based on field studies it has become evident that seismogenic slip along a fault is often manifested in zones that are only very few millimetres thick (Rice, 2006). As a result of seismogenic slip, thin layers of ultracataclasites can form in the fault core (Chester and Chester, 1998). The fault core can also be characterized by rock lenses or horses (Lindanger et al., 2007). Such lenses may be formed by a tip-line bifurcation or by the removal of an asperity (Childs et al., 1996). Between the lenses, high strain zones are located (Gabrielsen et al., 2016).

2.4.3 Pseudotachylites

Pseudotachylite is a glassy or very fine-grained material that forms on the principal slip surface. It often occurs not just along the slip surface, but also in veins and may contain inclusions of wall-rock fragments (Maddock, 1983; Trouw et al., 2010). Pseudotachylite is typically dark in colour and glassy in appearance. It received its name because it resembles the basaltic glass,

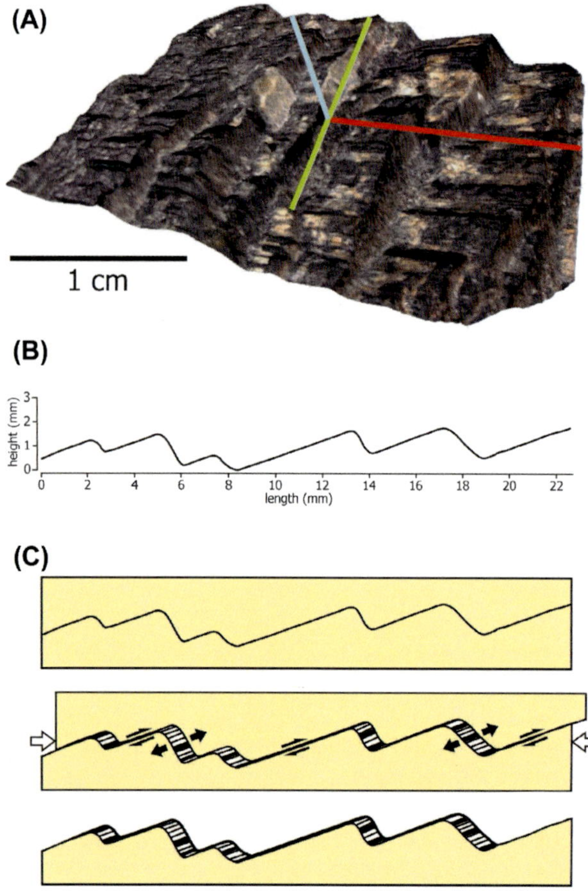

FIG. 2.18 Stair-step slickenside. (A) 3-D image of a slickenside. (B) Extracted stair-step topography. (C) Evolution of the slickenside, where jogs in the fault trace lead to the growth of mineral fibres.

tachylyte. The glass may also contain crystals with quench textures that formed via crystallization from a melt (Trouw et al., 2010). Pseudotachylites are also found in shock metamorphic rocks of impact structures. Because of this evidence, the common consensus is that they are caused by fast frictional sliding, such as rapid fault movement, associated with a seismic event (Sibson, 1975).

2.4.4 Strain hardening/strain softening of the fault core

Progressive deformation can result in strain hardening or strain softening of the deformed fault core material (Fig. 2.2). Changes in fault-core properties can have a direct impact on fault behaviour. In this context, strain hardening (the strength increase of the material with increasing displacement of the fault) has been often observed in experimental studies (Chapter 5 "Faulting in the laboratory"). In addition, the commonly recognised increase of fault zone thickness with increasing displacement has been attributed to strain hardening (Hull, 1988). This is also an important process during deformation-band formation, where a single band evolves into a zone of parallel bands. Such a behaviour may be an effect of strain hardening with a progressive shift of deformation into the parts of the host rock, which are still weak, consequently leading to an increase in the thickness of the deformation zone.

The development of the final slip surface might be an effect of strain softening within the crushed material (Aydin and Johnson, 1978). Hull (1988) described that the rate of brittle deformation zone widening can be a function of grain-size reduction. From a physical point of view, strain hardening is seen as a function of Hall-Petch hardening (Hull, 1988). Hall-Petch hardening (or grain boundary hardening) means the increase of the yield strength of a polycrystalline material due to grain-size reduction (Yu et al., 2018). The vast amount of observation of strain hardening processes in experiments and nature, leads to the question whether active faults will become stronger over time prior to failure (Morrow et al., 1982). However, fault behaviour is also found to be more complex. The work of Wojtal and Mitra (1986) shows that the deformation behaviour can change over time. During early stages of deformation, strain hardening can occur, whereas further increase of deformation may lead to strain softening. Water in the gouge material can lower the amount of strain hardening, but the gouge strength does not correlate with grain-size distribution or the mineral composition (Morrow et al., 1982). Other studies found evidence for a lithological control on the strain behaviour. Knott et al. (1996) showed that coarse sandstone produce wider fault zones and attributed this to strain hardening, while argillaceous material is prone to strain softening. We summarize; fault behaviour is complex, but fault evolution can often follow a path from initial strain hardening behaviour to later strain softening after reaching peak stress.

2.4.5 Fault surface geometry and roughness

Fault surfaces can have different geometries. Two end-member geometries exist, planar and listric faults. One possible explanation for listric fault surfaces was described in Section 2.3.5. Listric faults are curved in the dip direction, so that the dip angle decreases down-dip, but an along-strike curvature of fault surfaces is also common. This curvature can result in a low-amplitude oscillation of the fault surface. In addition, fault surfaces exposed in outcrops show surface roughness on different scales. These are reported to be formed by;

(1) drag of strong asperities across the fault surface,
(2) hierarchical linkage of fault segments during lateral fault expansion,
(3) damage formation near the crack tip, and
(4) both brittle and plastic deformation on the sliding interface during seismic slip (Renard and Candela, 2017).

The topography of fault surfaces has a strong impact on the frictional strength, fluid-flow, seismic behaviour of faults, and the formation of fault gouge and breccias, as summarized by Power and Tullis (1991). Along a fault surface, two rock volumes are in contact only at a small fraction of the entire surface (Goldsby and Tullis, 2011; see Section 2.5.1.2 "Barriers and asperities"). Consequently, there is the need to describe and classify the surface of fault surfaces. Over the years it has been often demonstrated that fractal geometry is a powerful tool to characterize surfaces irregularities in general (e.g., Majumdar and Tien, 1990; Persson, 2014). The key descriptor in fractal analysis is the fractal dimension, i.e., the D-value (Mandelbrot, 1967). Fractals are objects that do not have integer dimension, but rather have typical "broken" dimensions (e.g., 1.8, therefore they are named "fractals"; Mandelbrot, 1982). Fractal objects are scale-invariant, which means that their geometric characteristics do not change with scale. As fault surfaces have irregular surfaces in outcrops (Fig. 2.19), different fractal approaches have been applied over the last 30 years to analyse fault surfaces. Scholz and Aviles (1986) showed that the topography of fractures and faults can be represented by the Weierstraß-Mandelbrot fractal function (Fig. 2.20). This fractal is a graph that is continuous but not differentiable. The fractal approach is suitable to describe irregular topographies (Scholz and Aviles, 1986).

More recently, two different types of fractal models have been used to describe fault surfaces; self-similar and self-affine models. For a self-similar fault surface, a small area of the surface, if magnified isotropically, will be statistically identical to the entire fault surface. In the case of a self-affine fault surface, a magnified area of the fault surface is only statistically identical with the entire fault surface if different magnifications are utilized for the directions parallel and perpendicular to the fault surface (Power and Tullis, 1991). Detailed analyses has shown that fault surface roughness is anisotropic. Roughness amplitudes parallel to the slip direction are one magnitude lower than the amplitudes perpendicular to the slip direction, which means the fault surface is smoother in the slip direction (Power et al., 1987; Fig. 2.20).

FIG. 2.19 Field images of fault surfaces. Fault surfaces can be polished or characterized by step-like slickensides. (A) Polished fault surface of a strike-slip fault, Corona Heights, California (pylons for scale). (B) Fault surface with low amplitude wavy geometry, northern Alps. (C) Detail of a fault surface with slickenside mineral fibres, northern Alps (scale is in centimetres).

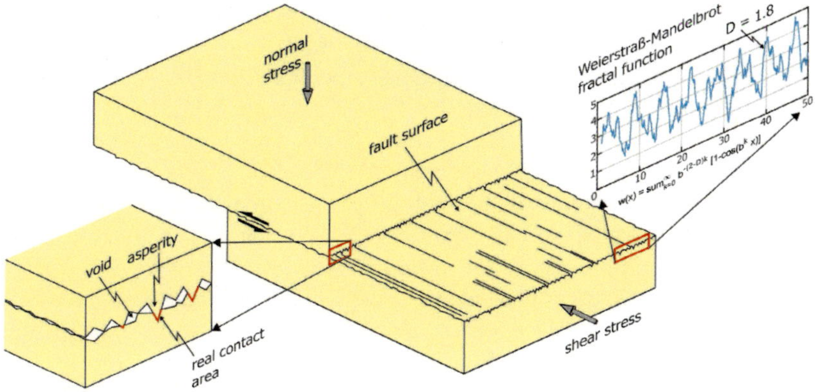

FIG. 2.20 Fault surface topography. Fault surfaces can be characterized by a roughness that is represented by the Weierstraß-Mandelbrot fractal. The fault surface is in general smoother parallel to the transport direction than perpendicular to the transport direction.

With the establishment of LiDAR techniques and photogrammetry, it has become possible to analyse fault surfaces in high resolution and three dimensions (Candela et al., 2009, 2011; 2012; Renard and Candela, 2017). Using these data, two mechanisms can be differentiated; mechanisms that create roughness during formation and lateral propagation, and mechanisms that modify the surface roughness during the slip along a pre-existing fault surface (Renard and Candela, 2017). Understanding the formation of fault surface roughness has important implications for seismology. Fault surface roughness influences the frictional properties of a fault and controls rupture processes (Renard and Candela, 2017). It can explain the great variety of slip velocities during earthquakes and the occurrence of earthquakes of different sizes and on the same fault (Fournier and Morgan, 2012). Experiments have shown that fault behaviour can change from stick-slip to stable sliding with a change in surface roughness and the growth of contact asperities (Voisin et al., 2007). The analysis of natural faults in the field suggests that fault-surface roughness changes over time and that the fault surface become smoother with increasing slip (Sagy et al., 2007; Brodsky et al., 2011).

2.4.6 The process zone

The process zone develops ahead of the tip line of a fault and contains the structures that form due to the propagation of the fault tip (Vermilye and Scholz, 1998; Fig. 2.12). At the tip of a fault, the displacement decreases to zero (e.g., Faulkner et al., 2011). Knowledge of the parameters that control the evolution and geometry of this region is important to understand fault nucleation and propagation (Reches and Lockner, 1994), for the extrapolation of faults in the subsurface and to better interpret seismic reflection data with its inherent resolution (Cowie and Shipton, 1998). Outcrop studies indicate that the width of the process zone is linearly proportional to the length of the fault and has a proportionality constant of $\sim 10^{-2}$. Microfractures in the process zone show a logarithmic decrease in density with distance from the principal fault slip surface (Vermilye and Scholz, 1998).

Further knowledge on the process zone has been gathered from experiments. Reches and Lockner (1994) showed, using triaxial experiments of granite samples, that an evolving fault propagates along its own plane, at an angle of 20–30 degrees to the maximum compression axis, within a process zone characterized by intense microcracking. Zang et al. (2000) performed a uniaxial compression test with granite samples. During these experiments, a shear fracture evolved with a process zone at the fracture tip. The experiments show that fast-propagating fractures have a wider process zone than slow-propagating fractures. The density of microcracks and acoustic emissions, measured during the experiment, increased towards the main fracture and the shear displacement shows linear scaling with the fracture length. The ratio of the process zone width to the length of the fault ranges from 0.01 to 0.1 (Zang et al., 2000). These results are consistent with observations made in outcrop studies. As the fault propagates through the rock volume, the fractures that initially formed in the processes zone will later be part of the fault-flanking damage zone.

2.4.7 Deformation bands

Deformation bands are structural elements that develop in porous rock material such as sandstones, tuffs and limestones and in unconsolidated granular material. They are restricted to the brittle upper crust (Fossen et al., 2007). They occur in a number of different tectonics settings (Ballas et al., 2015) and with various driving mechanisms, such as faulting and compaction (Du Bernard et al., 2002). Typical for deformation bands are their relatively small offsets and their tabular, sheet-like geometry with thicknesses that range from millimetres to centimetres (e.g., Aydin, 1978; Fossen et al., 2007) (Fig. 2.21). Internally, deformation bands can show a significant pore-space reduction compared to their host material, which results in their low hydraulic permeability (e.g., Hesthammer and Fossen, 2001; Shipton et al., 2005; Torabi and Fossen, 2009; Tindall, 2014). The pore-space loss has been suggested to be caused by compaction (Mollema and Antonellini, 1996) and/or cataclasis (Ballas et al., 2015).

Altogether four main types of deformation bands have been defined. Du Bernard et al. (2002) showed dilation bands that form due to pure extension that form as equivalents to mode I fractures, and compaction bands that result from load-induced compaction (Fig. 2.22). The compaction bands show a pore space reduction, whereas dilation bands are characterized by an increase in pore space. Shear-deformation bands show a distinct offset due to shear and are considered to be equivalent to mode II fractures. In shear-deformation

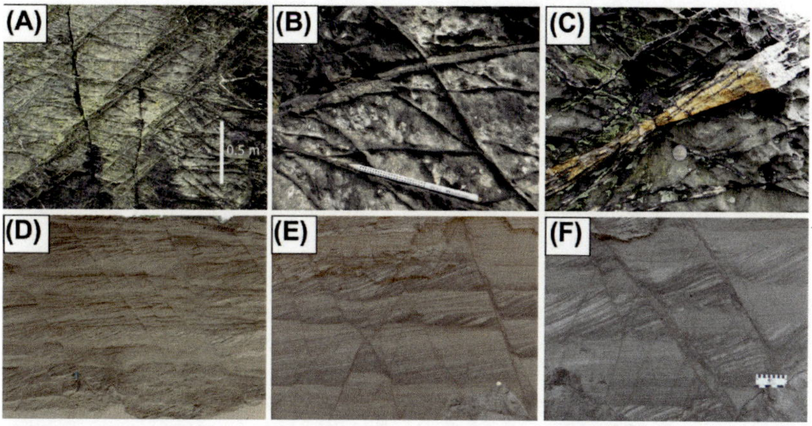

FIG. 2.21 Field images of shear deformation bands. (A and B) Shear-deformation bands developed in Cretaceous sandstones in northern Germany. Note the tabular geometry of the deformation bands. (C) Shear deformation band cluster developed in Cretaceous sandstones in northern Germany (2 Euro coin for scale). (D–F) Shear-deformation bands in unconsolidated Pleistocene sediments. The bands show the typical branching and merging and they enclose lenses of the host material.

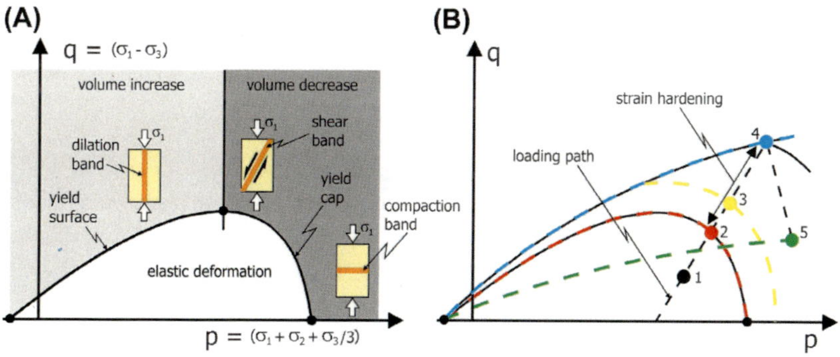

FIG. 2.22 Model for the evolution of deformation bands. (A) The q-p diagram. The yield surface separates the field of elastic deformation from the field of plastic deformation. On the positive slope side, volume increase occurs during deformation (e.g., dilational shear bands), whereas on the negative slope side volume decreases (e.g., compactional shear bands). (B) Evolution of deformation bands. During the formation of deformation bands, strain hardening can occur, which leads to the formation of deformation bands clusters that culminate in the formation of a fault. *Figures are compiled from and modified after Schultz and Siddhartan (2005), Fossen et al. (2007) and Torabi and Zarifi (2014).*

bands, the pore-space loss is likely a function of grain rolling and sliding during the shear process that leads to a reorganisation of the grain fabric and a related denser packing of the grains. A fourth type of deformation bands are so-called shear-enhanced compaction bands that accommodate both shear and compaction (Eichhubl et al., 2010). Deformation bands are analogues to low-displacement fractures or faults. In contrast to faults and fractures that appear as discrete surfaces in non-porous rocks, deformation bands have a pronounced tabular geometry with a distinct thickness. This different deformation behaviour is controlled by the lithology of the host rocks, with these bands most often forming in high-porosity sandstones.

Near-surface deformation bands often develop as so-called disaggregation bands. They form due to grain rolling and grain-boundary sliding processes. In greater depths, the grains within a deformation band were crushed. Such bands are called cataclastic bands. Due to grain orientation and cataclasis, the deformation bands often have a lower porosity than their host rock. Several parallel deformation bands are called clusters (Fossen et al., 2007). In an outcrop, deformation bands often form conjugate sets, with the acute angle enclosing the maximum principle stress. Individual bands can branch or merge down-dip as well as along strike (Brandes et al., 2018a). They can also enclose lenticular blocks of their host material (Fig. 2.21F). Continued deformation can lead to the formation of complex arrays of deformation bands, where en-échelon deformation bands can grow into an overlapping pattern by the linking of bands (Schultz and Balasko, 2003).

Because their low permeability affects fluid flow in hydrocarbon reservoirs, deformation bands have attracted much attention. Deformation bands show a close relationship to faults; they form in both the process zone of a fault (Ballas et al., 2015) and the damage zone of faults (Shipton and Cowie, 2001, 2003; Fossen et al., 2007). Consequently, deformation bands have been analysed with experimental studies (Mair et al., 2000) and are now incorporated into reservoir simulations (Qu and Tveranger, 2016; Zuluaga et al., 2016).

The formation of deformation bands has been a focus of research over the last two decades. In material science and soil science features like shear-deformation bands are named *shear bands*. Shear bands in solids attracted attention early on (e.g., Nadai, 1931). They are defined as thin layers between two sub-parallel material discontinuity surfaces (Vardoulakis and Sulem, 1995). Fundamental work on this topic was carried out by Thomas (1961) and Hill (1962). A mechanical model that is suitable to explain shear band evolution is now referred to as Thomas-Hill-Mandel shear-band model (e.g., Vardoulakis and Sulem, 1995; Guo and Stolle, 2013). In this model, shear band formation is treated as an equilibrium bifurcation from a homogeneous deformation (Vardoulakis and Sulem, 1995). Further studies have refined the knowledge on the mechanical evolution of shear bands and compaction bands (e.g., Rudnicki and Rice, 1975; Rice, 1976; Olsson, 1999) and were used to answer key questions, e.g., why the compaction localizes in form of a distinct band (Issen and Rudnicki, 2001). From mechanical models, it becomes evident that the key controlling factor for the formation of deformation bands are the material properties. The development of bands, with their characteristic tabular geometry, is determined by the porosity of their host material.

The standard Coulomb approach (which works for non-porous rocks) is not sufficient to describe the faulting of porous material (Schultz and Siddhartan, 2005). In porous material, deformation causes grain fabric reorganisation and grain-size reduction due to grain crushing. Because of this different mechanical behaviour, non-porous rocks develop discrete fractures, whereas porous material develops deformation bands (Schultz and Siddhartan, 2005). Shear band formation is described as an equilibrium bifurcation from a homogeneous deformation (Vardoulakis and Sulem, 1995). During the deformation process, the material properties bifurcate into the less deformed host material and the highly deformed band (Schultz and Siddhartan, 2005). The development of deformation bands can be visualized in a q-p diagram, a version of the Mohr diagram, where the horizontal axis shows p (the effective mean stress, $\sigma_1 + \sigma_2 + \sigma_3/3$) and the vertical axis q (the deviatoric stress, $\sigma_1 - \sigma_3$) (Wong et al., 1992) (Fig. 2.22A). A key element of the diagram is that the yield surface has a positive and a negative slope. The yield surface separates the field of elastic deformation from the field of plastic deformation. On the positive slope side, volume increase occurs during deformation (e.g., dilational shear bands), whereas on the negative slope side (called yield cap) volume decreases (e.g., compactional shear bands) (Schultz and Siddhartan, 2005; Klimczak et al., 2011) (Fig. 2.22A). The formation of the deformation bands is a consequence of yielding (either dilation and shear or compaction and shear). Consequently, either dilation bands, dilatant shear band, or shear bands can form. Along the negative part of the curve, compactional shear bands and compaction bands form (Schultz and Siddhartan, 2005; Fossen et al., 2007; Torabi and Zarifi, 2014). During the formation of deformation bands, strain hardening can occur, which leads to the formation of deformation bands clusters that culminate in the formation of a fault (Figs 2.22B and 2.23). The slip surface develops at the margin of the deformation bands zone (Aydin and Johnson, 1978), where the slip surfaces nucleate at the points that show high-intensity grain crushing (Shipton et al., 2005). However, there is a different view on how deformation bands develop into faults, as presented by Nicol et al. (2013). In their geometric model, jogs in the deformation bands will accumulate new deformation bands, which form a cluster that acts as an asperity. During the process of asperity removal a thoroughgoing slip surface is formed.

The more recent observation that deformation bands also develop in unconsolidated near-surface sediments along active faults during earthquakes (Cashman et al., 2007), allows the use of deformation bands to indicate paleo-earthquakes and can help to identify recently-active faults. This makes the analysis of deformation bands highly relevant to the field of seismology, paleoseismology, neotectonics, and Quaternary geology. Studies by Brandes and Tanner (2012) and Brandes et al. (2018a,b) show shear-deformation bands form in unconsolidated sediments in the process zone of regional faults. The strike of the deformation bands follows the trend of the faults in the subsurface, indicating their close connection.

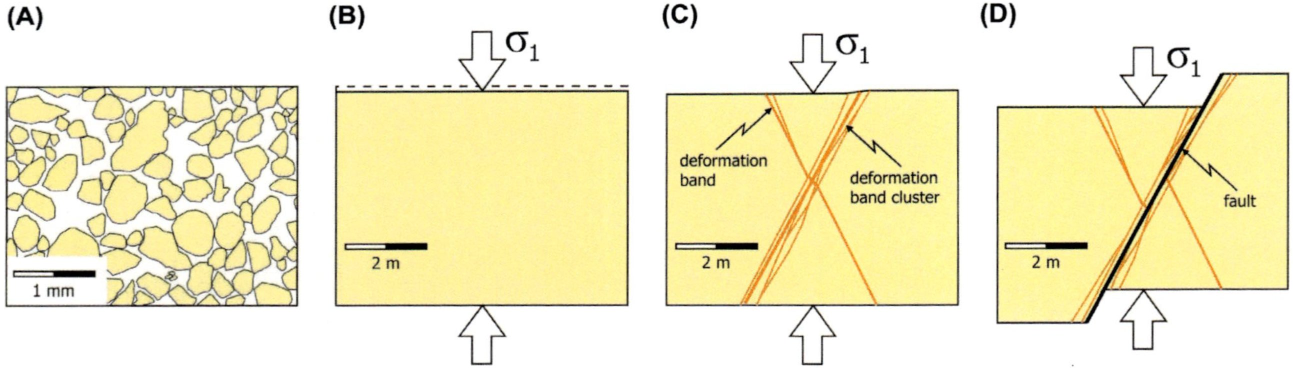

FIG. 2.23 Evolution of deformation bands and the subsequent development of a fault with a distinct slip surface from a deformation band cluster. (A) Undeformed granular material. (B) Material deforms elastically. (C) Strain localizes in the form of shear-deformation bands. (D) Strain hardening occurs and leads to the formation of a single fault surface, along which strain is manifested. *Modified after Aydin and Johnston (1978).*

2.4.8 Fault groups and their characterization

The previous part of this chapter describes single faults and their evolution (except for Section 2.3.4 "Stress state and dynamic fault classification of Anderson", which deals with conjugate faults). However, faults do tend to occur in groups, which are described with different terms. Schultz et al. (2010) use the terms *fault set*, *fault array*, *fault zone*, *fault system* and *fault population* to classify fault groups. A *fault set* has a common element such as the age, the length or the orientation. *Fault arrays* are composed of faults that are genetically linked. Schultz et al. (2010) use the term *fault zone* for a narrow fault array, where the faults have a similar strike. As shown above, the term fault zones is now often use to describe the sum of all structural-mechanical units of a fault (Chester and Logan, 1986) (see Section 2.4.1 "The fault zone"). A *fault system* is an extensive array, in which the faults show mechanical interaction and the term *fault population* describes a group of faults that have a wide range of lengths, spacing and displacements that result from a progressive deformation of an area (Schultz et al., 2010). Peacock et al. (2016) use the adjectives *geometric*, *topological, kinematic* and *mechanical* to describe faults. *Geometric* is used to describe the shape and pattern of faults, *topological* is used to describe the arrangement and relationships between faults (fault groups), *kinematic* is used for the description of the displacement and *mechanical* to describe the processes that formed a fault (Peacock et al., 2016). In the nomenclature of Peacock et al. (2016), the term *array* means a set of stepping or en-échelon faults and the term *network* describes a system of linked and interacting faults (Fig. 2.24). van der Pluijm and Marshak (2004) use the term *array* or *system* to describe a group of related faults and subdivide arrays into parallel arrays, anastomosing arrays, en-échelon arrays, relay arrays, conjugate arrays and random arrays. In fault systems, *isolated* and *segmented faults* occur. *Isolated faults* have no significant mechanical interactions with the neighbouring faults, whereas *segmented faults* consist of overlapping segments that show en-échelon arrangements (Schultz et al., 2010). Detailed information on fault growth and the arrangement of fault segments, related displacement patterns and fault interaction is given in Chapter 6, "The growth of faults".

2.4.8.1 Fault arrangement and fractal geometry

As faults can be arranged in complex systems and populations, there is the need for a quantitative description. Above that, fault populations contain a wide range of large-scale and small-scale faults, which can have different effects on the fluid flow, where small-scale faults can act as potential fluid conduits and large-scale faults as potential seals. This, together with the difficulty to visualize small-scale faults with geophysical methods, drives the demand for a predictive classification tool (Needham et al., 1996). The self-similarity of faults suggests that a fractal approach is a feasible tool to

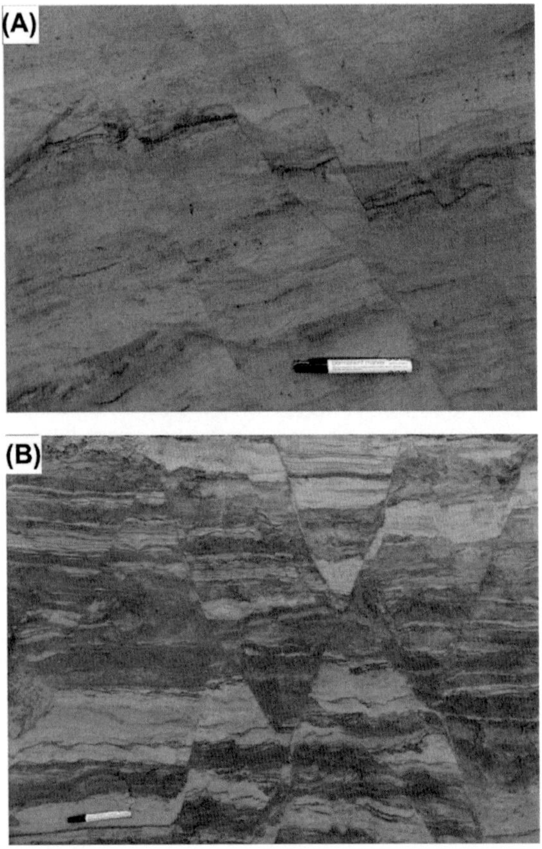

FIG. 2.24 Fault and deformation band groups. (A) Fault/deformation band *array* means a set of stepping or en-échelon faults/deformation bands (scale is in centimetres). (B) The term fault/deformation band *network* describes a system of linked and interacting faults/deformation bands, following the nomenclature of Peacock et al. (2016) (pen for scale).

predict the presence of faults (Walsh et al., 1991; Yielding et al., 1992; Needham et al., 1996). Fractal geometry has not only been used to characterize fault surfaces (see Section 2.4.5 "Fault surface geometry and roughness"), but it has also been applied to analyse the arrangement of faults in fault systems. Fig. 2.1B shows shear-deformation bands that branch and these branches split into smaller branches forming a self-similar "dendroid" pattern. The idea behind this approach is also supported by the fact that earthquakes are a fractal process (Kagan and Knopoff, 1980), and that there is a dependency between fault geometry and the mode of strain release (Okubo and Aki, 1987). Aviles et al. (1987) showed that fractal dimensions with D-values in the range of 1.0008–1.0191 characterize the trace of the San Andreas Fault system and that it is not possible to distinguish between the fault segments that creep, exhibit seismic slip and micro-seismic activity, on the basis of their fractal dimension. In contrast, another study from the San Andreas Fault showed larger fractal dimensions in a range of 1.1−1.4, and from this study, it became evident that a more complicated fault geometry is associated with larger D-values (Okubo and Aki, 1987). The larger D-values, with respect to the study of Aviles et al. (1987), are seen as the result of the inclusion of fault branches and different minor fault systems (Aviles et al., 1987). The results of Okubo and Aki (1987) are supported by comparable observations that were made for a fault system in Egypt where smooth unidirectional fault traces have low D-values, while irregular and heterogeneous fault patterns are represented by higher D-values. This is related to seismic behaviour, where a smooth fault trace with low D-values is interpreted to sustain less seismic energy before it slips than a fault with higher D-values (Arab et al., 1994). In addition, the correlation of larger D-values with more irregular fault geometry is reported from the Sumatra fault system (Sukmono et al., 1996). A more recent study from Taiwan on the 1999 Chi-Chi earthquake was also able to show that the D-values of the surface ruptures are related to the coseismic displacement, the fault slip, and the geometric complexity of the fault (Chang et al., 2007). The observation that faults are self-similar over a wide scale range, allows us to describe a fault population with a power-law frequency distribution:

$$N = aS^{-D}$$

Where N is the number of features with a size equal or greater to S and S is the fault length or displacement, a is a constant that depends on the sample size. D is the fractal dimension. Defining the population distribution based on, e.g., seismic-scale observations, allows extrapolation to the sub-seismic scale, and to predict the number of small faults that are below the resolution of the geophysical method (Watterson, 1986; Yielding et al., 1992, 1996; Walsh et al., 1994; Needham et al., 1996). Fractal geometry approaches have been also been applied to fracture networks and deliver a better description than other methods (Bonnet et al., 2001). The fracture density is often fractal and scale-invariant, which make the fractal dimension a powerful tool, as it can be calculated without knowing the size of the fractures (La Pointe, 1988). Park et al. (2010) showed that the fractal dimension of fracture networks increases with each fracturing event, but then tends to stabilize. The increase in fractal dimension is the consequence of formation of new fractures and the subsequent stabilization of the fractal dimension is interpreted as a switch to increased fracture reactivation instead of the formation of new ones. The formation of new fractures in mature fracture networks is localized at the fracture tips (Park et al., 2010).

2.4.9 Fault evolution with depth

At the beginning of this chapter, the question was asked, "How do fault zones behave from the Earth's surface to the base of the lithosphere?" Fault behaviour is directly controlled by the variation in the mechanical processes, the structural style of the fault zone and the fault rock lithology. Faults show a structural variability along-strike and along-dip. At first, a fault is often a complex feature that consists of a series of aligned to partly overlapping fault segments. Depending on the level of erosion, very different fabrics and different fault-related rocks can be exposed in outcrop. At the Earth's surface, an active fault may form a fault-scarp that represents a topographic step in the landscape. A fault-scarp directly indicates local coseismic slip on the fault. Close to the Earth's surface, a fault will often penetrate young, unconsolidated, soft-sediments (see Section 2.6 in this chapter "Faults in soft-sediments"). At greater depth, in more consolidated material, a fault will be characterized by the occurrence of fault breccias and gouges (Fig. 2.25). Often slickensides are developed in rocks that result from slip processes. Within the upper 10 km of the lithosphere, brittle deformation dominates and frictional processes occur along faults. This deformation behaviour is controlled by the material properties of quartz. Below 10–12 km, the temperature will be approximately 300 °C and quartz starts to behave plastically. Based on natural faults and on experiments it can be derived that the brittle-plastic transition at strain rates of natural faults is around 350 °C (Brace and Kohlstedt, 1980; Hirth et al., 2001). Depending on the temperature, a several kilometre-thick brittle-plastic transition zone can developed. Below approximately 20 km and at temperatures of approximately 480 °C, the plastic regime is fully established and faults pass into lower crustal shear zones, where dislocation creep is the dominating deformation mechanism. Under these conditions, mylonites develop.

This view on faults and shear zones has been extended in the last decades and it has been established that both brittle and ductile features can be developed along a fault/shear zones system throughout the whole lithosphere (Huntington and Klepeis with 66 community contributors, 2018). In this context, Frost et al. (2011) derived that ductile deformation along a fault can occur during slow interseismic slip, whereas brittle deformation takes place during earthquakes. This indicates a possible strain-rate dependent deformation behaviour, where brittle failure can also occur in the ductile regime within the lower part of the brittle-ductile transition. Recent seismological studies have shown deep earthquakes in the continental lithosphere (e.g., Inbal et al., 2015; Brandes et al., 2019), which demonstrate the possibility of rupture processes in the lower crust and around the Moho.

2.4.10 Fault-related folding

We previously described brittle deformation in the near-field of faults (i.e., the damage zone). Nevertheless, there is also a far-field effect of fault activity that

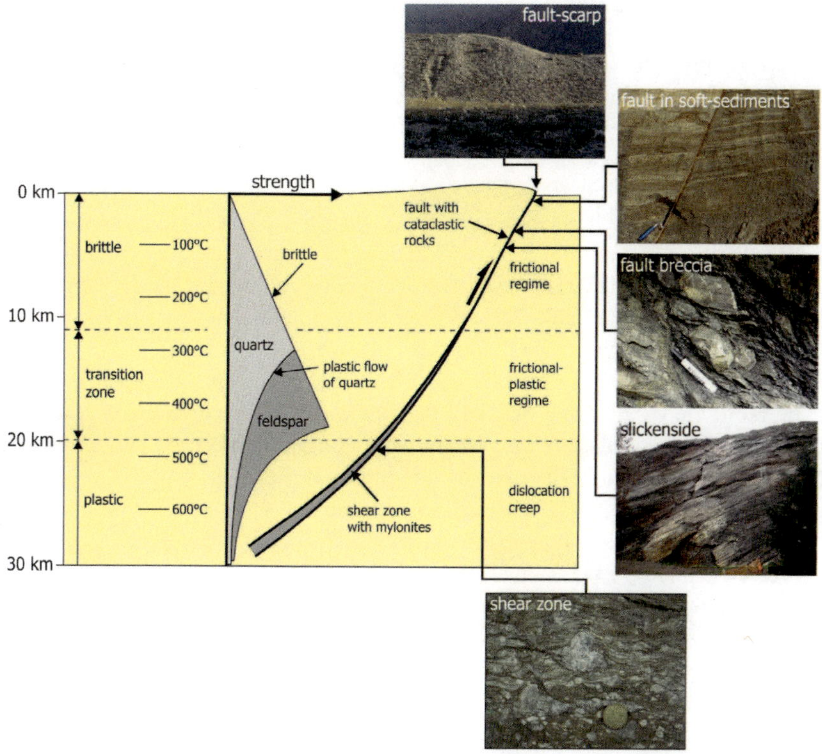

FIG. 2.25 Fault evolution with depth. Brittle failure that leads to fault formation occurs in the upper 10–12 km of the lithosphere in the frictional regime. Different features are developed along a fault depending on the depth. Fault-scarps can indicate active faults that reach and offset the Earth's surface. With increasing depth, the deformation varies from grain rolling/grain sliding processes in soft sediments to the formation of fault breccias and the development of slickensides. In the temperature range 300–400 °C the material becomes plastic and mylonites form. *Figure is based on Lin (2008), Fossen and Cavalcante (2017) and Huntington and Klepeis with 66 community contributors (2018).*

leads to the formation of folds that develop contemporaneous to the fault movement. This is named fault-related folding and has been in the focus of research for over a century (Buxtorf, 1916; Rich, 1934). Fault-related folding represents one of the most important deformation styles in the upper part of the lithosphere. However, its significance was recognized later on with the development of kinematic models for fault-related folds in the 1980s (e.g., Suppe, 1983; Suppe and Medwedeff, 1990; Erslev, 1991; Epard and Groshong, 1995), with a boost in the 1990s–2000s due to software/hardware implementation (Hardy and Ford, 1997; Allmendinger, 1998; Allmendinger et al., 2004; Cardozo and Aanonsen, 2009; Cardozo et al., 2011). A thorough treatment of fault-related folding is beyond the scope of this chapter. A summary and review of the kinematic models of fault-related folding can be found in Brandes and Tanner (2014).

2.5 Fault movement and seismicity

After a fault is fully established, it can further grow by the propagation of the tip line and increasing displacement along the fault surface. As summarized by Bürgmann et al. (1994), fault slip distribution is controlled by the geometry of the faults, the behaviour of the host rock as well as by remote boundary conditions and by the boundary conditions along the fault. In an elastic material, slip along a fault shows an elliptic distribution. In three dimensions, a fault is therefore an elliptic feature (Walsh and Watterson, 1987; see Chapter 6 "The growth of faults"). However, this ideal case can be modified by changes in the strength of the fault, the occurrence of spatial gradients in the stress field, inelastic deformation around the fault tip and variations of the elastic modulus of the host rock (Bürgmann et al., 1994). The presence of stiff material in the vicinity of a fault can flatten the slip profile in the area in which it is offset by the fault and the slip gradient along a fault can be influenced by the interaction of closely-related faults (Bürgmann et al., 1994).

Active faults can have contrasting behaviours. They rupture (slick-slip), show slow-slip, or they creep (Fig. 2.26). Analysing and differentiating these processes is an important step towards a better understanding of the kinematics of faulting and the related hazard potential. Houston (2001) stated "the way a rupture initiates, propagates and terminates is central for seismic faulting". In the introduction to this chapter, some of the grand challenges in structural geology and seismology are listed, in which half of them are related to understanding fault slip and rupture propagation. In this context, the question "what controls the transition from seismic rupturing to aseismic creep of a fault?" is of high relevance, as the same fault can show both seismic rupturing and aseismic creep.

FIG. 2.26 Fault rupture and fault creep. (A) Offset fence at the San Andreas Fault in the Olema Valley. The fence was offset by 5.5 m during the 1906 earthquake. (B) Offset along the Hayward Fault, California. The wall was built in the 1930s (Blueford and Craig, 2006). Lateral offset of the wall of 0.4 m in about 80 years.

2.5.1 Fault rupture

Rupture is the sudden failure process along a fault, which emits seismic waves (commonly referred to as an earthquake) and consequently has been in the research focus for decades. Stick-slip behaviour of faults has been recognized as a key factor for seismicity (Brace and Byerlee, 1966). Seismogenic fault rupture takes place in the so-called seismogenic zone and this zone is correlated with frictional behaviour, where shear motion is accommodated by discontinuous deformation that involves cataclasis and frictional sliding (Sibson, 1984). The lower limit of the seismogenic zone in continental crust is defined as the point at which a fault zone passes from the brittle into a quasi-plastic regime at temperatures of about 350 °C (see Section 2.4.9 Fault evolution with depth). Here, aseismic continuous shearing with mylonite formation dominates (Sibson, 1984). During an earthquake, potential energy is released and distributed both as the emitted seismic waves and as fault-related mechanical processes, such as fracture formation and the generation of frictional heat (Ammon et al., 2005). The area on a fault in which the earthquake rupture takes place is called the fault-rupture area (Harris, 2017). Earthquakes are related to mode II and mode III ruptures, where mode II ruptures predominantly occur at strike-slip faults and mode III ruptures are characteristic for large earthquakes at subduction zones (Bouchon et al., 2010). Seismic slip rates reach velocities of $0.1-2\ m\ s^{-1}$ and show re-occurrence intervals of 10^2-10^4 years (Sibson, 1984). In some cases the rupture speed can be higher than the shear-wave speed (~3 km/s) or even reach the speed of compressional waves (~6 km/s) (Das, 2015). This is called a "supershear" earthquake. Supershear earthquakes are often associated with mode II ruptures (Bouchon et al., 2010). A wide range of different trigger mechanisms have been discussed for earthquakes, such as fluid pressure variations (e.g., Costain et al., 1987), stress changes due to erosion processes at the earth surface (e.g., Van Arsdale et al., 2007; Calais et al., 2010), stress changes due to glacial isostatic adjustment (GIA; e.g., Wu and Hasegawa, 1996a,b) and even meteorological events (e.g., Hainzl et al., 2006) or tidal effects (Vergos et al., 2015).

The sliding motion on the fault initiates with a frictional instability, and the slip along the fault starts when the ratio of the shear stress to the normal stress reaches the static friction coefficient (Scholz, 1998) and the rupture front spreads across the fault. This takes place at a velocity that is usually slower than the ambient shear-wave speed (Ammon et al., 2005), but can be also higher (see supershear earthquakes above). Generally, the entire fault rupture process has a very short duration that takes seconds to a few minutes. Convers and Newman (2013) pointed out that the duration of the dynamic fault rupture is a crucial parameter for describing earthquake source processes, such as the rupture length, velocity and directivity. The rupture duration allows to rapidly distinguish between earthquakes with normal or slow ruptures (Convers and Newman, 2013). Rate and amount of slip are unevenly distributed along a fault. The three main rupture propagation directions are along-strike, up-dip and down-dip. In a study based on 96 earthquakes with magnitudes greater than 5.6–6, Chounet et al. (2018) suggest that along-strike rupture propagation is more frequent than along-dip propagation. This direction preference is likely due to the limited width of the seismogenic zone. The propagation direction is often controlled by material contrasts. For shallow subduction zone earthquakes, up-dip propagation is more frequent than down-dip propagation, because the rocks of the hanging-wall plate are more compliant than the oceanic crust in the footwall. At greater depth, where the slab reaches the mantle wedge, down-dip propagation is favoured (Chounet et al., 2018).

The physics of fault rupture is controlled by many different parameters. An important effect is the lithology of the fault blocks involved and the properties of the principle slip surface. Reches and Lockner (2010) point out that earthquake rupture requires a strength loss with the change from a static pre-slip friction to a dynamic coseismic friction. As summarized by Bullock et al. (2014), frictional behaviour is controlled by the normal stress, pore fluid pressure, temperature, pre-existing structures in the slip zone, the composition and distribution of the mineral phases in the slip zone and the rupture velocity. Friction can be time-, velocity- or displacement-dependent (Dieterich and Conrad, 1984). Especially the velocity-dependence of friction, which means that friction can vary in response to the speed of slip, has been in the research focus (e.g., Blanpied et al., 1991; Scruggs and Tullis, 1998; Perfettini and Ampuero, 2008). Velocity-strengthening is the increase of friction with increasing slip velocity, whereas velocity-weakening is the decrease of friction with increasing velocity. Faults where velocity-strengthening occurs show creep. Velocity-weakening behaviour leads to faster fault slip and is seen as a reason for stick-slip behaviour. At low velocities (up to a few hundreds of microns per second) many materials are characterized by velocity-weakening behaviour, but for higher velocities, velocity-strengthening occurs (Bar-Sinai et al., 2015).

2.5.1.1 The seismic cycle

From a temporal point of view, seismicity can be regarded as a cycle of stress build-up, sudden failure (accompanied with the emittance of seismic waves and a local stress drop), followed by a new phase of stress build-up (Fig. 2.27). This was first presented by Reid (1910). Over the years, this concept has been refined. The stress in the seismic cycle comes from the constantly moving tectonic plates. Around a fault, a locked zone exists, where the rocks do not move because of friction and stress can build up. The area near the fault accommodates elastic strain, because the material far away from the fault still moves with plate tectonic velocity. If the frictional strength of the fault is

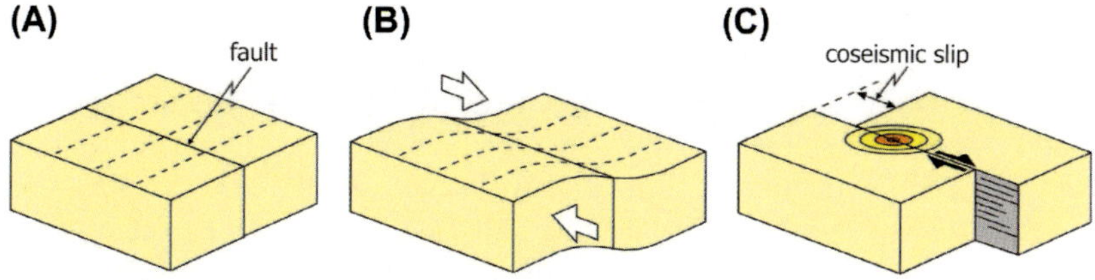

FIG. 2.27 Simple seismic cycle. (A) Fault without force. (B) Plate tectonic forces deform the rocks, due to a locked zone along the fault the volume around the fault cannot move and stress builds up. (C) Friction along the fault is overcome and the stress is released, with the emittance of seismic waves. *Based on Reid (1910).*

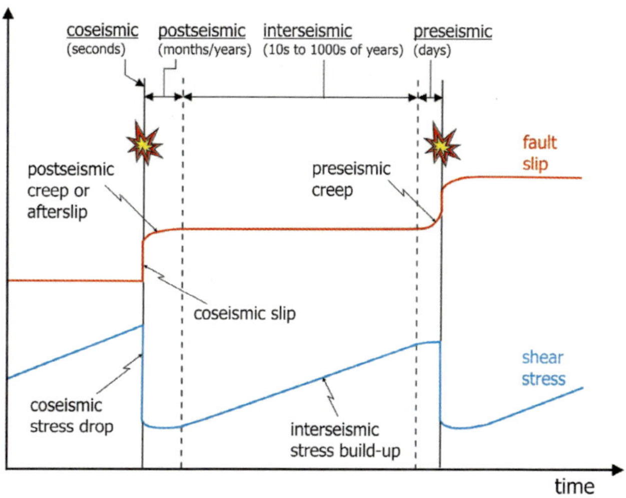

FIG. 2.28 The seismic cycle. The diagram shows the different stages of the seismic cycle, together with the stress build-up (*blue curve*) and stress release and the related fault slip over time (*red curve*). The time is shown on the horizontal axis. Over time, the shear stress increases until the fault strength is reached, which leads to fault movement and associated stress drop. *Figure is compiled and modified after Sibson (1986b, 1989) and Di Toro et al. (2012).*

exceeded, the fault moves and recovers the slip deficit that accumulated during the time the fault was locked. Following this, the stress drops and the elastic strain is reduced and the sudden fault movement emits seismic waves. After the earthquake, the fault locks again and new stress accumulates. This seismic cycle is now subdivided into a preseismic, a coseismic, a postseismic and an interseismic phase (Fig. 2.28). Sibson (1986b, 1989) used the terminology α-, β-, γ- and δ-phase for these periods. In the preseismic phase, a so-called preseismic creep may occur, which has been measured on some faults (e.g., Cohn et al., 1982). The preseismic phase has a duration of hours to days and besides the aseismic creep, it can be characterised by foreshocks (Jones, 1985; Reasenberg, 1999; Daskalaki et al., 2016), and/or show the emission of very high frequency (VHF) and very low frequency (VLF) electromagnetic signals (e.g., Kapiris et al., 2002). The coseismic phase, which takes seconds to minutes, is the main seismic event that produces fault rupture and the emittance of seismic waves. Subsequently the postseismic phase follows (Sibson, 1989) with deformation mostly located around the rupture zone (Helmstetter and Shaw, 2009). During this phase, aftershocks occur (e.g., Perfettini et al., 2005; Perfettini and Avouac, 2007) as well as postseismic creep (e.g., Cohn et al., 1982), that can continue for years (Hussain et al., 2016). This fault movement is also called afterslip and often contains creep events (Marone et al., 1991). Helmstetter and Shaw (2009) stated that the link between aseismic afterslip and aftershock activity is not clear. The last element of the seismic cycle is the subsequent, longer interseismic phase, in which new stress builds up, until the failure strength of the fault is reached again. During this phase, processes of interseismic healing of the rupture surface may occur, which are caused by pressure solution, recrystallization or hydrothermal cementation and lead to strengthening of the fault (Sibson, 1986).

Damaging earthquakes often occur in unexpected locations, where only small earthquakes were previously recorded or where historical reports are lacking (Chen and Bürgmann, 2017). Knowledge about the seismic cycle is therefore important and basic earthquake recurrence models have been proposed. Earthquake activity can be periodic, time-predictable or slip-predictable (e.g., Shimazaki and Nakata, 1980) (Fig. 2.29). Periodic activity means that stress drop and the coseismic slip are proportional and constant. In such a case, both the timing and the slip can be predicted (Fig. 2.29).

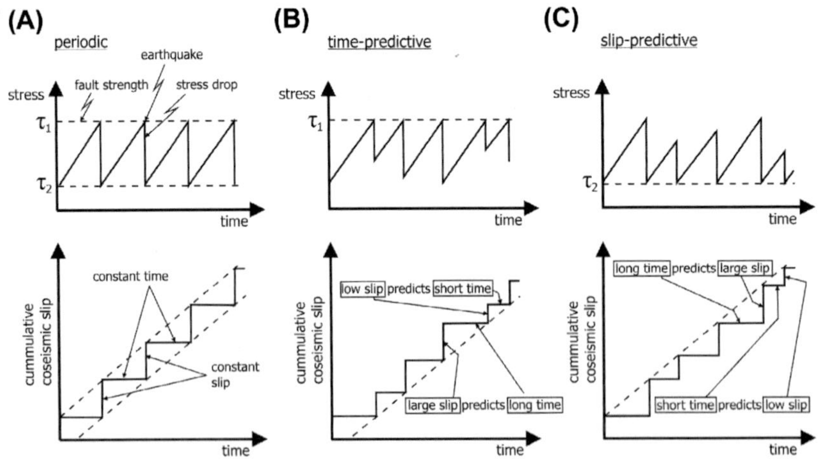

FIG. 2.29 Earthquake recurrence models. (A) Periodic behaviour where stress drop and coseismic slip are constantly proportional. (B) Time-predictive behaviour with a constant fault strength that allows to predict the timing of fault movements. (C) Slip-predictive model that allows to predict the slip along the fault based on the time between two earthquakes. *Figure is modified after Shimazaki and Nakata (1980).*

The Parkfield segment of the San Andreas Fault shows this pattern of activity (Bakun and Lindh, 1985). In the time-predictable case, slip occurs when fault strength is reached (τ_1 in Fig. 2.29). Due to a constant fault strength, an upper threshold for the stress of faulting is given and this regularity allows predicting the timing of fault movement and therefore this is called the time-predictable model (Shimazaki and Nakata, 1980). Low slip (small earthquake) predicts a short time interval to the next earthquake. Large slip (big earthquake) predicts a long time interval to the next earthquake. This behaviour was observed on the Calaveras Fault in California (Bufe et al., 1977). The slip-predicable model in contrast, means that the longer the stress build-up, the greater the future slip along the fault. In this model, a long time interval after an earthquake predicts a future large slip or big earthquake in the future, whereas a short time interval after an earthquake predicts a future low slip or small earthquake in the future. However, these are models, and have therefore limitations. In addition, they are endmembers and do not necessarily predict the behaviour of a particular fault. Discriminating between these models and improving our understanding of the seismic cycle requires information on the timing and slip for past earthquakes. Over the years, different approaches have been developed to determine fault timing and these will be discussed in Chapter 7, "Direct dating of fault movement".

Seismicity is mainly restricted to the seismogenic zone, where brittle processes dominate. The extent of the lower limit of the seismogenic zone is strongly controlled by the lithosphere temperature. For continental faults, the seismogenic zone is roughly located between 5 and 12 km depth (Fig. 2.30A). Above 5 km, also stick-slip behaviour dominates, but aseismic creep can occur (see Section 2.5.2, Fault creep). Below 12 km the temperature is high enough for quartz to deform in a ductile fashion.

The interface between a subducting plate and the overriding plate can be considered a fault (megathrust). It can be subdivided in different domains (Fig. 2.30B). In very shallow depths, aseismic behaviour or slow slip ruptures occur (domain A) (Lay et al., 2012). Further down-dip domain B is located, where stable sliding with relatively large slip occurs. Domain C is more heterogeneous, with smaller areas that show stable sliding. The deepest domain D is large characterized by aseismic sliding, with some small unstable slip patches (Lay et al., 2012).

The magnitude of an earthquake is proportional to the rupture area (Hanks and Kanamori, 1979; Wells and Coppersmith, 1994; Wesnousky, 2008). It is important to note that during an earthquake it is possible that only parts of the fault will move (the rupture surface), which means that the rupture surface is in general smaller than the entire fault. The rupture area is given by the length of the rupture surface (L) times the down-dip extent of the rupture surface (W) (Sibson, 2011; Fig. 2.31). During earthquakes, faults grow. The maximum amount a fault can typically grow during one earthquake is about 1% of its previous length (Cowie and Scholz, 1992b).

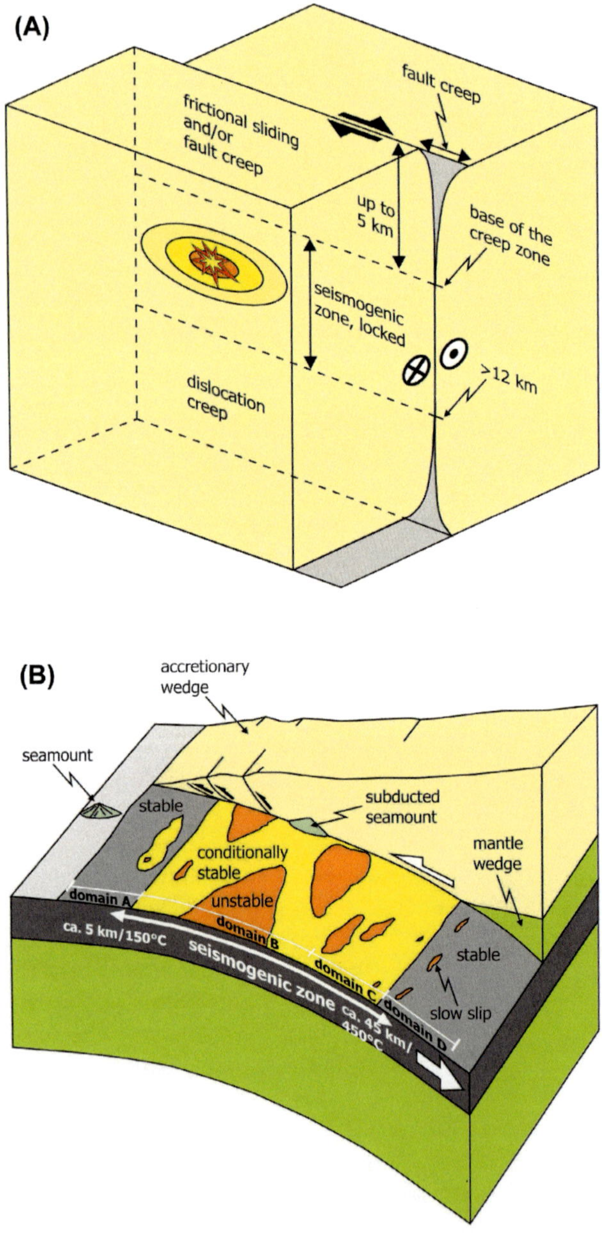

FIG. 2.30 (A) Seismogenic zone along a continental strike-slip fault. The seismogenic zone is at a depth between 5 and 12 km. Below 12 km dislocation creep occurs. Above 5 km frictional sliding with earthquakes or fault creep can occur. (B) Seismogenic zone along a subduction zone. *(A) Figure is compiled and modified after Hussain et al. (2016) and Jiang and Lapusta (2017). (B) Based on Lay et al., 2012.*

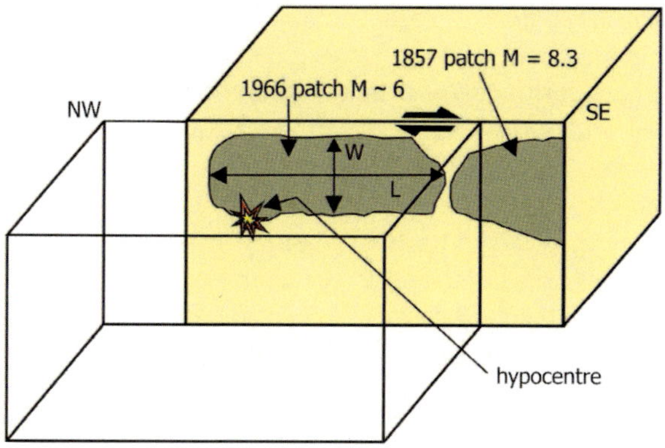

FIG. 2.31 Block model of the San Andreas Fault near Parkfield. The olive areas are the rupture surfaces in 1857 and 1966. *Modified after Stuart and Tullis (1995). The nomenclature is based on Sibson (2011).*

2.5.1.2 Barriers and asperities

Fault strength is not constant along strike. Patches referred to as barriers and asperities occur on faults. They are an important controlling factor for the rupture process. Barriers represent unbroken patches on a fault surface after an earthquake (Das and Aki, 1977; Aki, 1984). Initially the entire fault surface is stressed, and this stress is released during an earthquake. In case the stress release is not evenly distributed along the fault surface, patches occur where the stress has been released that are separated by areas that are still stressed and unbroken (the barriers; Fig. 2.32). Asperities are defined as areas between

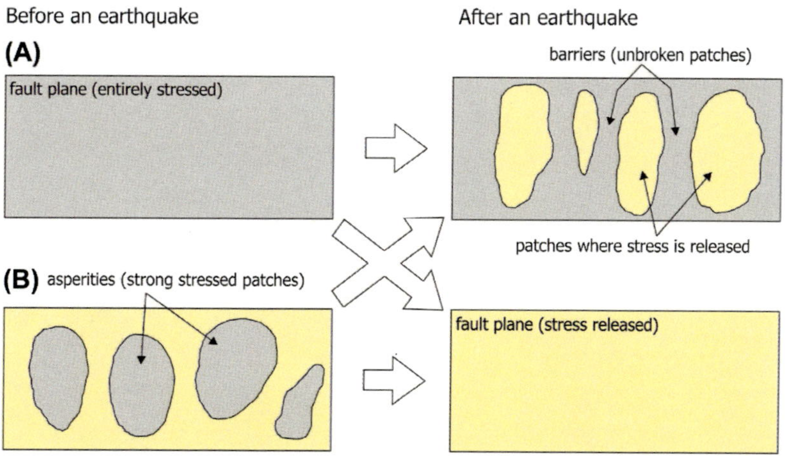

FIG. 2.32 Barriers and asperities. (A) Barriers represent unbroken patches on a fault surface after an earthquake. (B) Asperities are defined as areas between two fault blocks that have a high frictional strength, which leads to fault locking at these locations. *Figure is modified after Aki (1984).*

two fault blocks that have a high frictional strength, which leads to fault locking at these locations. They are strong patches on a fault surface that are stressed, which are surrounded by areas where the stress was already released (Lay and Kanamori, 1981). During an earthquake, asperities break and stress is released. Large fault surfaces, for example, the interface between the hanging wall plate and the down-going slab at a subduction zone (often called megathrusts), can have several asperities along-strike and along-dip. At these asperities, stress is high, compared to the average stress on the fault surface. When shear stress on the fault surface exceeds the yield stress of the asperities, localized slip occurs, which leads to an increase of stress at stronger asperities (Lay and Kanamori, 1981). The breakage of asperities is the onset of the rupture process. Frictional sliding requires the failure of asperities from grain-scale to crustal scale jogs of faults (Brantut et al., 2013). Due to the occurrence of both foreshocks and aftershocks, strong patches on a fault surface behave as barriers and others as asperities (Aki, 1984). One of the key questions in the introduction "How do faults slip?" is directly related to the breakage of asperities. The fault rupture process can be controlled by the failure of isolated asperities or neighbouring asperities can fail together and produce a larger rupture (Konca et al., 2008). Asperities occur on very different scales, from grain-scale to several tens of kilometre-size patches on the interface of a megathrust at a subduction zone (Cloos, 1992; Brantut et al., 2013). Meter-scale asperities can be observed in outcrops (Fig. 2.33). Asperities on the fault surface such as seamounts on a subducting plate can produce an interface between two fault blocks (Cloos, 1992; Yang et al., 2012) that is characterized by a strong lateral variation in seismic coupling (Lay and Kanamori, 1981). The subduction of seamounts may lead to a segmentation of the subduction zone and as a consequence may reduce the size of a megathrust earthquake (Singh et al., 2011).

Outcrop data can deliver valuable information about the geometry of asperities. Fig. 2.33 shows asperities developed on a small normal fault that

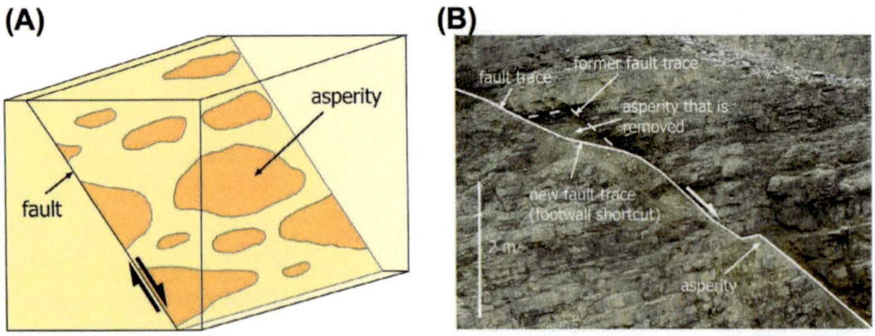

FIG. 2.33 Asperities. (A) Theoretical model of asperities along a normal fault. (B) Asperities in a field example of a normal fault. Due to a footwall shortcut, a jog in the fault surface (a type of asperity) has been sheared off. Another small asperity is still present.

cuts through thin-bedded limestone. One of the exposed asperities is intact and is represented by a jog or bend in the fault surface. The other asperity has already been removed. Fault movement has sheared-off the asperity by the development of a footwall shortcut (Fig. 2.33). The asperity is then transported down-dip along the fault and now represents a displaced rock volume that shows a characteristic triangular shape in cross-section. The increase of displacement during fault evolution can foster the progressive interaction of larger asperities and thus can lead to a width-increase of the fault-related damage zone (Faulkner et al., 2011).

During earthquakes, faults can undergo weakening by slip-related processes. Pore fluids can be thermally pressurized within and adjacent to the fault core, leading to a decrease in effective stress (see Section 2.3.2, The Coulomb failure criterion and the Mohr circle). Furthermore, frictional heating can reduce the friction coefficient by the production of pseudotachylyte or silica gel formation (Rice, 2006). Melt is generated at highly stressed asperities, which leads to a significant drop in friction and the promotion of fault slip (flash heating, flash weakening) (Beeler et al., 2008; Rempel and Weaver, 2008). These processes are summerized under the term fault lubrication and comprise gelification, decarbonation and dehydration reactions as well as melt formation (Di Toro et al., 2011). Velocity-weakening caused by flash heating is a potential explanation for earthquake ruptures that propagate as self-healing slip-pulses.

"How do ruptures stop?" was another of the grand challenges that were introduced in the introduction to this chapter. Knowledge about rupture arrest is important because it controls the size of an earthquake and, in addition, the deceleration of ruptures causes the radiation of high frequency energy that leads to strong ground motion (Sibson, 1985). The propagation of an earthquake rupture can be stopped by geometric barriers and inhomogeneous barriers. Geometric barriers are, e.g., edges where plate margins change direction, whereas inhomogeneous barriers are locations where ruptures stop without a geometric discontinuity. Inhomogeneous barriers are associated with volcanic features or anomalies in the seismic velocity (Aki, 1979).

Rupture processes are closely related to fault parameters. This allows linking the rupture area and the maximum earthquake magnitude that can occur along this fault (Wyss, 1979), where an increase of the rupture area leads to an increase of the magnitude (Wells and Coppersmith, 1994).

Over the last few decades, rupture behaviour has been successfully analysed with numerical simulations. In early simulations, earthquake fault motion was often modelled assuming uniform fault slip (or a uniform stress drop) over the entire fault surface (Aki, 1984), with the propagation controlled by friction laws (Ide, 2014). This view has changed and more heterogeneous fault models that cause irregular slip motion are now utilized (Aki, 1984). These models where further developed in recent years. Chen and Bai (2006) pointed out that fault growth processes cannot be described assuming a simple elastic crack, because this would have a physically unreasonable stress singularity at the tip of the crack. They recommend an elastic-plastic fracture model that has a zone of inelastic deformation at the fault tip. Shibazaki & Matsu'ura (1992) used a numerical approach to simulate the entire process from rupture nucleation over rupture propagation to rupture stop. They designed a fault surface that represents features typical of fault surfaces. The fault in their simulation was of uniform strength and characterized by a weak zone, a local strong asperity and is surrounded by strong barriers. With increasing external stress, a rupture nucleates quasi-statically at the weakest point. This process leads to stress increase at the asperity. Subsequently, dynamic failure occurs at the asperity, the rupture propagates and rupture ceases. After a critical level is reached, dynamic rupture propagation starts and the propagating rupture accelerates to S-wave velocity. When the rupture propagates into the barriers it decelerates and stops (Shibazaki & Matsu'ura, 1992). After the rupture stops cycle starts again with the build-up of high stress at the ends of the weak zone.

An important controlling factor for the rupture processes is friction. Friction is not constant and the static friction can be time-dependent that increases with the hold time. Dynamic friction is velocity dependent and can show a velocity-weakening or velocity-strengthening behaviour (Marone, 1998). In recent years, fault slip processes have been successfully analysed with experiments (e.g., Marone, 1998; Niemeijer et al., 2008, 2010; 2011). This topic is treated in Chapter 5 "Faulting in the laboratory".

2.5.2 Fault creep

Creeping faults were first described in the early 1940s (Louderback, 1942). Creeping is a contrasting behaviour to rupturing. As shown above, rupture processes are limited in duration to seconds or very few minutes, whereas creeping faults show continuous movement between earthquakes that can last for several decades (Harris, 2017). Fault creep along the central segment of the San Andreas Fault was measured at 0.1–10 μm/s, which is five or more orders of magnitude smaller than seismic faulting (King et al., 1973). However, it is probable that creep has both temporal and spatial variations. The best examples of this are subduction thrusts. Many GPS studies assume creep on faults at depth – this is typically assumed to be temperature dependent. Typically, fault creep can be gradual (constant over month/years) or episodic, with short so-called creep events in the range of hours to days (Lyons and Sandwell, 2003). Fault creep can occur continuously (permanent creep) and/or as a postseismic (afterslip) creep (Gratier et al., 2014).

As summarized by Avouac (2015), the key factors that promote aseismic slip are clay-rich fault rocks, high temperature and elevated pore-fluid pressure. Kaduri et al. (2017) summarized that three main deformation mechanisms used to describe fault creep under low temperature and moderate stress within the upper crust: brittle creep, pressure-solution, stress-driven, viscous creep and velocity-strengthening grain frictional sliding. Brittle creep is a time-dependent rock failure by static fatigue processes that occur below the short-term failure strength (Brantut et al., 2012, 2013). Pressure solution creep is a temperature- and pressure-dependent deformation mechanism that modifies the grain shape e.g., by diffusion processes along the grain boundaries (Gundersen et al., 2002). Velocity-strengthening frictional behaviour occurs in gouge layers, with a velocity-strengthening magnitude that varies inversely with normal stress and directly with the gouge thickness and surface roughness. It can be seen as a result of an increase in the dilatancy rate upon an increase in slip rate (Marone et al., 1990). Consequently, an important controlling factor for fault creep is the friction between the fault blocks; some rocks such as serpentinites and talc favour creep (e.g., Reinen et al., 1991). Phyllosilicate-rich gouges have a velocity-strengthening behaviour for low slip rates and therefore tend to show stable sliding/fault creep (Blanpied et al., 1991; Faulkner et al., 2003). Additional controlling factors for fault creep are the pore-fluid pressure and the near-surface presence of unconsolidated material. Elevated pore-fluid pressure is interpreted to favour creep (Scholz, 1998; Avouac, 2015), but also the opposite may occur, as described by Ougier-Simonin and Zhu (2013).

In continental regions, fault-creep was mainly documented at strike-slip faults and only seldom for normal faults and thrust faults. Chen and Bürgmann (2017) assume that this is partly due to the fact that creep is more difficult to recognize on dip-slip faults and it requires more data to be collected to better understand the controlling factors for fault creep in the different tectonic regimes. Creeping faults are, in general, regarded as less dangerous because their constant movement dissipates stress build up, which reduces the potential for large slip events (Chen and Bürgmann, 2017) and fault creep does not produce observable seismic waves (King et al., 1973). Nevertheless, it is still unclear why large earthquakes occur on faults that previously showed creeping behaviour and whether there is a common formulation that describes the controlling factors for the creep rate (Chen and Bürgmann, 2017). Several studies have shown that the same fault can show both creep and seismogenic rupture processes. Today, the Hayward Fault in California is characterized by creep with a rate in a range of 3.5−6.5 mm/year (Lienkaemper et al., 1991), but a magnitude 6.8 earthquake took place on this fault in 1868 (Harris, 2017). This implies that the kinematic behaviour of faults can change over time. Noda and Lapusta (2013) showed that dynamic weakening can change creeping fault segments into rupturing ones. Commonly, it is assumed that two types of fault segments occur, one with rate-weakening friction that is characterized by stick-slip behaviour, and one with stable rate-strengthening friction that shows creep. Noda and Lapusta (2013) developed a model that can explain the change from creeping to rupturing. Stable, rate-strengthening at low slip rates can turn to unstable slip behaviour with coseismic weakening caused by rapid shear heating of pore fluids.

2.5.3 Slow earthquakes

Slow earthquakes have been recognized since the 1990s (Ide et al., 2007). They represent fault movement between rupturing and creep. Slow earthquakes can have rupture processes over weeks to months (Kaproth and Marone, 2013), whereas creep can have a duration of decades (Harris, 2017). There are different types of slow earthquakes. Beroza and Ide (2009) categorized slow earthquakes into: (1) low-frequency earthquakes (duration <1 s), (2) very-low-frequency earthquakes (duration about 20 s) and (3) slow-slip events (duration days to months). Slow earthquakes occur at subduction zones, at shallower and deeper depths than the seismogenic zone (Obara and Kato, 2016).

The processes of slow earthquakes have been analysed with laboratory experiments. Based on such experiments, Ikari et al. (2013) proposed a slip-weakening mechanism as driver for slow earthquakes. However, the origin of slow earthquakes are not yet well understood (Kaproth and Marone, 2013). Hulbert et al. (2019) summerized the main questions, in the context of slow earthquakes, as: (1) What are the limiting mechanism(s) for fault slip-speed and rupture propagation speed during the quasi-dynamic, self-propagating rupture process? (2) Why do earthquakes and slow-slip events occur on the same fault segment? (3) How can slow-slip trigger dynamic rupture? (4) Is there a continuum of the different slip modes? An important point about slow earthquakes is that they occur in regions that neighbour megathrust seismogenic zones and some of the slow earthquakes show a similar style to megathrust earthquakes. Consequently, slow earthquakes can provide valuable information on rupture physics and they can help to better understand the megathrust rupture cycle (Obara and Kato, 2016).

2.5.4 The Cosserat theory as a concept to describe fault and deformation band behaviour

As shown above, field studies imply that faults are complex structural elements, which are composed of one or more principle slip surfaces located within the fault core. In addition, the fault core is often characterized by a wide range of different materials with varying textures such as fault breccias and gouge. Strain in a granular gouge zone may be accommodated by distributed strain in the gouge or by movement on localized slip surfaces (Scott et al., 1994). Alonso-Marroquín et al. (2006) showed that grain rotation is an important process in fault gouges. Mechanical models therefore should incorporate this complexity. Traditionally, fault movement is treated with standard continuum mechanic formulations. In such a case, faults are usually modelled as discontinuities within a velocity field. This oversimplification makes it difficult to determine e.g., fault- weakening processes (Veveakis et al., 2012). To overcome some of these limitations, a different approach, based on the Cosserat theory, has been proposed and applied.

The Cosserat theory was developed in the late 19th and early 20th Century by Eugène-Maurice-Pierre and François Cosserat (Cosserat and Cosserat, 1909; Schaefer, 1967) and is an extension of standard continuum mechanics. Commonly, individual material points within a continuum are characterized by three degrees of freedom that allow translation of each point. In a so-called Cosserat continuum, each material point is considered as rigid particle and has a total of six degrees of freedom (three for translation and three additional degrees of freedom that allow for rotational movements) (Mühlhaus, 1985; De Borst and Sluys, 1991; Papamichos, 2010). To each point of the deformable continuum, a rigid orthogonal tripod is assigned and consequently each point has a direction, which allows the Cosserat continuum to be considered as a polar medium (Schaefer, 1967). These tripods can rotate independently of the translation field (Besdo, 1974). The Cosserat theory is commonly used to solve engineering problems such as the analysis of surface instabilities (Mühlhaus, 1985), the bending of thin elastic shells (Gruttmann et al., 1989) or the calculation of borehole failure (Papanastasiou and Vardoulakis, 1992). Furthermore, it is particularly suitable to describe the mechanical behaviour of microstructured brittle material, where deformation processes involve the initiation, coalescence and propagation of micro-cracks (Addessi, 2014). The Cosserat approach has been successfully applied to model fault gouge (Alonso-Marroquín et al., 2006).

Using a micropolar approach to the Cosserat theory, Twiss et al. (1991) presented a description for brittle deformation. Their model treats the crust as granular material with individual particles representing rigid fault blocks within a fault zone, which means that the rock volume in the model is composed of an aggregate of blocks (Twiss and Unruh, 2007). In their model, two independent scales of deformation may act. Macromotion is defined by progressive simple shear, controlling the average motion of the mass centroids of the blocks and micromotion describes the individual rotation of the blocks around their mass centroids (Fig. 2.34). Twiss and Unruh (2007) analysed the structure and

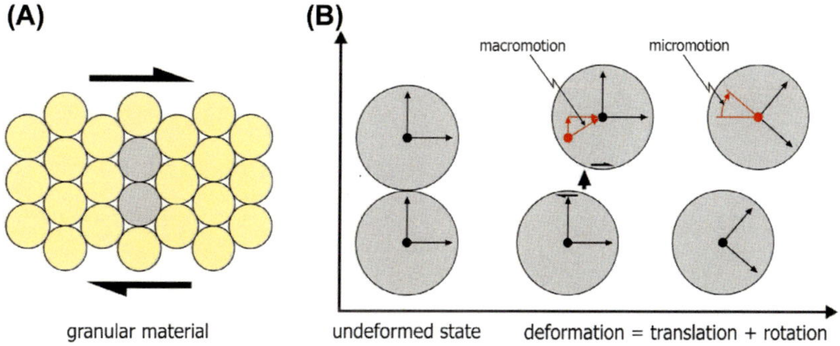

FIG. 2.34 Micropolar medium. Fault processes are described by the translation and rotation of particles. *Modified after Twiss and Unruh (2007).*

mechanical behaviour of the Loma Prieta Fault using this approach. The results of this study imply that a micropolar motion is valid for block rotation in a fault zone. The use of a micropolar description is a step towards a more realistic fault representation in numerical models and allows to better address the complex structure of fault zones where fault breccias and gouge occurs. Veveakis et al. (2012) used Cosserat continuum mechanics to model fault gouge and were able to show that thermal effects fosters the localization of deformation with the formation of very thin shear bands. Addessi (2014) pointed out that the Cosserat approach is also suitable to model brittle-like materials that have a strain-softening constitutive behaviour. This is important because a mature fault zone can undergo strain softening (Sibson, 1984).

The Cosserat theory is also the foundation for a new description of fault reactivation. In this model, fault reactivation is considered to be controlled by a symmetric macrostrain tensor and a skewed-symmetric relative microstrain tensor, where the large-scale deformation is described with the macrostrain tensor and local microrotations of blocks between faults by the microstrain tensor (Žalohar and Vrabec, 2010).

The fact that the Cosserat approach allows rotational movements makes it especially useful to describe fault creep, where particle rotation is a possible explanation for the gradual fault movement. In addition, the Cosserat theory is very suitable to model the behaviour of granular material (Chang and Ma, 1991) and can therefore utilized to describe the development of deformation bands in unconsolidated material (Alonso-Marroquín and Vardoulakis, 2005), which involves the rotation of grains as an important deformation mechanism (Brandes and Tanner, 2012). In this context, Alonso-Marroquín and Vardoulakis (2005) used a numerical simulation with a discrete element approach to model the evolution of shear-deformation bands. During deformation, the particles lose or gain contact with each other and the particle assembly undergoes restructuring by rolling and sliding of the individual particles. When loading starts, the frictional dissipation has a uniform distribution and progressively localizes into thin layers that culminate in the formation of shear bands, with a thickness of 6–8 particles. The simulation results show that the particles were arranged in large vorticity cells, between these cells rolling mechanisms dominate. Alonso-Marroquín and Vardoulakis (2005) ask the question whether this can explain the occurrence of seismic gaps along large faults.

2.5.5 Large overthrusts and the effect of fluid pressure

Large overthrusts (e.g., nappes and thrust sheets) have been well documented for more than one hundred years (e.g., Willis, 1894; Heim, 1921). They are key features of many fossil and active orogens, but their orientation is not in accordance with the theory of Anderson (Scholz, 2002). The movement of these thrust sheets has been regarded as a paradox, and has been treated in different studies during the last hundred years (e.g., Smoluchowski, 1909). Thrust sheets are very thin, compared to their lateral extent. Many are only very few kilometres thick, but have a lateral extent of several tens to hundred kilometres. The transport distance of foreland fold-and-thrust sheets can reach >100 km and their dimensions are up to >700 km (strike length) and >3 km (thickness) (Hatcher, 2004). This geometry made it very difficult to explain their kinematics.

Different stresses interact within the thrust sheet. Normal stress, σ_n, occurs, given by the thickness of the thrust sheets times the density of the rocks times the acceleration of gravity (Scholz, 2002). A horizontal stress σ_h is applied to the thrust sheet from one side, which results from plate motion leading to orogen shortening (Fig. 2.35). To move the thrust sheet this horizontal stress has to overcome the frictional resistance σ_f (see above) along the basal detachment. For a thrust sheet of several tens of kilometres of lateral extent, this stress would exceed the failure strength of the rock material, thus leading to a local (brittle) deformation on the side where the stress is applied, rather than moving the entire thrust sheet. Hubbert and Rubey (1959) proposed a fluid pressure model to explain the movement of large thrust sheets. In their

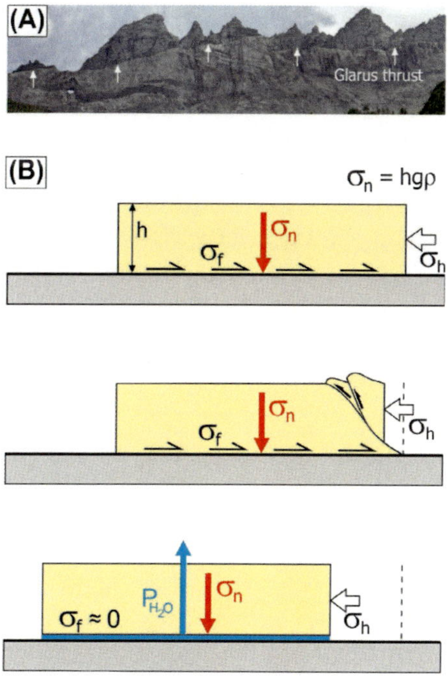

FIG. 2.35 The paradox of large overthrusts. (A) The Glarus Thrust in the Alps. Note the horizontal trace of the thrust. (B) Fluid pressure model. To move a thrust sheet, the horizontal stress has to overcome the normal stress and the friction at the basal detachment. The required horizontal stress to move a thrust sheet is in general higher than the failure strength of the rocks, thus leading to local deformation. Thrust movement becomes possible if high fluid pressure eliminates the normal stress and decouples the thrust sheet from the underlying rocks. *(B) is Modified after van der Pluijm and Marshak (2004).*

model, the fluid pressure decreases the effective normal stress (see Section 2.3.2, The Coulomb failure criterion and the Mohr circle), which locks the basal detachment. If the fluid pressure is high enough, the thrust sheet will be decoupled and can be moved under relatively low horizontal stress, without exceeding the failure strength of the rock material (Fig. 2.35).

This simple model is reasonable and over the years it has been applied and tested (Aydin and Engelder, 2014; Yue and Suppe, 2014), but also questioned (Gretener, 1972, 1981). It has been questioned, whether it is realistic that the high pore-fluid pressure is present along the entire detachment simultaneously and how this pressure can be maintained over the total area and the transportation distance (Gretener, 1981). Price (1988) pointed out that the total displacement along the basal detachment shows lateral variations and only a small part of the fault is active for a given time interval. The thrust sheet is not sliding over the entire detachment, but due to the propagation of individual dislocations that were controlled by the strength heterogeneity, anisotropy of the rock mass and by variations in the regional stress field (Price, 1988). Yue and Suppe (2014) tested the Hubbert-Rubey model for thrusts, based on examples from Taiwan. They were able to show that the fluid pressure is not important for explaining fault-weakening and that dynamic-weakening mechanisms are more important (Yue and Suppe, 2014). For the Pine Mountain Thrust in the Appalachians, Aydin and Engelder (2014) could show that fluid over-pressure was a likely driver for overthrust faulting, as predicted by the Hubbert-Rubey model. They concluded that the overpressuring was enhanced by the maturation of organic material in black shales. Such a maturity-controlled fluid pressure increase has been also discussed for other settings with pronounced basal detachments, such as the Niger Delta (Cobbold et al., 2004, 2009).

2.6 Faults in soft-sediments

Most studies consider faults in consolidated rocks. Nevertheless, unconsolidated sediments are also affected by faulting. In general, active faults that reach the Earth's surface penetrate soft-sediments. Many paleoseismological studies rely on trenching of such near-surface fault segments and consequently focus on faults and fault zones developed in soft-sediments. The deformation of soft-sediment in the hanging wall of a basement fault was already analysed with analogue models in early studies (Rettger, 1935). Benchmark work on the deformation of soft-sediments was published by Maltman (1994).

An important difference to solid rocks is that unlithified sediments show a non-linear stress-strain relationship. In rocks, elastic deformation takes place at the scale of atomic bonds, whereas in soft-sediments, the deformation induces grain-sliding with a very low elastic component and thus cannot be restored (Jones, 1994). The deformation mechanisms of unconsolidated sand are a function of porosity and depth. For near-surface sand, cataclasis and particulate flow are the relevant processes (Fig. 2.36). At depths above 1 km,

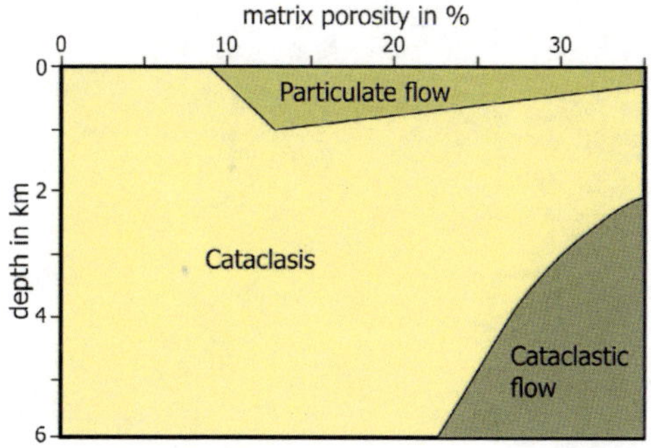

FIG. 2.36 Deformation mechanisms in unconsolidated sediments. *Modified after Fulljames et al. (1997) and Bense et al. (2003).*

particulate flow is the dominant deformation mechanism, where the grains roll. The grain fabric disaggregates and the pores within the deformation zone tend to dilate (Bense et al., 2003). At depths below 1 km, cataclasis can occur, where grains are crushed. However, both processes can act at the same time. Cataclastic flow occurs at greater depths and is not relevant for the deformation of near-surface sediments (Bense et al., 2003).

Faults in soft-sediments can be represented by very different structural elements that depend on the material properties of the sediment. In fine-grained material with low porosities such as clay and silt, more discrete fault surfaces are developed, whereas in porous sands, often tabular deformation bands occur (see Section 2.4.7, "Deformation bands") (Fig. 2.37). In gravel-dominated sediments, faults are diffuse and can show a preferred orientation of the long axes of the pebbles, which indicate the fault trace (Kim et al., 2004).

In soft-sediments (especially if the fault offsets a layered sequence with a sand-clay intercalation), the host material is often incorporated into the fault surface. The term 'clay smear' describes different processes and structures that can occur along normal faults, which offset a sedimentary sequence prone of shale intercalations, where clay is sheared into the fault (e.g., Vrolijk et al., 2016). Clay smear processes can produce faults that act as effective seals. Details on clay smear processes and the resulting fault sealing effects are described in Chapter 8 "Fault seal". Shearing can also incorporate unconsolidated sand into the fault zone (Lewis et al., 2002) (Fig. 2.37), which can lead to higher permeability that turns the fault zone into a fluid conduit. Fault surfaces in soft-sediments can be also the root of hydrofractures that evolve due to high fluid pressure during an earthquake. These hydrofractures can propagate as mode I fractures into the host sediment (Fig. 2.37) and are suitable indicators of paleoseismic events.

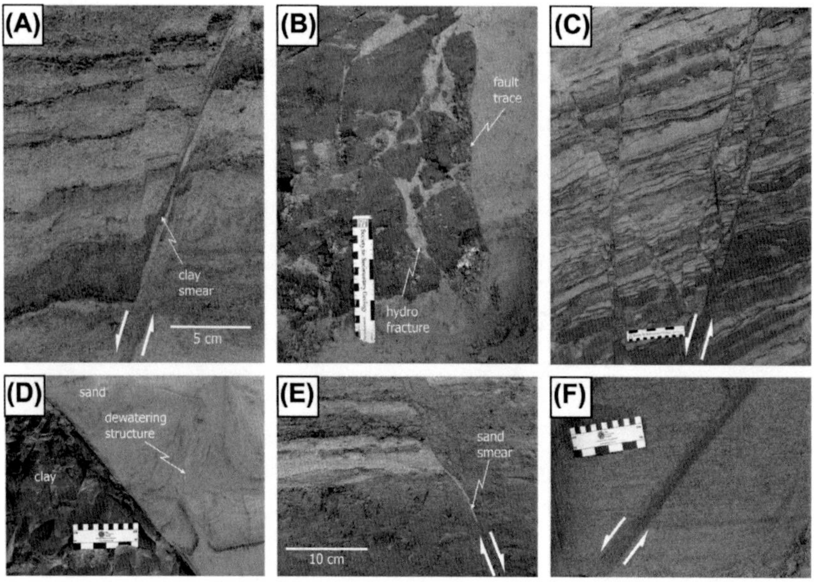

FIG. 2.37 Faults in soft-sediments. (A) Normal fault with clay smear, northern Denmark. (B) Hydrofractures that propagate from a fault surface into to host sediment, northern Denmark (scale in centimetres and inches). (C) Deformation bands, northern Denmark (scale in centimetres and inches). (D) Thrust fault, northern Denmark (scale in centimetres and inches). (E) Sand smear, northern Denmark. (F) Deformation band, northern Germany (scale in centimetres and inches).

References

Addessi, D., 2014. A 2D Cosserat finite element based on a damage-plastic model for brittle materials. Comput. Struct. 135, 20–31. https://doi.org/10.1016/j.compstruc.2014.01.003.

Aki, K., 1979. Characterization of barriers on an earthquake fault. J. Geophys. Res. 84 (B11), 6140–6148.

Aki, K., 1984. Asperities, barriers, characteristic earthquakes and strong motion prediction. J. Geophys. Res. 89 (B7), 5867–5872.

Al-Busaidi, A., Hazzard, J.F., Young, R.P., 2005. Distinct element modeling of hydraulically fractured Lac du Bonnet granite. J. Geophys. Res. 110, B06302. https://doi.org/10.1029/2004JB003297.

Allmendinger, R.W., Zapata, T., Manceda, R., Dzelalija, F., 2004. Trishear kinematic modeling of structures, with examples from the Neuquén Basin, Argentina. In: McClay, K.R. (Ed.), Thrust Tectonics and Hydrocarbon Systems, vol. 82. AAPG Mem, pp. 356–371.

Allmendinger, R.W., 1998. Inverse and forward numerical modeling of trishear fault-propagation folds. Tectonics 17, 640–656.

Alonso-Marroquín, F., Vardoulakis, I., 2005. Micromechanics of shear bands in granular media. In: García-Rojo, R., Herrmann, H., McNamara, S. (Eds.), Powders and Grains 2005. AMMG. A.A. Balkema, Leiden, The Netherlands, pp. 701–704.

Alonso-Marroquín, F., Vardoulakis, I., Herrmann, H.J., Weatherley, D., Mora, P., 2006. Effect of rolling on dissipation in fault gouges. Phys. Rev. E74. https://doi.org/10.1103/PhysRevE.74.031306.

Ammon, C.J., Ji, C., Thio, H.-K., Robinson, D., Ni, S., Hjorleifsdottir, V., Kanamori, H., Lay, T., Das, S., Helmberger, D., Ichinose, G., Polet, J., Wald, D., 2005. Rupture process of the 2004 Sumatra-Andaman earthquake. Science 308, 1133–1139.

Anderson, E.M., 1905. The dynamics of faulting. Trans. Edinb. Geol. Soc. 8 (3), 387–402.

Anderson, E.M., 1951. The Dynamics of Faulting and Dyke Formation with Application to Britain, second ed. Oliver and Boyd, Edinburgh. 206 p.

Angelier, J., Mechler, P., 1977. Sur une méthode graphique de recherche des constraintes principales également utilisable en tectonique et en séismologie: la méthode des dièdres droits. Bull. Soc. Géol. Fr. 7, 1309–1318. https://doi.org/10.2113/gssgfbull.S7-XIX.6.1309.

Angelier, J., Bergerat, F., Chu, H.T., Juang, W.S., Lu, C.Y., 1990. Paleostress analysis as a key to margin extension: the Penghu islands, South China Sea. Tectonophysics 183, 161–176.

Arab, N., Kazi, A., Rieke, H.H., 1994. Fractal geometry of faults in relation to the 12 October 1992 Cairo earthquake. Nat. Hazards 10, 221–233.

Atkinson, B.K., 1982. Subcritical crack propagation in rocks: theory, experimental results and applications. J. Struct. Geol. 4/1, 41–56.

Aviles, C.A., Scholz, C.H., Boatwright, J., 1987. Fractal analysis applied to characteristic segments of the San Andreas Fault. J. Geophys. Res. 92 (B1), 331–344.

Avouac, J.-P., 2015. From geodetic imaging of seismic and aseismic fault slip to dynamic modeling of the seismic cycle. Annu. Rev. Earth Planet. Sci. 43, 233–271.

Aydin, M.G., Engelder, T., 2014. Revisiting the Hubbert-Rubey pore pressure model for overthrust faulting: inferences from bedding-parallel detachment surfaces within Middle Devonian gas shale, the Appalachian Basin, USA. J. Struct. Geol. 69, 519–537.

Aydin, A., Johnson, A.M., 1978. Development of faults as zones of deformation bands and as slip surfaces in sandstones. Pure Appl. Geophys. 116, 931–942.

Aydin, A., 1978. Small faults formed as deformation bands in sandstone. Pure Appl. Geophys. 116, 913–930.

Bakun, W.H., Lindh, A.G., 1985. The Parkfield, California, earthquake prediction experiment. Science 229, 619–624.

Ballas, G., Fossen, H., Soliva, R., 2015. Factors controlling permeability of cataclastic deformation bands and faults in porous sandstone reservoirs. J. Struct. Geol. 76, 1–21. https://doi.org/10.1016/j.jsg.2015.03.013.

Bar-Sinai, Y., Spatschek, R., Brener, E.A., Bouchbinder, E., 2015. Velocity-strengthening friction significantly affects interfacial dynamics, strength and dissipation. Sci. Rep. 5 https://doi.org/10.1038/srep07841.

Beeler, N.M., Tullis, T.E., Goldsby, D.L., 2008. Constitutive relationships and physical basis of fault strength due to flash heating. J. Geophys. Res. 113, B01401. https://doi.org/10.1029/2007JB004988.

Bense, V.F., Gleeson, T., Loveless, S.E., Bour, O., Scibek, J., 2013. Fault zone hydrogeology. Earth Sci. Rev. 127, 171–192.

Bense, V.F., Van den Berg, E.H., Van Balen, R.T., 2003. Deformation mechanisms and hydraulic properties of fault zones in unconsolidated sediments; the Roer Valley Rift System. The Netherlands. Hydrogeol. J. 11, 319–332. https://doi.org/10.1007/s10040-003-0262-8.

Ben-Zion, Y., Sammis, C.G., 2003. Characterization of fault zones. Pure Appl. Geophys. 160, 677–715.

Beroza, G.C., Ide, S., 2009. Deep tremors and slow quakes. Science 324/5930, 1025–1026. https://doi.org/10.1126/science.1171231.

Besdo, D., 1974. Ein Beitrag zur nichtlinearen Theorie des Cosserat-Kontinuums. Acta Mech. 20, 105–131.

Blanpied, M.L., Lockner, D.A., Byerlee, J.D., 1991. Fault stability inferred from granite sliding experiments at hydrothermal conditions. Geophys. Res. Lett. 18, 609–612.

Blau, P., 2013. Amonton's law of friction. In: Wang, C. (Ed.), Encyclopedia of Tribology. Springer, p. 71.

Blueford, J., Craig, M., 2006. Tracing Hayward Fault. A Potential Disaster Area. Field Guide. Northern California Geological Society, 37 p.

Bonnet, E., Bour, O., Odling, N.E., Davy, P., Main, I., Cowie, P., Berkowitz, B., 2001. Scaling of fracture systems in geological media. Rev. Geophys. 39, 347–383.

Bott, M.H.P., 1959. The mechanics of oblique slip faulting. Geol. Mag. 96, 109–117.

Bouchon, M., Karabulut, H., Bouin, M.-P., Schmittbuhl, J., Vallée, M., Archuleta, R., Das, S., Renard, F., Marsan, D., 2010. Faulting characteristics of supershear earthquakes. Tectonophysics 493, 244–253.

Brace, W.F., Byerlee, J.D., 1966. Stick-slip as a mechanism for earthquakes. Science 153, 990–992.

Brace, W.F., Kohlstedt, D.L., 1980. Limits on lithospheric stress imposed by laboratory experiments. J. Geophys. Res. 85, 6248–6252.

Brace, W.F., 1960. An extension of the Griffith theory of fracture to rocks. J. Geophys. Res. 65/10, 3477–3480.

Brandes, C., Tanner, D.C., 2012. Three-dimensional geometry and fabric of shear deformation-bands in unconsolidated Pleistocene sediments. Tectonophysics 518–521, 84–92.

Brandes, C., Tanner, D.C., 2014. Fault-related folding: a review of kinematic models and their application. Earth Sci. Rev. 138, 352–370.

Brandes, C., Schmidt, C., Tanner, D.C., Winsemann, J., 2013. Paleostress pattern and salt tectonics within a developing foreland basin (north-western Subhercynian Basin, northern Germany). Int. J. Earth Sci. 102, 2239–2254.

Brandes, C., Igel, J., Loewer, M., Tanner, D.C., Lang, J., Müller, K., Winsemann, J., 2018a. Visualisation and analysis of shear-deformation bands in unconsolidated Pleistocene sand using ground-penetrating radar: implications for paleoseismological studies. Sediment. Geol. 367, 135–145.

Brandes, C., Steffen, H., Sandersen, P.B.E., Wu, P., Winsemann, J., 2018b. Glacially induced faulting along the NW segment of the Sorgenfrei-Tornquist Zone, northern Denmark: implications for neotectonics and Lateglacial fault-bound basin formation. Quat. Sci. Rev. 189, 149–168.

Brandes, C., Plenefisch, T., Tanner, D.C., Gestermann, N., Steffen, H., 2019. Evaluation of deep crustal earthquakes in northern Germany – possible tectonic causes. Terra Nova 31, 83–93. https://doi.org/10.1111/ter.12372.

Brantut, N., Baud, P., Heap, M.J., Meredith, P.G., 2012. Micromechanics of brittle creep in rocks. J. Geophys. Res. 117, B08412. https://doi.org/10.1029/2012JB009299.

Brantut, N., Heap, M.J., Meredith, P.G., Baud, P., 2013. Time-dependent cracking and brittle creep in crustal rocks: a review. J. Struct. Geol. 52, 17–43. https://doi.org/10.1016/j.jsg.2013.03.007.

Brodsky, E.E., Gilchrist, J.J., Sagy, A., Collettini, C., 2011. Faults smooth gradually as a function of slip. Earth Planet. Sci. Lett. 302, 185–193.

Bufe, C.G., Harsh, P.W., Burford, R.O., 1977. Steady-state seismic slip – a precise recurrence model. Geophys. Res. Lett. 4/2, 91–94.

Bullock, R.J., De Paola, N., Holdsworth, R.E., Trabucho-Alexandre, J., 2014. Lithological controls on the deformation mechanisms operating within carbonate-hosted faults during the seismic cycle. J. Struct. Geol. 58, 22–42.

Bürgmann, R., Pollard, D.D., Martel, S.J., 1994. Slip distributions on faults: effects of stress gradients, inelastic deformation, heterogeneous host-rock stiffness, and fault interaction. J. Struct. Geol. 16/12, 1675–1690.

Buxtorf, A., 1916. Prognosen und Befunde beim Hauensteinbasis- und Grenchenbergtunnel und die Bedeutung der letztern für die Geologie des Juragebirges. Abh. Naturforsch. Ges. Basel 27, 184–254.

Cacace, M., Blöcher, G., Watanabe, N., Moeck, I., Börsing, N., Scheck-Wenderoth, M., Kolditz, O., Huenges, E., 2013. Modelling of fractured carbonate reservoirs: outline of a novel technique via a case study from the Molasse Basin, southern Bavaria, Germany. Environ. Earth Sci. 70, 3585–3602.

Caine, J.S., Evans, J.P., Forster, C.B., 1996. Fault zone architecture and permeability structure. Geology 24/11, 1025–1028.

Calais, E., Freed, A.M., Van Arsdale, R., Stein, S., 2010. Triggering of New Madrid seismicity by late-Pleistocene erosion. Nature 466, 608–611. https://doi.org/10.1038/nature09258.

Candela, T., Renard, F., Bouchon, M., Brouste, A., Marsan, D., Schmittbuhl, J., Voisin, C., 2009. Characterization of fault roughness at various scales: implications of three-dimensional high resolution topography measurements. Pure Appl. Geophys. 166, 1817–1851.

Candela, T., Renard, F., Schmittbuhl, J., Bouchon, M., Brodsky, E.E., 2011. Fault slip distribution and fault roughness. Geophys. J. Int. 187, 959–968.

Candela, T., Renard, F., Klinger, Y., Mair, K., Schmittbuhl, J., Brodsky, E.E., 2012. Roughness of fault surfaces over nine decades of length scales. J. Geophys. Res. 117, B08409. https://doi.org/10.1029/2011JB009041.

Cardozo, N., Aanonsen, S., 2009. Optimized trishear inverse modeling. J. Struct. Geol. 31, 546–560.

Cardozo, N., Jackson, C.A.-L., Whipp, P.S., 2011. Determining the uniqueness of best-fit trishear models. J. Struct. Geol. 33, 1063–1078.

Cashman, S.M., Baldwin, J.N., Cashman, K.V., Swanson, K., Crawford, R., 2007. Microstructures developed by coseismic and aseismic faulting in near-surface sediments, San Andreas Fault, California. Geology 35, 611–614.

Célérier, B., Etchecopar, A., Bergerat, F., Vergely, P., Arthaud, F., Laurent, P., 2012. Inferring stress from faulting: from early concepts to inverse methods. Tectonophysics 581, 206–219.

Chang, C.S., Ma, L., 1991. A micromechanical-based micropolar theory for deformation of granular solids. Int. J. Solids Struct. 28, 67–86.

Chang, Y.-F., Chen, C.-C., Liang, C.-Y., 2007. The fractal geometry of the surface ruptures of the 1999 Chi-Chi earthquake, Taiwan. Geophys. J. Int. 170, 170–174.

Chen, Z., Bai, W., 2006. Fault creep growth model and its relationship with occurrence of earthquakes. Geophys. J. Int. 165, 272–278.

Chen, K.H., Bürgmann, R., 2017. Creeping faults: good news, bad news? Rev. Geophys. 55, 282–286. https://doi.org/10.1002/2017RG000565.

Chester, F.M., Chester, J.S., 1998. Ultracataclasite structure and friction processes of the Punchbowl fault, San Andreas system, California. Tectonophysics 295, 199–221.

Chester, F.M., Logan, J.M., 1986. Implications for mechanical properties of brittle faults from observations of the Punchbowl fault zone, California. Pure Appl. Geophys. 124, 79–106.

Childs, C., Watterson, J., Walsh, J.J., 1996. A model for the structure and development of fault zones. J. Geol. Soc. Lond. 153, 337–340.

Childs, C., Manzocchi, T., Walsh, J.J., Bonson, C.G., Nicol, A., Schöpfer, M.P.J., 2009. A geometric model of fault zone and fault rock thickness variations. J. Struct. Geol. 31, 117–127.

Choi, J.-H., Edwards, P., Ko, K., Kim, Y.-S., 2016. Definition and classification of fault damage zones: a review and new methodological approach. Earth Sci. Rev. 152, 70–87.

Chounet, A., Vallée, M., Causse, M., Courboulex, F., 2018. Global catalog of earthquake rupture velocities shows anticorrelation between stress drop and rupture velocity. Tectonophysics 733, 148–158.

Clausen, J.A., Gabrielsen, R.H., Johnsen, E., Korstgård, J.A., 2003. Fault architecture and clay smear distribution. Examples from field studies and drained ring-shear experiments. Norw. J. Geol. 83, 131–146.

Cloos, M., 1992. Thrust-type subduction-zone earthquakes and seamount asperities: a physical model for seismic rupture. Geology 20, 601–604.

Cobbold, P.R., Mourgues, R., Boyd, K., 2004. Mechanism of thin-skinned detachment in the Amazon Fan: assessing the importance of fluid overpressure and hydrocarbon generation. Mar. Pet. Geol. 21, 1013–1025.

Cobbold, P.R., Clarke, B.J., Løseth, H., 2009. Structural consequences of fluid overpressure and seepage forces in the outer thrust belt of the Niger Delta. Petrol. Geosci. 15, 3–15.

Cohn, S.N., Allen, C.R., Gilman, R., Goulty, N.R., 1982. Preearth-quake and postearthquake creep on the Imperial Fault and the Brawley Fault zone. In: The Imperial Valley, California, earthquake of October 15, 1979, 1254. U.S. Geol. Surv. Prof. Pap., pp. 161–167

Convers, J.A., Newman, A.V., 2013. Rapid earthquake rupture duration estimates from teleseismic energy rate, with application to real-time warning. Geophys. Res. Lett. 40, 5844–5848.

Cooke, M.L., Madden, E.H., 2014. Is the Earth lazy? A review of work minimization in fault evolution. J. Struct. Geol. 66, 334–346.

Cosserat, E., Cosserat, F., 1909. Théorie des corps déformables. Hermann & Sons, Paris, p. 226.

Costain, J.K., Bollinger, G.A., Speer, J.A., 1987. Hydroseismicity — a hypothesis for the role of water in the generation of intraplate seismicity. Geology 15, 618–621. https://doi.org/10.1130/0091-7613.

Coulomb, C.A., 1776. Essai sur une application des règles des maximis et minimis à quelques problèmes de statique, relatifs à l'architecture. Mem. Acad. r. Sci. Paris 7, 343–387.

Cowie, P.A., Scholz, C.H., 1992a. Physical explanation for the displacement-length relationship of faults using a post-yield fracture mechanics model. J. Struct. Geol. 14, 1133–1148.

Cowie, P.A., Scholz, C.H., 1992b. Growth of faults by accumulation of seismic slip. J. Geophys. Res. 97 (B7), 11085–11095.

Cowie, P.A., Shipton, Z.K., 1998. Fault tip displacement gradients and process zone dimensions. J. Struct. Geol. 20/8, 983–997.

Das, S., Aki, K., 1977. Fault plane with barriers: a versatile earthquake model. J. Geophys. Res. 82, 5658–5670.

Das, S., 2015. Supershear Earthquake Ruptures – Theory, Methods, Laboratory Experiments and Fault Superhighways: An Update. In: Ansal, A. (Ed.), Perspectives on European Earthquake Engineering and Seismology Volume 2, Geotechnical, Geological and Earthquake Engineering, p. 455. https://doi.org/10.1007/978-3-319-16964-4.

Daskalaki, E., Spiliotis, K., Siettos, C., Minadakis, G., Papadopoulos, G.A., 2016. Foreshocks and short-term hazard assessment of large earthquakes using complex networks: the case of the 2009 L'Aquila earthquake. Nonlinear Process Geophys. 23, 241–256.

De Borst, R., Sluys, L.J., 1991. Localisation in a Cosserat continuum under static and dynamic loading conditions. Comput. Methods Appl. Mech. Eng. 90, 805–827.

Delvaux, D., Moeys, R., Stapel, G., Melnikov, A., Ermikov, V., 1995. Palaeostress reconstructions and geodynamics of the Baikal region, Central Asia, part I. Palaeozoic and Mesozoic pre-rift evolution. Tectonophysics 252, 61–101. https://doi.org/10.1016/0040-1951(95)00090-9.

Di Toro, G., Han, R., Hirose, T., De Paola, N., Nielsen, S., Mizoguchi, K., Ferri, F., Cocco, M., Shimamoto, T., 2011. Fault lubrication during earthquakes. Nature 471, 494–498. https://doi.org/10.1038/nature09838.

Di Toro, G., Mittempergher, S., Ferri, F., Mitchell, T.M., Pennacchioni, G., 2012. The contribution of structural geology, experimental rock deformation and numerical modelling to an improved understanding of the seismic cycle. Preface to the special volume "Physico-chemical processes in seismic faults". J. Struct. Geol. 38, 3–10.

Dieterich, J.H., Conrad, G., 1984. Effect of humidity on time- and velocity-dependent friction in rocks. J. Geophys. Res. 89 (B6), 4196–4202.

Dor, O., Rockwell, T.K., Ben-Zion, Y., 2006. Geological observations of damage asymmetry in the structure of the San Jacinto, San Andreas and Punchbowl Faults in Southern California: a possible indicator for preferred rupture propagation direction. Pure Appl. Geophys. 163, 301–349.

Du Bernard, X., Eichhubl, P., Aydin, A., 2002. Dilation bands: a new form of localized failure in granular media. Geophys. Res. Lett. 29, 29-1–29-4.

Dyer, R., 1988. Using joint interactions to estimate paleostress ratios. J. Struct. Geol. 10/7, 685–699.

Dyskin, A.V., Germanovich, L.N., 1993. A model of crack growth in microcracked rock. Int. J. Rock Mech. 30/7, 813–820.

Eichhubl, P., Hooker, J.N., Laubach, S.E., 2010. Pure and shear-enhanced compaction bands in Aztec Sandstone. J. Struct. Geol. 32, 1873–1886.

Engelder, T., Geiser, P., 1980. On the use of regional joint sets as trajectories of paleostress fields during the development of the Appalachian Plateau, New York. J. Geophys. Res. 85 (B11), 6319–6341.

Epard, J.-L., Groshong, R.H., 1995. Kinematic model of detachment folding including limb rotation, fixed hinges and layer-parallel strain. Tectonophysics 247, 85–103.

Erslev, E.A., 1991. Trishear fault-propagation folding. Geology 19, 617–620.

Evans, J.P., 1990. Thickness-displacement relationships for fault zones. J. Struct. Geol. 12/8, 1061–1065. https://doi.org/10.1016/0191-8141(90)90101-4.

Faulkner, D.R., Lewis, A.C., Rutter, E.H., 2003. On the internal structure and mechanics of large strike-slip fault zones: field observations of the Carboneras fault in southeastern Spain. Tectonophysics 367, 235–251.

Faulkner, D.R., Mitchell, T.M., Jensen, E., Cembrano, J., 2011. Scaling of fault damage zones with displacement and the implications for fault growth processes. J. Geophys. Res. 116, B05403. https://doi.org/10.1029/2010JB007788.

Flekkøy, E.G., Malthe-Sørenssen, A., Jamveit, B., 2002. Modeling hydrofracture. J. Geophys. Res. 107 (B8), 2151. https://doi.org/10.1029/2000JB000132.

Forsyth, D.W., Lay, T., Aster, R.C., Romanowicz, B., 2009. Grand challenges for seismology. EOS 90, 361–362.

Fossen, H., Cavalcante, G.C.G., 2017. Shear zones – a review. Earth Sci. Rev. 171, 434–455.

Fossen, H., Schultz, R.A., Shipton, Z.K., Mair, K., 2007. Deformation bands in sandstone: a review. J. Geol. Soc. 164, 755–769.

Fossen, H., 2010. Structural Geology. Cambridge University Press, 463 p.

Fournier, T., Morgan, J., 2012. Insights to slip behavior on rough faults using discrete element modeling. Geophys. Res. Lett. 39, L12304. https://doi.org/10.1029/2012GL051899.

Friedman, M., 1975. Fracture in rock. Rev. Geophys. Space Phys. 13, 352–358.

Frost, E., Dolan, J., Ratschbacher, L., Hacker, B., Seward, G., 2011. Direct observation of fault zone structure at the brittle-ductile transition along the Salzach-Ennstal-Mariazell-Puchberg fault system, Austrian Alps. J. Geophys. Res. 116, B02411. https://doi.org/10.1029/2010JB007719.

Fulljames, J.R., Zijerveld, L.J.J., Franssen, R.C.M.W., 1997. Fault seal processes: systematic analysis of fault seals over geological and production time scales. In: Møller-Pedersen, P., Koestler, A.G. (Eds.), Hydrocarbon Seals — Importance for Exploration and Production, vol.7. Norwegian Petroleum Society, Oslo, pp. 51–59. NPF, Special Publication.

Fyfe, W.S., Price, N.J., Thompson, A.B., 1978. Fluids in the Earth's Crust: Their Significance in Metamorphic, Tectonic, and Chemical Transport Processes. Elsevier Scientific Pub. Co., Amsterdam, p. 401 p.

Gabrielsen, R.H., Braathen, A., Kjemperud, M., Valdresbråten, M.L.R., 2016. The geometry and dimensions of fault-core lenses. In: Childs, C., Holdsworth, R.E., Jackson, C.A.-L., Manzocchi, T., Walsh, J.J., Yielding, G. (Eds.), The Geometry and Growth of Normal Faults, vol. 439. Geological Society, London, Special Publications, pp. 249–269. https://doi.org/10.1144/SP439.4.

Ghani, I., Koehn, D., Toussaint, R., Passchier, C.W., 2013. Dynamic development of hydrofracture. Pure Appl. Geophys. 170, 1685–1703.

Gillespie, P.A., Howard, C.B., Walsh, J.J., Watterson, J., 1993. Measurement and characterisation of spatial distributions of fractures. Tectonophysics 226, 113–141.

Goldsby, D.L., Tullis, T.E., 2011. Flash heating leads to low frictional strength of crustal rocks at earthquake slip rates. Science 334, 216–218.

Goodman, R.E., 1980. Introduction to Rock Mechanics. John Wiley & Sons, New York, 478 p.

Gratier, J.P., Renard, F., Vial, B., 2014. Postseismic pressure solution creep: evidence and time-dependent change from dynamic indenting experiments. J. Geophys. Res. 119, 2764–2779. https://doi.org/10.1002/2013JB010768.

Gretener, P.E., 1972. Thoughts on overthrust faulting in a layered sequence. Bull. Can. Petrol. Geol. 20, 583–607.

Gretener, P.E., 1981. Pore pressure, discontinuities, isostasy and overthrusts. In: McClay, K.R., Price, N.J. (Eds.), Thrust and Nappe Tectonics. Geological Society of London Special Publications 9, pp. 33–39.

Griffith, A.A., 1921. The phenomena of rupture and flow in solids. Philos. Trans. R. Soc. Lond. A 221, 163–198.

Gruttmann, F., Stein, E., Wriggers, P., 1989. Theory and numerics of thin elastic shells with finite rotations. Ing. Arch. 59, 54–67.

Gudmundsson, A., Brenner, S.L., 2001. How hydrofractures become arrested. Terra Nova 13, 456–462.

Gudmundsson, A., 2013. Great challenges in structural geology and tectonics. Front. Earth Sci. 1 https://doi.org/10.3389/feart.2013.00002.

Gundersen, E., Renard, F., Dysthe, D.K., Bjørlykke, K., Jamtveit, B., 2002. Coupling between pressure solution creep and diffusive mass transport in porous rocks. J. Geophys. Res. 107, B11. https://doi.org/10.1029/2001JB000287.

Guo, P., Stolle, D.F.E., 2013. Coupled analysis of bifurcation and shear band in saturated soils. Soils Found. 53/4, 525–539.

Hainzl, S., Kraft, T., Wassermann, J., Igel, H., Schmedes, E., 2006. Evidence for rainfall-triggered earthquake activity. Geophys. Res. Lett. 33, L19303. https://doi.org/10.1029/2006GL027642.

Hajiabdolmajid, V., Kaiser, P.K., Martin, C.D., 2002. Modelling brittle failure of rock. Int. J. Rock Mech. Min. Sci. 39 (6), 731–741. https://doi.org/10.1016/S1365-1609(02)00051-5.

Hancock, P.L., Kadhi, A., 1978. Analysis of mesoscopic fractures in the Dhruma-Nisah segment of the central Arabian graben system. J. Geol. Soc. 135, 339–347.

Hancock, P.L., 1985. Brittle microtectonics: principles and practice. J. Struct. Geol. 7, 437–457.

Hanks, T.C., Kanamori, H., 1979. A moment magnitude scale. J. Geophys. Res. 84 (B5), 2348–2350.

Hardy, S., Ford, M., 1997. Numerical modelling of trishear fault propagation folding. Tectonics 16, 841–854.

Harris, R.A., 2017. Large earthquakes and creeping faults. Rev. Geophys. 55, 169–198. https://doi.org/10.1002/2016RG000539.

Hatcher, R.D., 2004. Properties of thrusts and upper bounds for the size of thrust sheets. In: McClay, K.R. (Ed.), Thrust Tectonics and Hydrocarbon Systems, vol. 82. AAPG Memoir, pp. 18–29.

Heim, A., 1921. Geologie der Schweiz. die Schweizer Alpen 1. Hälfte, Band 2. Tauchnitz, Leipzig.

Helmstetter, A., Shaw, B.E., 2009. Afterslip and aftershocks in the rate-and-state friction law. J. Geophys. Res. 114, B01308. https://doi.org/10.1029/2007JB005077.

Hesthammer, J., Fossen, H., 2001. Structural core analysis from the Gullfaks area, northern North Sea. Mar. Pet. Geol. 18, 411–439.

Higgins, M.W., 1971. Cataclastic rocks. In: Geological Survey Professional Paper, vol. 687. United States Government Printing Office, Washington, 97 p.

Hill, R., 1962. Acceleration waves in solids. J. Mech. Phys. Solids 10, 1–16.

Hirth, G., Teyssier, C., Dunlap, W.J., 2001. An evaluation of quartzite flow laws based on comparisons between experimentally and naturally deformed rocks. Int. J. Earth Sci. 90, 77–87.

Hobbs, B.E., Means, W.D., Williams, P.F., 1976. An Outline of Structural Geology. John Wiley & Sons, 571 p.

Hoek, E., Martin, C.D., 2014. Fracture initiation and propagation in intact rock - a review. J. Rock Mech. Geotech. Eng. 6, 287–300.

Houston, H., 2001. Influence of depth, focal mechanism, and tectonic setting on the shape and duration of earthquake source time functions. J. Geophys. Res. 106 (B6), 11137–11150.

Hubbert, M.K., Rubey, W.W., 1959. Role of fluid pressure in mechanics of overthrust faulting. 1. Mechanics of fluid-filled porous solids and its application to overthrust faulting. Bull. Geol. Soc. Am. 70, 115–166.

Hubbert, M.K., Willis, D.G., 1957. Mechanics of hydraulic fracturing. AIME Pet. Transac. 210, 153–168.

Hulbert, C., Rouet-Leduc, B., Johnson, P.A., Ren, C.X., Rivière, J., Bolton, D.C., Marone, C., 2019. Similarity of fast and slow earthquakes illuminated by machine learning. Nat. Geosci. 12, 69–74. https://doi.org/10.1038/s41561-018-0272-8.

Hull, J., 1988. Thickness-displacement relationships for deformation zones. J. Struct. Geol. 10/4, 431–435.

Huntington, K.W., Klepeis, K.A., With 66 Community Contributors, 2018. Challenges and Opportunities for Research in Tectonics: Understanding Deformation and the Processes that Link Earth Systems, from Geologic Time to Human Time. A Community Vision Document Submitted to the U.S. National Science Foundation. University of Washington. https://doi.org/10.6069/H52R3PQ5, 84 p.

Hussain, E., Wright, T.J., Walters, R.J., Bekaert, D., Hooper, A., Houseman, G.A., 2016. Geodetic observations of postseismic creep in the decade after the 1999 Izmit earthquake, Turkey:

implications for a shallow slip deficit. J. Geophys. Res. Solid Earth 121, 2980–3001. https://doi.org/10.1002/2015JB012737.

Husseini, M.I., 1977. Energy balance for motion along a fault. Geophys. J. R. Astron. Soc. 49, 699–714.

Ide, S., Beroza, G.C., Shelly, D.R., Uchide, T., 2007. A scaling law for slow earthquakes. Nature 447, 76–79. https://doi.org/10.1038/nature05780.

Ide, S., 2014. Modeling fast and slow earthquakes at various scales. Proc. Jpn. Acad. Ser. B 90, 259–277.

Ikari, M.J., Marone, C., Saffer, D.M., Kopf, A.J., 2013. Slip weakening as a mechanism for slow earthquakes. Nat. Geosci. 6, 468–472. https://doi.org/10.1038/ngeo1818.

Inbal, A., Ampuero, J.P., Clayton, R.W., 2016. Localized seismic deformation in the upper mantle revealed by dense seismic arrays. Science 354, 88–92. https://doi.org/10.1126/science.aaf1370.

Issen, K.A., Rudnicki, J.W., 2001. Theory of compaction bands in porous rock. Phys. Chem. Earth A 26, 95–100.

Jiang, J., Lapusta, N., 2017. Connecting depth limits of interseismic locking, microseismicity, and large earthquakes in models of long-term fault slip. J. Geophys. Res. Solid Earth 122, 6491–6523. https://doi.org/10.1002/2017JB014030.

Jones, L.M., 1985. Foreshocks and time-dependent earthquake hazard assessment in southern California. Bull. Seismol. Soc. Am. 75, 1669–1679.

Jones, M., 1994. Mechanical principles of sediment deformation. In: Maltman, A. (Ed.), The Geological Deformation of Sediments. Chapman and Hall, pp. 37–72.

Kaduri, M., Gratier, J.-P., Renard, F., Çakir, Z., Lasserre, C., 2017. The implications of fault zone transformation on aseismic creep: example of the North Anatolian Fault, Turkey. J. Geophys. Res. Solid Earth 122, 4208–4236. https://doi.org/10.1002/2016JB013803.

Kagan, Y.Y., Knopoff, L., 1980. Spatial distribution of earthquakes: the two-point correlation function. Geophys. J. R. Astron. Soc. 62, 303–320.

Kapiris, P., Polygiannakis, J., Peratzakis, A., Nomicos, K., Eftaxias, K., 2002. VHF-electromagnetic evidence of the underlying pre-seismic critical stage. Earth Planets Space 54, 1237–1246.

Kaproth, B.M., Marone, C., 2013. Slow earthquakes, preseismic velocity changes, and the origin of slow frictional stick-slip. Science 341, 1229–1232.

Killick, A.M., 2003. Fault rock classification: an aid to structural interpretation in mine and exploration geology. S. Afr. J. Geol. 106, 395–402.

Kim, Y.-S., Sanderson, D.J., 2010. Inferred fluid flow through fault damage zones based on the observation of stalactites in carbonate caves. J. Struct. Geol. 32, 1305–1316.

Kim, Y.-S., Peacock, D.C.P., Sanderson, D.J., 2004. Fault damage zones. J. Struct. Geol. 26, 503–517.

King, C.-Y., Nason, R.D., Tocher, D., 1973. Kinematics of fault creep. Phil. Trans. Roy. Soc. Lond. Ser. A Math. Phys. Sci. 274, 355–360.

Klimczak, C., Soliva, R., Schultz, R.A., Chéry, J., 2011. Sequential growth of deformation bands in a multilayer sequence. J. Geophys. Res. 116, B09209. https://doi.org/10.1029/2011JB008365.

Knipe, R.J., Jones, G., Fisher, Q.J., 1998. Faulting, fault sealing and fluid flow in hydrocarbon reservoirs: an introduction. In: Jones, G., Fisher, Q.J., Knipe, R.J. (Eds.), Faulting, Fault Sealing and Fluid Flow in Hydrocarbon Reservoirs, vol. 147. Geological Society, London, pp. vii–xxi. Special Publications.

Knott, S.D., Beach, A., Brockbank, P.J., Brown, J.L., McCallum, J.E., Welbon, A.I., 1996. Spatial and mechanical controls on normal fault populations. J. Struct. Geol. 18, 359–372.

Konca, A.O., Avouac, J.P., Sladen, A., Meltzner, A.J., Sieh, K., Fang, P., Li, Z., Galetzka, J., Genrich, J., Chlieh, M., Natawidjaja, D.H., Bock, Y., Fielding, E.J., Ji, C., Helmberger, D.V., 2008. Partial rupture of a locked patch of the Sumatra megathrust during the 2007 earthquake sequence. Nature 456, 631–635. https://doi.org/10.1038/nature07572.

La Pointe, P.R., 1988. A method to characterize fracture density and connectivity through fractal geometry. Int. J. Rock Mech. Min. Sci. Geomech. Abstr. 25, 421–429.

Lacombe, O., 2012. Do fault slip data inversions actually yield 'paleostresses' that can be compared with contemporary stresses? A critical discussion. C.R. Geosci. 344, 159–173.

Lawn, B., 1993. Fracture of brittle solids. In: Cambridge Solid State Science Series, second ed. Cambridge University Press, Cambridge. 378 p.

Lay, T., Kanamori, H., 1981. An asperity model of large earthquake sequences. Earthquake prediction: an international review. Maurice Ewing Ser. 4, 579–592.

Lay, T., Kanamori, H., Ammon, C.J., Koper, K.D., Hutko, A.R., Ye, L., Yue, H., Rushing, T.M., 2012. Depth-varying rupture properties of subduction zone megathrust faults. J. Geophys. Res. 117, B04311. https://doi.org/10.1029/2011JB009133.

Lay, T. (Ed.), 2009. Seismological Grand Challenges in Understanding Earth's Dynamic Systems. Report to the National Science Foundation, IRIS Consortium, 76 p.

Legarth, B., Huenges, E., Zimmermann, G., 2005. Hydraulic fracturing in a sedimentary geothermal reservoir: results and implications. Int. J. Rock Mech. Min. Sci. 42, 1028–1041.

Lewis, G., Knipe, R.J., Li, A., 2002. Fault seal analysis in unconsolidated sediments: a field study from Kentucky. In: Koestler, A.G., Hunsdale, R. (Eds.), Hydrocarbon Seal Quantification, vol. 11. NPF, Special Publication, Elsevier B. V., Amsterdam, pp. 243–253.

Lienkaemper, J.J., Borchardt, G., Lisowski, M., 1991. Historic creep rate and potential for seismic slip along the Hayward Fault, California. J. Geophys. Res. 96 (B11), 18261–18283.

Lin, J., Parmentier, E.M., 1988. Quasistatic propagation of a normal fault: a fracture mechanics model. J. Struct. Geol. 10/3, 249–262.

Lin, A., 2008. Seismic slip in the lower crust inferred from granulite-related pseudotachylyte in the Woodroffe thrust, Central Australia. Pure Appl. Geophys. 165, 215–233.

Lindanger, M., Gabrielsen, R.H., Braathen, A., 2007. Analysis of rock lenses in extensional faults. Norw. J. Geol. 87, 361–372.

Lisle, R.J., 2013. A critical look at the Wallace-Bott hypothesis in fault-slip analysis. Bull. Soc. Géol. Fr. 184, 299–306. https://doi.org/10.2113/gssgfbull.184.4-5.299.

Lockner, D., Byerlee, J.D., 1977. Hydrofracture in Weber Sandstone at high confining pressure and differential stress. J. Geophys. Res. 82/14, 2018–2026.

Louderback, G.D., 1942. Faults and earthquakes. Bull. Seismol. Soc. Am. 32, 305–330.

Lyons, S., Sandwell, D., 2003. Fault creep along the southern San Andreas from interferometric synthetic aperture radar, permanent scatterers, and stacking. J. Geophys. Res. 108/B1, 2047. https://doi.org/10.1029/2002JB001831.

Maddock, R.H., 1983. Melt origin of fault-generated pseudotachylites demonstrated by textures. Geology 11 (6), 105–108.

Maerten, L., Maerten, F., Lejri, M., Gillespie, P., 2016. Geomechanical paleostress inversion using fracture data. J. Struct. Geol. 89, 197–213. https://doi.org/10.1016/j.jsg.2016.06.007.

Mair, K., Main, I., Elphick, S., 2000. Sequential growth of deformation bands in the laboratory. J. Struct. Geol. 22, 25–42.

Majumdar, A., Tien, C.L., 1990. Fractal characterization and simulation of rough surfaces. Wear 136, 313–327.

Maltman, A., 1994. Introduction and overview. In: Maltman, A. (Ed.), The Geological Deformation of Sediments. Chapman and Hall, pp. 1–35.

Mandelbrot, B., 1967. How long is the coast of Britain? Statistical self-similarity and fractional dimension. Science 156/3775, 636–638.

Mandelbrot, B., 1982. The Fractal Geometry of Nature. Freeman and Co., 460 p.

Marone, C., Raleigh, C.B., Scholz, C.H., 1990. Frictional behavior and constitutive modeling of simulated fault gouge. J. Geophys. Res. 95, 7007–7025.

Marone, C.J., Scholz, C.H., Bilham, R., 1991. On the mechanics of earthquake afterslip. J. Geophys. Res. 96 (B5), 8441–8452.

Marone, C., 1998. Laboratory-derived friction laws and their application to seismic faulting. Annu. Rev. Earth Planet. Sci. 26, 643–696.

Marrett, R., Allmendinger, R.W., 1990. Kinematic analysis of fault-slip data. J. Struct. Geol. 12, 973–986.

Mohr, O., 1900. Welche Umstände bedingen die Elastizitätsgrenze und den Bruch eines Materials. Z. Des. Vereines Dtsch. Ingenieure 44, 1524–1530 und 45, 1572–1577.

Mollema, P.N., Antonellini, M.A., 1996. Compaction bands: a structural analog for anti-mode I cracks in aeolian sandstone. Tectonophysics 267, 209–228.

Morrow, C.A., Shi, L.Q., Byerlee, J.D., 1982. Strain hardening and strength of clay-rich fault gouges. J. Geophys. Res. 87 (B8), 6771–6780.

Mühlhaus, H.-B., 1985. Oberflächen-Instabilität bei geschichtetem Halbraum mit Biegesteifigkeit. Ing. Arch. 55/5, 388–400.

Nadai, A., 1931. Plasticity: A Mechanics of the Plastic State of Matter. McGraw-Hill Book Co., 349 p.

Needham, T., Yielding, G., Fox, R., 1996. Fault population description and prediction using examples from the offshore U.K. J. Struct. Geol. 18 (2–3), 155–167.

Nelson, R.A., 1979. Natural fracture systems: description and classification. Am. Assoc. Pet. Geol. Bull. 63/12, 2214–2232.

Nelson, R.A., 2001. Geologic Analysis of Naturally Fractured Reservoirs, second ed. Gulf Professional Publishing, Woburn, Massachusetts. 332 p.

Nemčok, M., Schamel, S., Gayer, R., 2005. Thrustbelts. Structural Architecture, Thermal Regimes, and Petroleum Systems. Cambridge University Press, Cambridge, 541 p.

Nicol, A., Childs, C., Walsh, J.J., Schafer, K.W., 2013. A geometric model for the formation of deformation band clusters. J. Struct. Geol. 55, 21–33.

Niemeijer, A.R., Spiers, C.J., Peach, C.J., 2008. Frictional behaviour of simulated quartz fault gouges under hydrothermal conditions: results from ultra-high strain rotary shear experiments. Tectonophysics 460, 288–303.

Niemeijer, A.R., Marone, C., Elsworth, D., 2010. Frictional strength and strain weakening in simulated fault gouge: competition between geometrical weakening and chemical strengthening. J. Geophys. Res. 115, B10207. https://doi.org/10.1029/2009JB000838.

Niemeijer, A.R., Di Toro, G., Nielsen, S., Di Felice, F., 2011. Frictional melting of gabbro under extreme experimental conditions of normal stress, acceleration, and sliding velocity. J. Geophys. Res. 116, B07404. https://doi.org/10.1029/2010JB008181, 074044/1.

Noda, H., Lapusta, N., 2013. Stable creeping fault segments can become destructive as a result of dynamic weakening. Nature 493, 518–521. https://doi.org/10.1038/nature11703.

Obara, K., Kato, A., 2016. Connecting slow earthquakes to huge earthquakes. Science 353, 253–257. https://doi.org/10.1126/science.aaf1512.

Okubo, P.G., Aki, K., 1987. Fractal geometry in the San Andreas Fault system. J. Geophys. Res. 92 (B1), 345–355.

Olsson, W.A., 1999. Theoretical and experimental investigation on compaction bands in porous rock. J. Geophys. Res. 104 (B4), 7219–7228.

Ougier-Simonin, A., Zhu, W., 2013. Effects of pore fluid pressure on slip behaviors: an experimental study. Geophys. Res. Lett. 40, 2619–2624. https://doi.org/10.1002/grl.50543.

Papamichos, E., 2010. Continua with microstructure: Cosserat theory. Eur. J. Environ. Civil Eng. 14 (8–9), 1011–1029. https://doi.org/10.1080/19648189.2010.9693277.

Papanastasiou, P.C., Vardoulakis, I.G., 1992. Numerical treatment of progressive localization in relation to borehole stability. Int. J. Numer. Anal. Methods Geomech. 16, 389–424.

Park, S.-I., Kim, Y.-S., Ryoo, C.-R., Sanderson, D.J., 2010. Fractal analysis of the evolution of a fracture network in a granite outcrop, SE Korea. Geosci. J. 14/2, 201–215.

Pascal, C., 2002. Interaction of faults and perturbation of slip: influence of anisotropic stress states in the presence of fault friction and comparison between Wallace–Bott and 3D distinct element models. Tectonophysics 356, 307–322.

Peacock, D.C.P., Nixon, C.W., Rotevatn, A., Sanderson, D.J., Zuluaga, L.F., 2016. Glossary of fault and other fracture networks. J. Struct. Geol. 92, 12–29.

Peacock, D.C.P., 2001. The temporal relationship between joints and faults. J. Struct. Geol. 23, 329–341.

Perfettini, H., Ampuero, J.-P., 2008. Dynamics of a velocity strengthening fault region: implications for slow earthquakes and postseismic slip. J. Geophys. Res. 113, B09411. https://doi.org/10.1029/2007JB005398.

Perfettini, H., Avouac, J.-P., 2007. Modeling afterslip and aftershocks following the 1992 Landers earthquake. J. Geophys. Res. 112, B07409. https://doi.org/10.1029/2006JB004399.

Perfettini, H., Avouac, J.-P., Ruegg, J.-C., 2005. Geodetic displacements and aftershocks following the 2001 $M_w = 8.4$ Peru earthquake: implications for the mechanics of the earthquake cycle along subduction zones. J. Geophys. Res. 110, B09404. https://doi.org/10.1029/2004JB003522.

Persson, B.N.J., 2014. On the fractal dimension of rough surfaces. Tribol. Lett. 54, 99–106.

Pollard, D.D., Saltzer, S.D., Rubin, A.M., 1993. Stress inversion methods: are they based on faulty assumptions? J. Struct. Geol. 15, 1045–1054.

Power, W.L., Tullis, T.E., 1991. Euclidean and fractal models for the description of rock surface roughness. J. Geophys. Res. 96 (B1), 415–424.

Power, W.L., Tullis, T.E., Brown, S.R., Boitnott, G.N., Scholz, C.H., 1987. Roughness of natural fault surfaces. Geophys. Res. Lett. 14/1, 29–32.

Price, N.J., Cosgrove, J.W., 1990. Analysis of Geological Structures. Cambridge University Press, Cambridge, 502 p.

Price, R.A., 1988. The mechanical paradox of large overthrusts. GSA Bull. 100, 1898–1908.

Qu, D., Tveranger, J., 2016. Incorporation of deformation band fault damage zones in reservoir models. Am. Assoc. Pet. Geol. Bull. 100, 423–443.

Ramsey, J.M., Chester, F.M., 2004. Hybrid fracture and the transition from extension fracture to shear fracture. Nature 428, 63–66.

Reasenberg, P.A., 1999. Foreshock occurrence before large earthquakes. J. Geophys. Res. 104 (B3), 4755–4768.

Reches, Z., Lockner, D.A., 1994. Nucleation and growth of faults in brittle rocks. J. Geophys. Res. 99 (B9), 18159–18173.

Reches, Z., Lockner, D.A., 2010. Fault weakening and earthquake instability by powder lubrication. Nature 467, 452–455. https://doi.org/10.1038/nature09348.

Reid, H.F., 1910. The California Earthquake of April 18, 1906: Report of the State Earthquake Investigation Commission. In: The Mechanics of the Earthquake, vol. 2. Carnegie Institution of Washington, Washington, DC, 192 p.

Reinen, L.A., Weeks, J.D., Tullis, T.E., 1991. The frictional behavior of serpentinite: implications for aseismic creep on shallow crustal faults. Geophys. Res. Lett. 18/10, 1921–1924.

Rempel, A.W., Weaver, S.L., 2008. A model for flash weakening by asperity melting during high-speed earthquake slip. J. Geophys. Res. 113, B11308. https://doi.org/10.1029/2008JB005649.

Renard, F., Candela, T., 2017. Scaling of fault roughness and implications for earthquake mechanics. In: Thomas, M.Y., Mitchell, T.M., Bhat, H.S. (Eds.), Fault Zone Dynamic Processes: Evolution of Fault Properties during Seismic Rupture, vol. 227. Geophysical Monograph, pp. 197–215.

Rettger, R.E., 1935. Experiments on soft-rock deformation. Am. Assoc. Pet. Geol. Bull. 19, 271–292.

Rice, J.R., 1976. The localization of plastic deformation. In: Koiter, W.T. (Ed.), Theoretical and applied mechanics. Proceedings of the 14th International Congress on Theoretical and Applied Mechanics, vol. 1. North Holland Publishing Co., Delft, pp. 207–220.

Rice, J.R., 2006. Heating and weakening of faults during earthquake slip. J. Geophys. Res. 111, B05311. https://doi.org/10.1029/2005JB004006.

Rich, J.L., 1934. Mechanics of low-angle overthrust faulting as illustrated by Cumberland Thrust block, Virginia, Kentucky, and Tennessee. Am. Assoc. Pet. Geol. Bull. 18, 1584–1596.

Riley, P.R., Goodwin, L.B., Lewis, C.J., 2010. Controls on fault damage zone width, structure, and symmetry in the Bandelier Tuff, New Mexico. J. Struct. Geol. 32, 766–780.

Robertson, E.C., 1983. Relationship of fault displacement to gouge and breccia thickness. Am. Inst. Min. Eng. Trans. 274, 1426–1432.

Romano, V., De'Haven Hyman, J., Karra, S., Valocchi, A.J., Battaglia, M., Bigi, S., 2017. Numerical modeling of fluid flow in a fault zone: a case study from Majella Mountain (Italy). Energy Procedia 125, 556–560. https://doi.org/10.1016/j.egypro.2017.08.191.

Rotevatn, A., Fossen, H., 2011. Simulating the effect of subseismic fault tails and process zones in a siliciclastic reservoir analogue: implications for aquifer support and trap definition. Mar. Pet. Geol. 28, 1648–1662. https://doi.org/10.1016/j.marpetgeo.2011.07.005.

Rotevatn, A., Sandve, T.H., Keilegavlen, E., Kolyukhin, D., Fossen, H., 2013. Deformation bands and their impact on fluid flow in sandstone reservoirs: the role of natural thickness variations. Geofluids 13/3, 359–371. https://doi.org/10.1111/gfl.12030.

Rudnicki, J.W., Rice, J.R., 1975. Conditions for localization of deformation in pressure-sensitive dilatant materials. J. Mech. Phys. Solids 23, 371–394.

Sagy, A., Brodsky, E.E., Axen, G.J., 2007. Evolution of fault-surface roughness with slip. Geology 35, 283–286.

Sato, K., 2012. Fast multiple inversion for stress analysis from fault-slip data. Comput. Geosci. 40, 132–137.

Schaefer, H., 1967. Das Cosserat Kontinuum. Z. Angew. Math. Mech. 47/8, 485–498.

Scholz, C.H., Aviles, C.A., 1986. The fractal geometry of faults and faulting. Geophys. Monogr. Ser. 37, 147–155.

Scholz, C.H., 1987. Wear and gouge formation in brittle faulting. Geology 15, 493–495.

Scholz, C.H., 1998. Earthquakes and friction laws. Nature 391, 37–42.

Scholz, C.H., 2002. The Mechanics of Earthquakes and Faulting, second ed. Cambridge University Press, Cambridge. 471 p.

Schultz, R.A., Balasko, C.M., 2003. Growth of deformation bands into echelon and ladder geometries. Geophys. Res. Lett. 30 https://doi.org/10.1029/2003GL018449.

Schultz, R.A., Siddharthan, R., 2005. A general framework for the occurrence and faulting of deformation bands in porous granular rocks. Tectonophysics 411, 1–18.

Schultz, R.A., Soliva, R., Okubo, C.H., Mège, D., 2010. Fault populations. In: Watters, T.R., Schultz, R.A. (Eds.), Planetary Tectonics. Cambridge University Press, pp. 457–510.

Schultz, R.A., 1999. Understanding the process of faulting: selected challenges and opportunities at the edge of the 21st century. J. Struct. Geol. 21, 985–993.

Schulz, S.E., Evans, J.P., 1998. Spatial variability in microscopic deformation and composition of the Punchbowl Fault, southern California: implications for mechanisms, fluid–rock interaction, and fault morphology. Tectonophysics 295, 223–244.

Scott, D.R., Marone, C.J., Sammis, C.G., 1994. The apparent friction of granular fault gouge in sheared layers. J. Geophys. Res. 99 (B4), 7231–7246.

Scruggs, V.J., Tullis, T.E., 1998. Correlation between velocity dependence of friction and strain localization in large displacement experiments on feldspar, muscovite and biotite gouge. Tectonophysics 295, 15–40.

Secor, D.T., 1965. Role of fluid pressure in jointing. Am. J. Sci. 263, 633–646.

Shibazaki, B., Matsu'ura, M., 1992. Spontaneous processes for nucleation, dynamic propagation, and stop of earthquake rupture. Geophys. Res. Lett. 19/12, 1189–1192.

Shimazaki, K., Nakata, T., 1980. Time-predictable recurrence model for large earthquakes. Geophys. Res. Lett. 7/4, 279–282.

Shipton, Z.K., Evans, J.P., Thompson, L.B., 2005. The geometry and thickness of deformation-band fault core and its influence on sealing characteristics of deformation-band fault zones. In: Sorkhabi, R., Tsuji, Y. (Eds.), Faults, Fluid Flow, and Petroleum Traps, vol. 85. AAPG Memoir, pp. 181–195.

Shipton, Z.K., Cowie, P.A., 2001. Damage zone and slip-surface evolutionover μm to km scales in high-porosity Navajo sandstone, Utah. J. Struct. Geol. 23, 1825–1844.

Shipton, Z.K., Cowie, P.A., 2003. A conceptual model for the originof fault damage zone structures in high-porosity sandstone. J. Struct. Geol. 25, 333–345.

Shipton, Z.K., Soden, A.M., Kirkpatrick, J.D., Bright, A.M., Lunn, R.L., 2006. How thick is a fault? Fault displacement-thickness scaling revisited. Earthquakes: radiated energy and the physics of faulting. Geophys. Monogr. Ser. 170, 193–198. https://doi.org/10.1029/170GM19.

Sibson, R.H., 1975. Generation of pseudotachylyte by ancient seismic faulting. Geophys. J. R. Astron. Soc. 43, 775–794.

Sibson, R.H., 1977. Fault rocks and fault mechanisms. J. Geol. Soc. Lond. 133, 191–213.

Sibson, R.H., 1984. Roughness at the base of the seismogenic zone: contributing factors. J. Geophys. Res. 89 (B7), 5791–5799.

Sibson, R.H., 1985. Stopping of earthquake ruptures at dilational fault jogs. Nature 316, 248–251.

Sibson, R.H., 1986a. Brecciation processes in fault zones: inferences from earthquake rupturing. Pure Appl. Geophys. 124 (1–2), 159–175.

Sibson, R.H., 1986b. Earthquakes and rock deformation in crustal fault zones. Annu. Rev. Earth Planet. Sci. 14, 149–175.

Sibson, R.H., 1989. Earthquake faulting as a structural process. J. Struct. Geol. 11, 1–14.

Sibson, R.H., 2011. The scope of earthquake geology. In: Fagereng, A., Toy, V.G., Rowland, J.V. (Eds.), Geology of the Earthquake Source: A Volume in Honour of Rick Sibson, vol. 359. Geological Society, London, pp. 319–331. https://doi.org/10.1144/SP359.18. Special Publication.

Shipton, Z.K., Cowie, P.A., 2001. Damage zone and slip-surface evolution over μm to km scales in high-porosity Navajo sandstone, Utah. J. Struct. Geol. 23, 1825–1844.

Shipton, Z.K., Cowie, P.A., 2003. A conceptual model for the origin of fault damage zone structures in high-porosity sandstone. J. Struct. Geol. 25, 333–345.
Simpson, R.W., 1997. Quantifying Anderson's fault types. J. Geophys. Res. 102 (B8), 17909–17919.
Singh, S.C., Hananto, N., Mukti, M., Robinson, D.P., Das, S., Chauhan, A., Carton, H., Gratacos, B., Midnet, S., Djajadihardja, Y., Harjono, H., 2011. Aseismic zone and earthquake segmentation associated with a deep subducted seamount in Sumartra. Nat. Geosci. 4, 308–311. https://doi.org/10.1038/NGEO1119.
Smoluchowski, M., 1909. Some remarks on the mechanics of overthrusts. Geol. Mag. 6, 204–205.
Stuart, W.D., Tullis, T.E., 1995. Fault model for preseismic deformation at Parkfield, California. J. Geophys. Res. 100 (B12), 24079–24099.
Sukmono, S., Zen, M.T., Kadir, W.G.A., Hendrajaya, L., Santoso, D., Dubois, J., 1996. Fractal geometry of the Sumatra active fault system and its geodynamical implications. J. Geodyn. 22 (1–2), 1–9.
Suppe, J., Medwedeff, D.A., 1990. Geometry and kinematics of fault-propagation folding. Eclogae Geol. Helv. 83, 409–454.
Suppe, J., 1983. Geometry and kinematics of fault-bend folding. Am. J. Sci. 283, 684–721.
Terzaghi, K., Peck, R.B., 1967. Soil Mechanics in Engineering Practice. John Wiley & Sons, New York, 729 p.
Terzaghi, K., 1923. Die Berechnung der Durchlässigkeitsziffer des Tones aus dem Verlauf der hydrodynamischen Spannungserscheinungen. Österreichische Akademie der Wissenschaften in Wien. Mathematisch Naturwissenschaftliche Kl. 132, 125–138.
Thomas, T.Y., 1961. Plastic Flow and Fracture in Solids. Academic Press, New York, 267 p.
Tindall, S.E., 2014. Simple calculations of fluid flow across jointed cataclastic deformation bands. Mar. Pet. Geol. 57, 152–159.
Torabi, A., Fossen, H., 2009. Spatial variation of microstructure and petrophysical properties along deformation bands in reservoir sandstones. Am. Assoc. Pet. Geol. Bull. 93, 919–938.
Torabi, A., Zarifi, Z., 2014. Energy release rate of propagating deformation bands and their hosted cracks. Int. J. Rock Mech. Min. Sci. 67, 184–190. https://doi.org/10.1016/j.ijrmms.2013.10.007.
Trouw, R.A.J., Passchier, C.W., Wiersma, D.J., 2010. Atlas of Mylonites- and Related Microstructures. Springer-Verlag, Berlin, Germany, ISBN 978-3-642-03607-1, 322 p.
Twiss, R.J., Unruh, J.R., 2007. Structure, deformation, and strength of the Loma Prieta fault, northern California, USA, as inferred from the 1989-1990 Loma Prieta aftershock sequence. GSA Bull. 119/9–10, 1079–1106. https://doi.org/10.1130/B25976.1.
Twiss, R.J., Protzman, G.M., Hurst, S.D., 1991. Theory of slickenline patterns based on the velocity gradient tensor and microrotation. Tectonophysics 186, 215–239.
Van Arsdale, R., Bresnahan, R., McCallister, N., Waldron, B., 2007. Upland Complex of the central Mississippi River valley: its origin, denudation, and possible role in reactivation of the New Madrid seismic zone. In: Stein, S., Mazzotti, S. (Eds.), Continental Intraplate Earthquakes: Science, Hazard, and Policy Issues. Geol. Soc. Am. Spec. Pap., 425, pp. 177–199.
van der Pluijm, B.A., Marshak, S., 2004. Earth Structure. Norton and Company, 656 p.
Vardoulakis, I.G., Sulem, J., 1995. Bifurcation Analysis in Geomechanics. Taylor & Francis, 462 p.
Vergos, G., Arabelos, D.N., Contadakis, M.E., 2015. Evidence for tidal triggering on the earthquakes of the Hellenic Arc, Greece. Phys. Chem. Earth 85–86, 210–215.
Vermilye, J.M., Scholz, C.H., 1998. The process zone: a microstructural view of fault growth. J. Geophys. Res. 103 (B6), 12223–12237.

Veveakis, E., Sulem, J., Stefanou, I., 2012. Modeling of fault gouges with Cosserat Continuum Mechanics: influence of thermal pressurization and chemical decomposition as coseismic weakening mechanism. J. Struct. Geol. 38, 254–264.

Voisin, C., Renard, F., Grasso, J.-R., 2007. Long term friction: from stick-slip to stable sliding. J. Geophys. Res. 34, L13301. https://doi.org/10.1029/2007GL029715.

Vrolijk, P.J., Urai, J.L., Kettermann, M., 2016. Clay smear: review of mechanisms and applications. J. Struct. Geol. 86, 95–152.

Wallace, R.E., 1951. Geometry of shearing stress and relation to faulting. J. Geol. 59, 118–130.

Walsh, J.J., Watterson, J., 1987. Distributions of cumulative displacement and seismic slip on a single normal fault surface. J. Struct. Geol. 9, 1039–1046.

Walsh, J.J., Watterson, J., Yielding, G., 1991. The importance of small-scale faulting in regional extension. Nature 351, 391–393.

Walsh, J.J., Watterson, J., Yielding, G., 1994. Determination and interpretation of fault size populations: procedures and problems. In: Aasen, J.O., et al. (Eds.), North Sea Oil and Gas Reservoirs III. Kluwer Academic, pp. 141–155.

Watterson, J., 1986. Fault dimensions, displacements and growth. Pure Appl. Geophys. 124 (1–2), 365–373.

Wells, D.L., Coppersmith, K.J., 1994. New Empirical relationships among magnitude, rupture length, rupture width, rupture area, and surface displacement. Bull. Seismol. Soc. Am. 84, 974–1002.

Wesnousky, S.G., 2008. Displacement and geometrical characteristics of earthquake surface ruptures: issues and implications for seismic-hazard analysis and the process of earthquake rupture. Bull. Seismol. Soc. Am. 98, 1609–1632.

Wibberley, C.A.J., Yielding, G., Di Toro, G., 2008. Recent advances in the understanding of fault zone internal structure: a review. In: Wibberley, C.A.J., Kurz, W., Imber, J., Holdsworth, R.E., Collettini, C. (Eds.), The Internal Structure of Fault Zones: Implications for Mechanical and Fluid-Flow Properties, vol. 299. Geological Society of London Special Publication, pp. 5–33. https://doi.org/10.1144/SP299.2.

Willis, B., 1894. The Mechanics of Appalachian Structure. Department of the Interior —U.S. Geological Survey, 217–281.

Wojtal, S., Mitra, G., 1986. Strain hardening and strain softening in fault zones from foreland thrusts. GSA Bull. 97, 674–687.

Wong, T.-F., Szeto, H., Zhang, J., 1992. Effect of loading path and porosity on the failure mode of porous rocks. Appl. Mech. Rev. 45/8, 281–293.

Woodcock, N.H., Mort, K., 2008. Classification of fault breccias and related fault rocks. Geol. Mag. 145, 435–440.

Wu, P., Hasegawa, H.S., 1996a. Induced stresses and fault potential in eastern Canada due to a disc load: a preliminary analysis. Geophys. J. Int. 125, 415–430. https://doi.org/10.1111/j.1365-246X.1996.tb00008.x.

Wu, P., Hasegawa, H.S., 1996b. Induced stresses and fault potential in Eastern Canada due to a realistic load: a preliminary analysis. Geophys. J. Int. 127, 215–229. https://doi.org/10.1111/j.1365-246X.1996.tb01546.x.

Wyss, M., 1979. Estimating expectable maximum magnitude of earthquakes from fault dimensions. Geology 7, 336–340.

Yamaji, A., 2000. The multiple inverse method: a new technique to separate stresses from heterogeneous fault-slip data. J. Struct. Geol. 22, 441–452.

Yang, H., Liu, Y., Lin, J., 2012. Effects of subducted seamounts on megathrust earthquake nucleation and rupture propagation. J. Geophys. Res. 39, L24302. https://doi.org/10.1029/2012GL053892.

Yew, C.H., 1997. Mechanics of Hydraulic Fracturing. Gulf Publishing Company, Houston Texas, 183 p.

Yielding, G., Walsh, J.J., Watterson, J., 1992. The prediction of small-scale faulting in reservoirs. First Break 10/12, 449–460.

Yielding, G., Needham, T., Jones, H., 1996. Sampling of fault populations using sub-surface data: a review. J. Struct. Geol. 18, 135–146.

Yu, H., Xin, Y., Wang, M., Liu, Q., 2018. Hall-Petch relationship in Mg alloys: a review. J. Mater. Sci. Technol. 34, 248–256. https://doi.org/10.1016/j.jmst.2017.07.022.

Yue, L.-F., Suppe, J., 2014. Regional pore-fluid pressures in the active western Taiwan thrust belt: a test of the classic Hubbert-Rubey fault-weakening hypothesis. J. Struct. Geol. 69, 493–518.

Žalohar, J., Vrabec, M., 2007. Paleostress analysis of heterogeneous fault-slip data: the Gauss method. J. Struct. Geol. 29, 1798–1810.

Žalohar, J., Vrabec, M., 2010. Kinematics and dynamics of fault reactivation: the Cosserat approach. J. Struct. Geol. 32, 15–27.

Zang, A., Wagner, F.C., Stanchits, S., Janssen, C., Dresen, G., 2000. Fracture process zone in granite. J. Geophys. Res. 105 (B10), 23651–23661.

Zuluaga, L.F., Rotevatn, A., Keilegavlen, E., Fossen, H., 2016. The effect of deformation bands on simulated fluid flow within fault-propagation fold trap types: lessons from the San Rafael monocline, Utah. Am. Assoc. Pet. Geol. Bull. 100, 1523–1540.

Chapter 3

Fault detection

David C. Tanner[a], Hermann Buness[a], Jan Igel[a], Thomas Günther[a], Gerald Gabriel[a], Peter Skiba[a], Thomas Plenefisch[b], Nicolai Gestermann[b], Thomas R. Walter[c]

[a]*Leibniz Institute for Applied Geophysics (LIAG), Hannover, Germany;* [b]*Federal Institute for Geosciences and Natural Resources, Hannover, Germany;* [c]*GFZ German Research Centre for Geosciences, Potsdam, Germany*

Chapter outline

Understanding Faults. https://doi.org/10.1016/B978-0-12-815985-9.00003-5

3.1 Introduction

Because the fault core is often more prone to weathering than the host rock, faults are seldom present in outcrop. However, even when they are, their occurrence is often limited to a cross-section, or the surface of the fault itself, without marker beds that show the displacement, etc. Furthermore, such outcrops are commonly only exposed in either quarries, or areas of high erosion or high seismic activity. On the contrary, in urban areas and intraplate regions, where the societal relevance of understanding the faults is higher, faults very rarely outcrop. This is where geophysics can play an important role.

Rocks possess a range of petrophysical properties (e.g., density, porosity, seismic velocity, and resistivity) that depend primarily on their mineralogical and fluid composition, depth in the crust, and state of deformation. These properties can be utilised by the various geophysical methods to image the structure of the Earth. Each method has its own advantages and disadvantages, mostly in terms of spatial resolution, cost of acquisition and processing, and interpretation (Table 3.1). However, they all typically cost much less than, for instance, drilling a borehole. Each geophysical method has the possibility to image a fault in a certain manner. Moreover, a particular survey acquisition can be planned to observe the same structure at different resolutions or scales.

We can broadly differentiate the geophysical methods according to their application in the coseismic/interseismic cycle of fault movement (Table 3.1). For instance, remote sensing of landforms is useful only directly before and after a seismic event, whereas seismological monitoring can only recognise seismic events, if and when they happen; faults in an interseismic stage do not show up with this method. Other methods, such as active seismics, ground-penetrating radar, geoelectrics, gravimetry and magnetics, require a fault, i.e., a structural discontinuity, to be already present in the rock and/or that movement has taken place to cause offset of the rock masses. In addition, under certain circumstances, the methods can actually obtain information from the fault core itself (e.g., seismic impedance, electrical conductivity).

In the following, each method will be discussed in detail, and then examples are shown to demonstrate the ability and/or the disadvantages of the method in different geological situations.

TABLE 3.1 Details of geophysical methods and remote sensing for detecting faults.

	Geophysical methods						
	Active seismics	**Ground-penetrating radar**	**Electrical resistivity tomography**	**Gravimetry**	**Magnetics**	**Seismology**	**Remote sensing**
State of fault in the seismic cycle	Inter-seismic	Inter-seismic	Inter-seismic	Inter-seismic	Inter-seismic	coseismic	inter- and coseismicmic
Is an active source required?	yes	yes	yes	no	no	no	no/yes
Depth of detection	m - crustal scale	cm - 10s m	cm — 500 m	m - crustal scale	10s m - crustal scale	10s m - planetary scale	cm - crustal scale
Acquisition time	days - months	hours - days	hours - days	days - weeks	days - weeks	continuous	days - weeks
Processing time	days - weeks	hours - days	hours - days	days	days	minutes - hours	hours - days
Interpretation time	days - weeks	hours - days	hours - days	days - weeks	days - weeks	hours - days	hours - days
Unusable for:	vertical-dipping faults without vertical offset of beds, or in scattering environments (e.g., lava)	shallow conductive layers (e.g., clay or saltwater)	same electrical properties on both sides of the fault or for the fault core and the host rock	shallow or flat-lying faults, or where there is no density contrast over the fault	shallow or flat-lying faults, or where there is no potential contrast over the fault	low magnitude events	inactive blind faults
Especially useful for:	defining the geometry of a fault, especially in 3-D	shallow faults and their damage zones (2D/3D), neotectonics	faults with electrically conductive cores (metals, graphite)	steep faults that offset basement against sediments	steep faults that offset basement against sediments	deep crustal/lithospheric faults	difficult to reach locations, determining fault offsets and scales

3.2 Active seismics

Among the geophysical methods, active seismics is the most often used tool for exploration of sub-surface structures, especially faults. It is able to deliver detailed structural information from the first metres of the subsurface to deep into the crust. This method utilizes a seismic source to generate elastic waves that travel into the subsurface stratum. They return, either due to a positive velocity gradient or by reflection at layer boundaries, where elastic properties change. The former case is exploited with the diving wave or refraction seismic approach, the latter case is the most commonly used approach, namely reflection seismics.

3.2.1 Seismic method

Historically, the development of active seismic methods originated from the use of natural sources, such as earthquakes. Active sources evolved on land from the use of mainly explosives in the early days, to predominantly vibroseis (a vibrating seismic source, either mechanical or electrical) nowadays. In the marine environment, airguns are the source of choice. For the near surface region, additional sources are available, such as sledgehammer, weight drop, guns, etc. (e.g., Miller et al., 1994; van der Veen et al., 2000).

The recording of seismic waves is traditionally conducted with geophones (devices that translate ground movement into voltage) that are specifically adapted to the source signal frequencies used. Recently, broadband MEMS (microelectromechanical systems) sensors have advanced into the market. The number of channels (each channel records one geophone or geophone group) has increased over time, from tens in the 1930s to several tens of thousands of channels nowadays used in 3D surveys.

3.2.2 Resolution

The ability of elastic waves to sense a fault in the subsurface depends on the variation of seismic impedance, which is the product of wave velocity and rock density. In theory, all impedance variations that exist in the subsurface could be sampled with seismic methods. However in practice, limitations occur. Besides the constrained illumination of the interior of the Earth with sources and sensors at the surface, the most severe limitation is the wavelength used, which depends, in turn, on wave velocity and frequency. An appropriate seismic source can generate a range of frequencies. Unfortunately, elastic waves also suffer from attenuation (reduction of the amplitude) as they pass through the substratum, whereby the higher frequencies are more strongly attenuated. This imposes an upper limit on the highest frequency that can be used in practice. On the other hand, low frequencies are constrained by the kind and size of the seismic source.

In practice, a seismic frequency range of 10–100 Hz is common for target depths from several hundred metres to several kilometres depth. Assuming a commonly-achievable dominant frequency of 50 Hz, and a rock velocity of 4000 m s^{-1} (typical for a large variety of deep-seated sediments, except carbonates, which show higher velocities), results in a wavelength of around 80 m. Vertical seismic resolution, according to the Rayleigh-criterion (Geldart and Sheriff, 1995), is at best a quarter of this wavelength, i.e., 20 m. However, this does not mean that smaller structures (such as, fault throw, fault core, etc.) cannot be detected, since in this definition resolution only means the minimum resolvable distance of two objects and not the detectability of objects in a homogenous medium. However, the reflection amplitude of objects smaller than a wavelength (this is called diffraction) decreases with the object's size, and they may not be detectable in the presence of noise and/or other reflections.

Horizontal resolution depends on the Fresnel region (frequency and depth-dependent part of a reflector from which most of the seismic energy is returned and therefore will be detected as a single arrival. Named after Augustin-Jean Fresnel (1788–1827)). Perfect migration (see below) improves resolution up to the Rayleigh criterion, although, in practice, only half of this value is regarded as realistic (Geldart and Sheriff, 1995).

In the near-surface region, recoverable frequencies, and thus resolution has meanwhile improved up to a wavelength of 4–5 m (e.g., Musmann and Buness, 2010). Recently, the use of shear waves (S-wave) instead of pressure waves (P-wave) allows an even better resolution, with wavelengths of about 1 m (Ghose et al., 2013; Krawczyk et al., 2013). This is possible because of the extraordinary low velocities of S-waves (relative to the P-wave) in unconsolidated sediments.

3.2.3 Seismic imaging of faults

Faults are predominantly imaged indirectly, usually by imaging the offset of sedimentary bedding or unconformities (Fig. 3.1). Processing methodologies of reflection seismic prefer planar, only gently-dipping layers. Fault zones are often steeply inclined and thus difficult to image, as one can infer from simple geometric considerations, bearing in mind that both sources and receivers are at the surface. Moreover, the traditional processing schemes were developed for planar horizontal layer boundaries. Another difficulty results from the impedance contrast: if the footwall and hanging-wall of the fault have the same lithology, the only impedance contrast that can exist must come from the fault itself, i.e., the fault core and/or damage zone. Fault zones can constitute zones of lower impedance due to grain size reduction (i.e., gouge) or zones of higher impedance due to mineralisation. Sometimes both phenomena appear at the same time (Lüschen et al., 2015). Fig. 3.2 shows an example of this in a

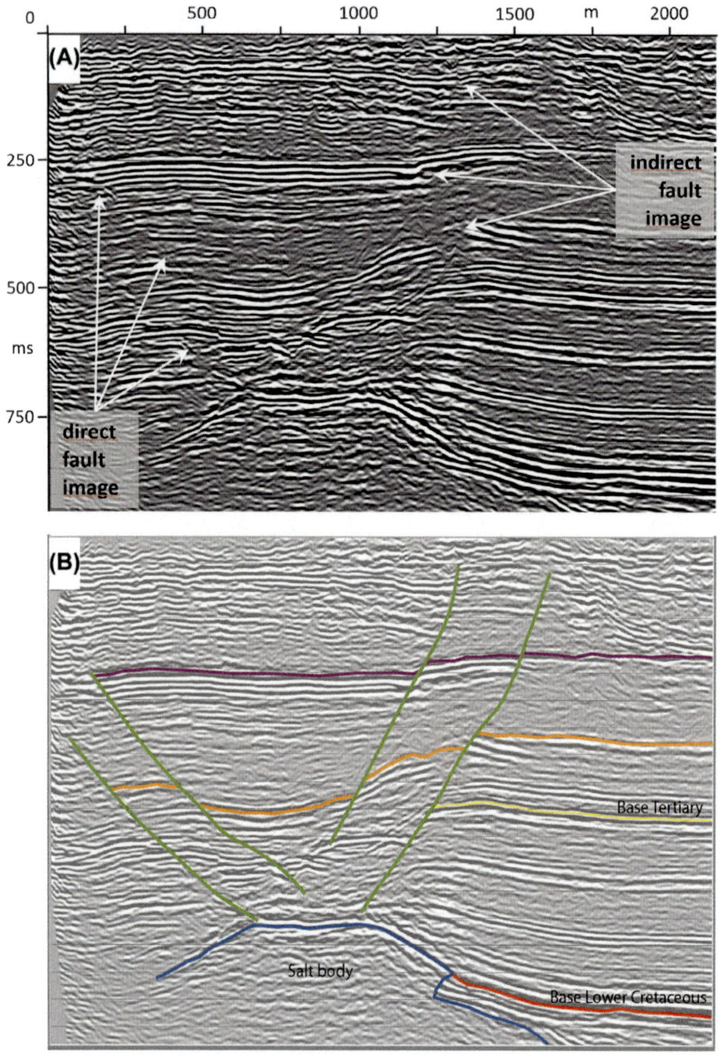

FIG. 3.1 Faults above a salt diapir in northern Germany, (A) without and (B) with interpretation. Faults originate at the top of the diapir and can be traced though the sediment pile. Due to the decreasing fault inclination with depth, they are recognizable as fault planes in the lower part but only as offset of sedimentary strata above.

granite pluton, where bedding does not exist. In this case, the fault zone must be thick enough relative to the seismic wavelength to ensure detectability. Geldart and Sheriff (1995) mention a practical limit of 1/20 to 1/30 of a wavelength.

If faults can be recognised by the offset of sedimentary strata, it offers the possibility to derive the kinematic history of the fault, providing the

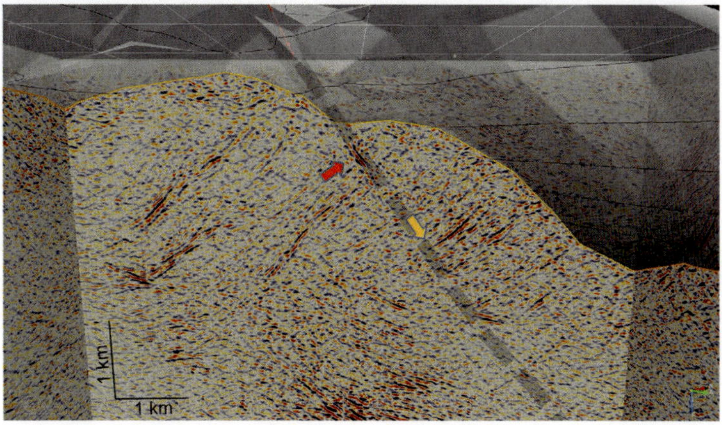

FIG. 3.2 Faults in crystalline rock: reflections inside a late-Variscan granite pluton in the Erzgebirge in Saxony, Germany. The transparent grey surface marks the upper boundary of the pluton. A major regional normal fault (*dashed grey line*) that can be traced from the surface (*red line*) is partly imaged directly (*red arrow*) and by a truncation of reflections with opposing dip (*orange arrow*). Note the different impedance contrasts of the normal faults (lower impedance inside the fault, blue-red-blue signature) and the one with opposing dip (higher impedance inside the fault, red-blue-red signature). Further details in Lušchen et al. (2015).

stratigraphy is known. Mapping the throw at all stratigraphic levels, i.e., using juxtaposition maps of throw on the fault, greatly helps to understand fault growth, development, and sealing potential (see Chapters 6 and 8).

Imaging steeply-inclined faults often pose a problem, since they are linked with sharp and small velocity changes, e.g., in the fault damage zone. These velocity variations are far too small to be resolved by standard processing. They can cause distortions (i.e., velocity pull ups/downs) of the underlying strata that are easily misinterpreted as apparent splitting of the fault with increasing depth (Couples et al., 2007; Rappin and Schnitzler, 2007).

A key processing step, which becomes increasingly important if the dip of an imaged fault is steeper, is *migration*. This process repositions reflections to their true subsurface locations, either after (poststack) or before (prestack) stacking. In the history of seismic imaging, there have been numerous techniques developed, mainly according to the progress in computing power. Starting with poststack time migration in the 1970s, subsequently dip moveout processing, poststack depth migration, and prestack time and depth migration have evolved (Jones et al., 2008). A number of special modifications exist, e.g., reverse time migration (Baysal et al., 1983) or Fresnel volume migration (Buske et al., 2009). These techniques were developed to handle increasingly complex structures (Fig. 3.3).

Another recent approach to visualise steeply-inclined faults is to elaborate seismic diffractions instead of specular reflections (Moser and Howard, 2008). Diffractions originate at objects that are smaller than a wavelength, e.g., at the edges of disrupted layers or the fault zone itself. This can be seen as

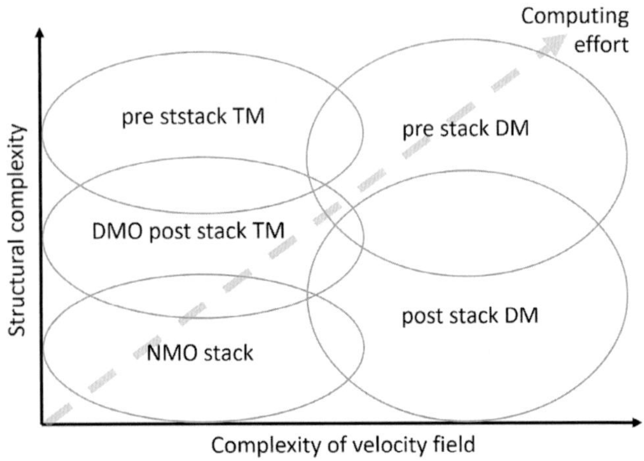

FIG. 3.3 Principal migration techniques. TM - time migration, DM - depth migration, NMO - normal moveout, DMO - dip moveout. The more advances techniques do not only need more computing effort, they may sometimes fail if an accurate velocity model cannot be established.

complementary to standard processing, since a geological/tectonic interpretation without specular reflections is unfeasible (Bréthaut et al., 2018).

3.2.4 Imaging of faults — 2-D and 3-D

Faults are often inferred by the absence of reflections, the reason being the disruption of otherwise reflecting stratigraphic interfaces. This holds true also for pure strike-slip faults, even if they do not offset the interfaces. However, these indications of faulting have to be taken with care, especially in 2-D data. P-waves can also be damped by gas accumulations, scattered by small (less than a wavelength) objects, or image other structures, e.g., impact, sinkholes, or even structures nearby the profile. Nevertheless, in 3-D seismic, a linear trace of a disrupted stratigraphic interface is a strong indication of a fault. 3-D seismic data offer further advantages in respect to 2-D:

(1) faults can be viewed, not only in one direction (as in 2-D seismic, giving typically an apparent dip angle) but also in their dip direction, by choosing appropriate interpretation lines. Nevertheless, the vast majority of 3-D seismic interpretation is still done only on inlines and crosslines.

(2) by the effective use of seismic attributes. A number of attributes can improve fault recognition. The most commonly used are dip magnitude and azimuth, curvature and coherency (Fig. 3.4). Especially coherency (equivalent attributes are variance, similarity, chaos), which describes the similarity of neighbouring seismic traces in a local window, is routinely used in fault interpretation (e.g., Chopra and Marfurt, 2007). In order to

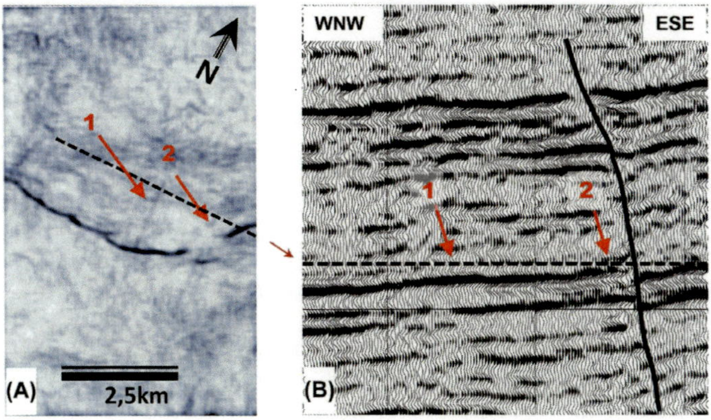

FIG. 3.4 Faults shown by a (A) time slice through a variance volume and (B) seismic section through molasses sediments in southern Germany. *Dashed lines* indicate the position of the seismic section (B) in (A) and vice versa. A large normal fault is clearly recognizable in (A) and (B), whereas some very faint E-W striking faults (red arrows) are seen only in the variance display. They would not be recognized in a 2-D seismic survey.

avoid smearing out of faults, edge- or lineament-preserving filters (Luo et al., 2002) should be applied during data processing to the original seismic amplitude data as well as to the attribute volumes (Marfurt, 2018). A number of algorithms exist for fault enhancement (e.g., ant-tracking, Randen et al., 2001) or sharpening (e.g., sceletonization, Li et al., 1997).

A drawback of 3-D seismic is that an appropriate coverage of the near-surface region is prohibitively expensive and most 3-D surveys cannot image the uppermost ~300–400 m. A supplementary survey using near-surface seismic methods is necessary, if faults should be imaged at these depths. An example from the Upper Rhine Graben is shown in Fig. 3.5. Using P-waves, an upper imaging limit of 20–40 m below the surface is usually achieved. S-wave imaging raises the limit to ~1–2 m below the ground (Inazaki, 2004; Pugin et al., 2009). For even shallower structural images, other techniques such as GPR (see Section 3.3) should be applied.

3.2.5 Fracture detection

As faults become smaller, they may not be visible anymore as single faults. However, they may be still recognizable in seismic data, if they form a set or band of fractures (see Chapter 2). These structures alter the elastic parameter and can be described by anisotropic equivalent media with continuous properties. If the fractures are vertically oriented, they exhibit a horizontal transverse anisotropy (HTI).

HTI can be recognized favourably in broad azimuth 3D seismic data by an azimuth-dependant processing of various seismic attributes, either derived

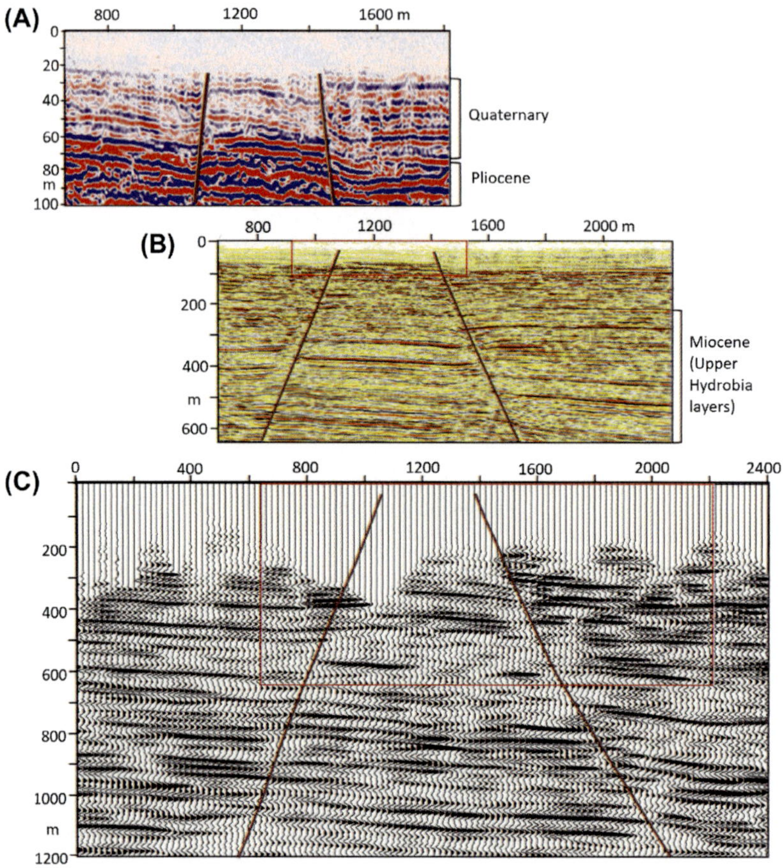

FIG. 3.5 Different seismic resolutions, illustrated in an example from the Upper Rhine Valley: (A and B) high resolution 2D seismic and (C) detail of a larger 3D seismic volume section. In (A), exaggeration is 4, (B and C) are without exaggeration. The *red boxes* in (B and C) indicates the positon of the profile in (A and B), respectively. Source frequencies in the 2D line were 30–300 Hz, in the 3D survey 12–96 Hz and CMP distances 1 m and 20 m, respectively. *Brown lines* indicate faults of a horst system. In (A), faults can be traced from only 20 m depth and fault activity in Quaternary times could be inferred. Details in Musmann and Buness (2010).

from reflections at the top of the fractured layer (e.g., amplitude, coherence), or from reflections at the bottom or further below (e.g., interval traveltime, velocity analysis; Liu and Martinez, 2012). Especially the amplitude versus offset (AVO) technique is suited to derive azimuth-controlled anisotropy (known as AzAVO).

S-waves are more sensitive to anisotropy than P-waves. They show a phenomena called S-wave splitting or birefringence (Crampin, 1985): an incident S-wave propagating through a fractured layer is split into two

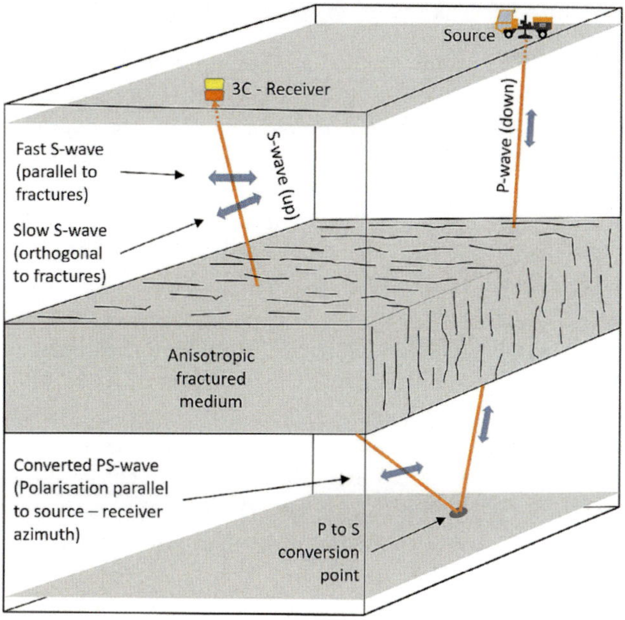

FIG. 3.6 Principle of S-wave splitting using converted waves. Blue arrows indicate polarisations. A P-wave is excited by a vertical source and penetrates the fractured medium. Below this medium, the P-wave converts to a reflected S-wave polarised in the source-receiver azimuth. As this S-wave passes upwards through the fractured medium, it is split into a fast S-wave polarized parallel to the fractures and a slow S-wave orthogonal to it. The fractures define a so-called natural coordinate system.

polarisation planes that are parallel and perpendicular to the fracture plane (Fig. 3.6). The former travels faster than the latter, the difference gives a measure of the fracture density. The data can be rotated, e.g., via the Alford rotation (Alford, 1986), into a coordinate system aligned orthogonally to the fracture system. Direct excitation of S-waves in seismic exploration is extremely rare. Instead, converted waves, detected using multicomponent recording, are more commonly used.

3.3 Ground-penetrating radar (GPR)

Ground-penetrating radar (GPR) is a geophysical field method that provides a high spatial resolution and is widely used for near-surface investigation, e.g., to reveal structures in igneous rocks, sediments and soils, to explore salt deposits, ice sheets and glaciers, to map soil moisture and the groundwater table, or for utility detection. The method utilises electromagnetic waves that are emitted into the subsurface and measures the returning reflected echo from the ground.

3.3.1 Principle

The principle of GPR is similar to that of seismics, but electromagnetic (EM) waves are used instead of elastic waves (Annan, 2005; Jol, 2009; Bristow and Jol, 2003). In air, EM waves propagate with the speed of light ($c_0 = 300\,000\ \mathrm{km\ s^{-1}}$) and in the subsurface with approx. one-third of c_0 ($\approx 100\,000\ \mathrm{km\ s^{-1}}$). The pulses are emitted by a transmitting antenna on the ground surface and propagate into the ground. If the wave hits an interface or object that is characterised by a change in EM wave impedance, a part of the energy is reflected while the other part is refracted. The impedance is a function of the magnetic and electric soil properties, i.e., magnetic permeability, electrical conductivity and dielectric permittivity. A receiving antenna records the reflected waves that reach the ground surface. In most cases, a constant-offset measurement is used, in which receiving and transmitting antenna, in a fixed setup, are towed across the ground surface, continuously recording radar signals (Fig. 3.7) that are stitched together in to a radargram. The radargram represents an image of the spatial changes of electromagnetic subsurface properties, which correlates to the subsurface architecture. A survey can be carried out along a profile to provide a 2D subsurface model, or several 2D lines can be combined to provide a 3D survey (Grasmueck, 1996).

While propagating through the subsurface, radar waves suffer from attenuation, which is caused by scattering and intrinsic attenuation by the subsurface material. Material with low electrical conductivity, such as hard rock, ice, sand and gravel, have low attenuation and depths of 10 m up to hundreds of meters can be investigated. In contrast, material with high electrical conductivity, such as clay, silt and salt water, strongly restrict the investigation depth or even render GPR inapplicable in extreme cases. The depth of investigation is also influenced by the frequency used: the lower the frequency the higher the penetration depth, but the lower the resolution.

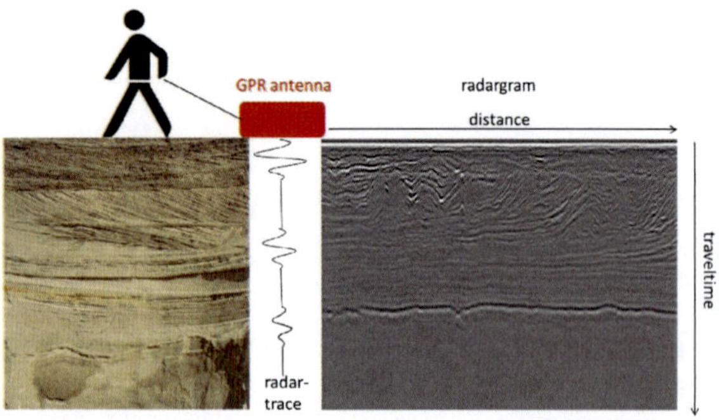

FIG. 3.7 Schematic illustration of constant-offset GPR principle: EM pulses are emitted into the ground and the signals reflected at interfaces are recorded (radar trace). The red box contains a transmitting and receiving antenna. By towing the antenna along the surface, single radar traces are recorded and stitched together in to a radargram, thus providing a subsurface image.

For geological investigations, frequencies between 20 MHz and 1 GHz are typically employed, which correspond to wavelength of 5 m−10 cm in the ground. The vertical resolution is equivalent to a quarter of the pulse width (Annan, 2005; Jol, 2009). This does not mean thinner layers cannot be detected, but defines the minimum distance between two interfaces that can be separated in the radargram. Lateral resolution is closely related to the Fresnel zone and is a function of the centre frequency and the depth to the target. Objects in the sub-wavelength range act as diffractors, which can be identified as such but no information on the size and shape of the diffractors can be retrieved. Two objects or interfaces that are closer than the resolution cannot be resolved as individual objects in the radargram.

Water plays an important role in GPR exploration because it has by far the highest relative dielectric permittivity ($\varepsilon_r \approx 80$) compared to other subsurface material, including minerals ($\varepsilon_r = 3 - 8$), and air ($\varepsilon_r = 1$). The water content influences wave velocity and impedance. Thus, not only changes of rock material can be mapped using GPR, but also subsurface water distribution, especially in porous material. The water content, however, is a function of the host material's porosity and pore-size distribution. This is one of the reasons why GPR is a powerful tool to reveal the structure of sandy aeolian or fluviatile sediments: the bedding planes are characterised by, although minor, changes in grain-size distribution and compaction, which changes the water content and therefore cause distinct GPR reflections. Another impact is that water in the subsurface contains mobile ions and contributes to electrical conductivity and thus increases intrinsic attenuation. This is especially the case for material with high cation exchange capacity (i.e., clay) and in saline environments, and thus both are not suited for GPR investigation.

3.3.2 Imaging of faults

A fault can be imaged by GPR either directly or indirectly. For direct detection, the fault itself needs to be characterised by a change of EM impedance, which causes a reflection, while for indirect detection, a particular strata must be observed on both sides of the fault. A direct image is possible, for example in the case of an open fracture or damage zone in hard rock that is characterised by a high porosity (Grasmueck, 1996). This may work in the unsaturated environment, where the fault has lower dielectric permittivity due to the increased air fraction compared to the intact rock material. In saturation, the fault may be characterised by higher permittivity due to the higher water content. Both will cause a distinct reflection when the EM wave reaches the fault. Also in sediments, faults, such as deformation bands, may be directly imaged due to their different porosity and pore-size distribution compared to the host medium (see Fig. 3.8 and Brandes et al., 2018). In contrast to seismics, GPR can image faults independent of their dip angle so that even steep structures, such as near-vertical faults, can be directly mapped (Grasmueck, 1996).

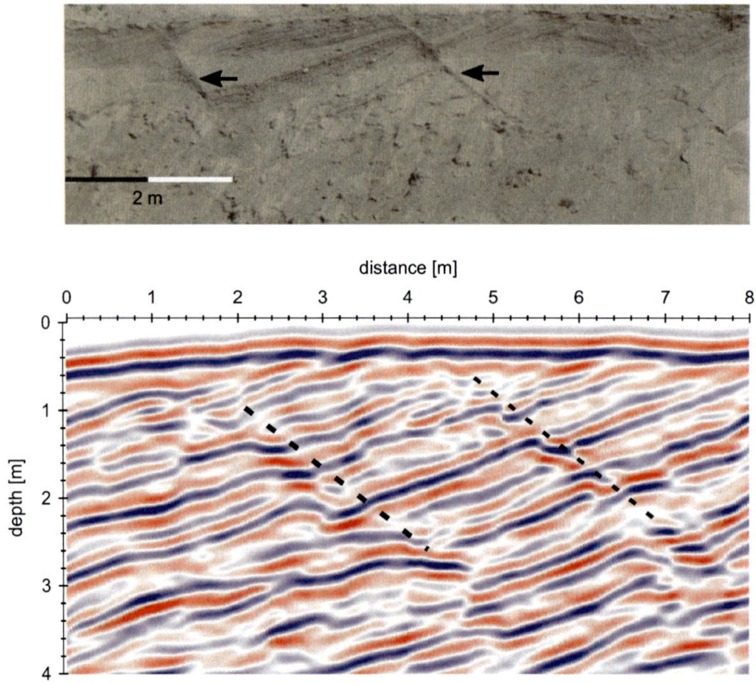

FIG. 3.8 Top: Outcrop photo in a sand pit showing sedimentary bedding and deformation bands (marked with *arrows*). Bottom: 400 MHz GPR section collected on top of the outcrop face showing reflections of sedimentary beds dipping leftwards. The deformation bands dipping right can be recognised by a reflection and by an offset of the bedding reflection (*dashed lines*).

In many cases, faults can be imaged indirectly. A precondition for this is a structured host medium comprising reflectors with high spatial correlation. Such media are layered sediments or rocks, such as, e.g., fluviatile sands and gravel, dune sands or banded sand- and limestones that are typically characterised by sub-horizontal bedding. Faults in these materials show as offsets of bedding reflectors on both sides of the fault (see Fig. 3.9 and Christie et al., 2009; McClymont et al., 2009; Brandes et al., 2018). Besides the location of the fault plane, it is also possible to determine the total displacement of the bedding reflectors (Christie et al., 2009) and to deduce mean deformation rates if information about the time of sedimentation is available.

Faults can be mapped by single 2D GPR lines, by combining individual 2D lines in to 3D interpretation (pseudo-3D GPR, e.g., Brandes et al., 2018; Christie et al., 2009) or by densely scanning the surface by 2D lines to provide a full-3D data cube that can be processed and interpreted in 3D (see Fig. 3.9, Grasmueck, 1996). Such surveys are time expensive and limited to small areas, but provide maximum information, even in a complex environment. 3D data also offer the advantage of effectively using attribute analysis, which can emphasis the signatures of faults (McClymont et al., 2008).

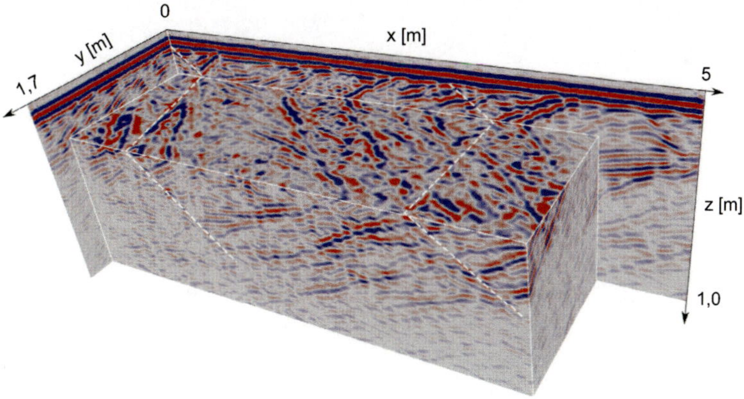

FIG. 3.9 Visualisation of a high-resolution (1.5 GHz) full-3D GPR data cube. Most reflections stem from main and subordinate bedding planes of the glaciofluvial sandy sediments. The *dashed lines* indicate the planes of two deformation bands that are characterised by either reflecting GPR waves and/or causing a shift of the bedding reflectors.

3.3.3 Examples

Shear-deformation bands developed in unconsolidated glaciofluvial sand deposits were investigated by means of 2D and 3D GPR (Brandes et al., 2018). It was postulated that these deformation bands are the process zone of deeper lying faults in the basement (Brandes et al., 2018). Fig. 3.8 shows a photo of an outcrop and a 2D GPR section collected on top of the slope, close to the edge. Both outcrop photo and GPR section show similar structures: bedding surfaces (delta clinoforms) dipping to the left and a number of deformation bands that dip to the right. In the radargram, the deformation bands can be visualised by the offset of reflections and party they cause a direct reflection. The differences between photo and GPR image are probably caused by the offset of 0.5−1 m between GPR section and outcrop face. Fig. 3.9 is the result of a high-resolution full-3D survey collected at the same location with high-frequency GPR (1.5 GHz centre frequency). Data were collected on top of the face on an area of 5 m × 1.6 m, with spatial sampling of 2 cm in both x and y directions. The data provide a high-resolution 3D image of the complex subsurface structures consisting of bedding planes, which is interspersed by planar deformation bands.

Many topographic scarps offset the basal surface of the glacial Quaternary deposits, on top of Devonian sandstones, on the northern coast of Kerry Head, County Kerry, Ireland. Outcrop analysis of the coast show that the scarp was caused by a fault in the Devonian sandstones. A pseudo-3D 80 MHz GPR survey was carried out over an area of 60 m × 40 m, with a line spacing of 1 m, in order to investigate the inland extension of the scarp trace. Fig. 3.10 shows a 2D line from the centre of the area. The upper part of the radargram is

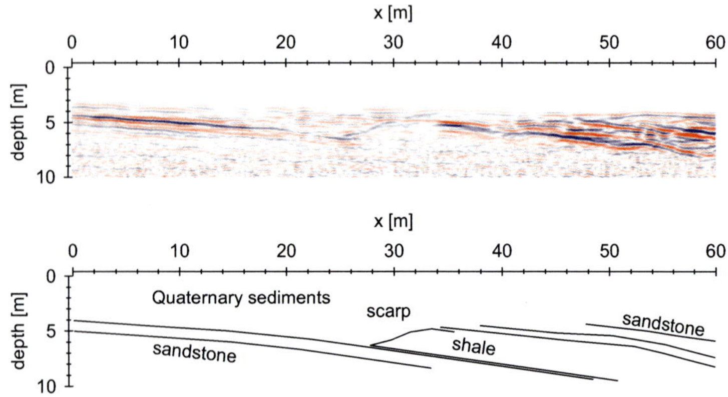

FIG. 3.10 Top: 80 MHz GPR section across a fault scarp, the early arrivals, including direct waves, have been muted. Bottom: Geological model based on GPR data and outcrop analysis.

dominated by antenna crosstalk and so has been muted. Therefore the first dominant signals are shallow reflections of the interface between Quaternary sediments and Devonian sandstones, and bedding planes inside the sandstone. The left- and right-hand sandstone facies are separated by the fault at approx. x = 30 m. The fault appears as a quite smooth slope in a shale layer, which was probably weathered before it was covered by Quaternary sediments. Due to the high electrical conductivity of clay and the resulting high attenuation of EM waves, no structural information can be obtained from below this layer. From the 3-D data (Fig. 3.11), information about the strike and dip of the

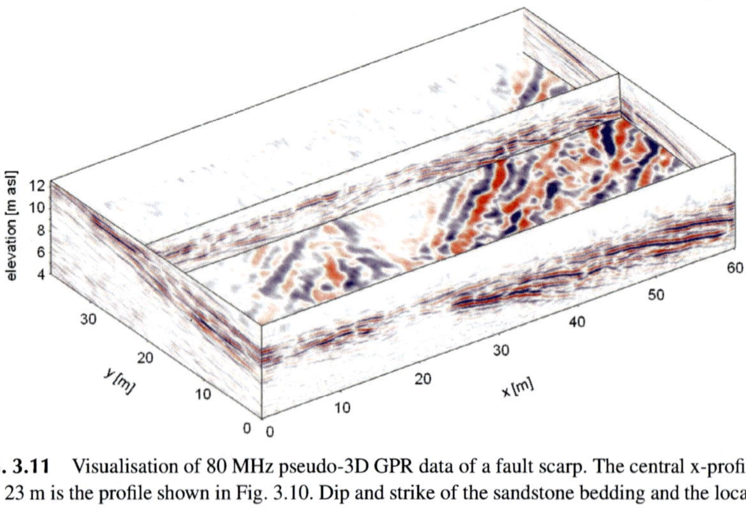

FIG. 3.11 Visualisation of 80 MHz pseudo-3D GPR data of a fault scarp. The central x-profile at y = 23 m is the profile shown in Fig. 3.10. Dip and strike of the sandstone bedding and the location of the scarp can be imaged.

bedding planes in the Devonian can be deduced as well as the paleotopography and the inland progression of the fault scarp from the outcrop at the coastline.

3.4 Electrical resistivity tomography (ERT)

The method of Electrical Resistivity Tomography is a widespread technique to explore the subsurface by means of injecting galvanic currents and measuring potentials, since the electrical resistivity is sensitive to changes in lithology (e.g., clay content) but also in the pore fluids (salinity). It does not have the resolution of reflection methods like seismics or GPR, but it can characterize the geological units if a material contrast is present. This makes it a valuable tool for fault characterization, particularly if clay or graphite is mineralized in the contact zones.

3.4.1 Background

The electrical resistivity of the subsurface is an important proxy for both the lithology (e.g., clay content, porosity), but also for indicating fluid type and saturation. Single measurements (such as deployed in the laboratory) use a four-point array, i.e., a current is injected by two electrodes and voltage is measured through two other electrodes. For a two-dimensional imaging of the subsurface, one needs many of these quadrupole measurements (Fig. 3.12A). This is most efficiently carried out by means of so-called multi-electrode instruments that connect a central unit by multi-core cables to a range of preinstalled (mostly equidistantly spread) electrodes that can be used for both current and voltage by relay switches. Modern instruments are able to acquire several thousand quadrupole data per hour. There are several arrays, such as the Wenner-Schlumberger-gradient type (potential dipole inside a longer current bipole) or dipole-dipole type (separated current and potential, one of each can be far away, leading to pole-dipole or pole-pole arrays), the latter of which is shown here subsequently. All data are used in an inversion process, where the resistivity distribution is numerically determined. Due this tomographic process, the method is called electrical resistivity tomography (ERT).

The depth of investigation mainly depends on the electrode separation that is determined by the number of electrodes and the electrode distance, but also from the signal-to-noise ratio. However, unlike electromagnetic methods, ERT is very robust against anthropogenic noise and can be applied in almost any condition, e.g., near settlements or across rough topographic undulations. Many ERT codes have been traditionally working with Finite Difference methods based on regular grids describing the subsurface. However, this is complicated for significant topography, as typically present at faults. Demirci et al. (2012) first used triangular grids to image a fault in Turkey, however still with a finite difference approach. On the contrary, the open-source code BERT

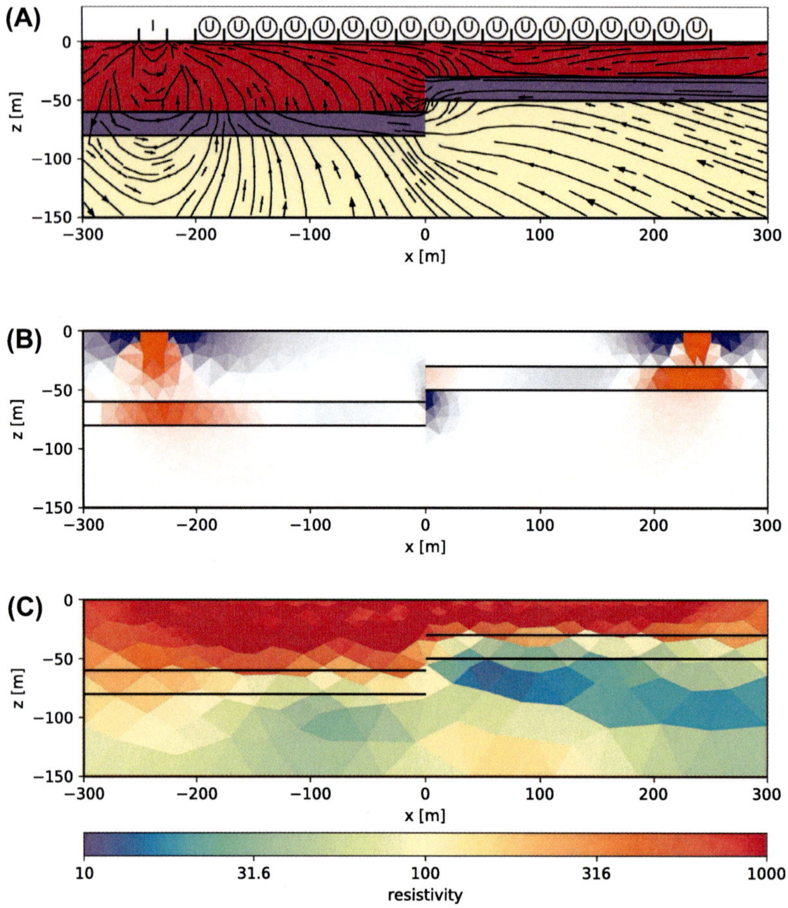

FIG. 3.12 Exemplary investigation of a fault zone using a dipole-dipole ERT experiment: (A) principle sketch of current injection (I) and voltage registration (U) along with current flowlines (*black lines*) for a synthetic fault in three stratigraphic layers, (B) sensitivity distribution for the largest separation between current and voltage dipoles, (C) inversion result of the synthetic data for the model (A).

(Günther et al., 2006) uses finite element solutions on irregular triangular meshes (or tetrahedra in 3D).

Fig. 3.12 illustrates an exemplary ERT experiment measuring the retrieved subsurface resistivity distribution around a simplified, synthetic fault of a medium-scale size. Three layers with typically occurring resistivities (from top to bottom; 1000 Ωm for a dry hardrock, 10 Ωm for a shale or marl, 100 Ωm for a saturated sandstone) are assumed to be displaced vertically by an offset of 30 m. 21 electrodes, spaced by 25 m, defines an electrode spread of 500 m, which is typically sufficient for a penetration of at least 100 m. Injection of a current I into the first dipole leads to current flowlines (black lines) that are strongly affected by the conductive second layer, which leads to sub-horizontal current vectors (Fig. 3.12A). This also influences the electric field that is measured at potential dipoles of adjacent electrodes (U). Fig. 3.12B shows the subsurface sensitivity of the measurement at the last dipole, i.e., the longest dipole spacing of 450 m. This measurement is mostly affected by the area directly beneath and next to current and potential dipoles, but also by the subsurface down to a depth beyond 150 m. The sensitivity increases in the good conductor due to the vertical current channeling and at the fault position where the resistivity changes laterally.

Subsequently, current is injected into all dipoles and potentials are measured between all remaining dipoles excluding the current electrodes so that a total number of 342 measurement is obtained. By comparison of reciprocal (i.e., exchanged current and potential dipoles) data, one can assess measurement errors. The inversion process reconstructs the subsurface resistivity, typically using a regularization that leads to the smoothest model that is able to fit the data. Fig. 3.12C shows the result of the synthetic data from the model in Fig. 3.12A, illustrating the features and limitations of ERT. The first layer is very well reconstructed, including its lower boundary, so that the fault position and displacement can be seen. Below, the good conducting layer is imaged on the right-hand side where it is shallower, however at a greater depth. Accordingly, the third layer shows up at even greater depth but with correct resistivity. On the left-hand side, the good conductor can only be estimated in the center of the profile and the third layer cannot be imaged. This is due to the current channeling by the good conductor that limits the penetration depth (Fig. 3.12B). In total, ERT can image the subsurface resistivity, but with an inherent image blurring. The latter can be counteracted by additional information, e.g., structural constraints from seismics or other electromagnetic data.

Among the publications using ERT for imaging fault zones, there is a domination of authors from the southeastern Europe (Italy, the Carpathian and Balkan countries, Turkey), as they have dominant tectonics with a lot fault zones and active research groups in the field of resistivity. Of these numerous publications, only a representative number can be given here. Štěpančíková et al. (2011) provide an impressive example of detailed imaging a very shallow fault zone in conjunction with gravity data. Both the different geological units and the extension of the individual fault zones are clearly visible. In a recent paper, Drahor and Berge (2017) give a very good overview on integrated geophysical fault imaging and provide an exemplary fault zone investigation using seismic refraction tomography (SRT), ground-penetrating radar (GPR), ERT and VLF electromagnetics, magnetics and self-potential. They clearly demonstrate on a strike-slip fault how the individual methods can complement each other and lead to an improved interpretation. Particularly, ERT reveals vertical structures and a conductive zone in the central part of the main fault.

Most recently, Chalupa et al. (2018) present a rather large-scale study where they combine ERT results with LiDAR measurements. They specifically note the lack of ERT studies at the scale of mountain ridges with penetration depths greater than 150 m. One of the rare larger-scale publications is that of Pucci et al. (2016) who image the fault zones at the region of the L'Aquila earthquake with investigation depths of up to 400 m. The largest example ever published is that of Storz et al. (2000) who imaged a graphitic fault zone in Germany on a profile length of 20 km with 3–4 km penetration depth. From these examples one can see that ERT can be applied in a wide range of scales, from meter-scale to kilometer-scale profile lengths. We present two examples describing a rather small-scaled imaging of an outcropping fault zone, and a large-scale investigation over a normal fault. In both examples, ERT is combined with other reflection methods, GPR and seismics (see corresponding sections).

ERT was applied to image the fault scarp on the coastline of Kerry Head, in addition to GPR (see Section 3.2) at the same profile, as shown in Fig. 3.13. We used a standard low-cost instrument (4-point light 10 W by LGM) with 148 electrodes, spaced 0.5 m. An optimized dipole-dipole array with jointly increasing dipole lengths and separations was used, resulting in 2850 quadrupoles measured in 1 h and 40 min. Fig. 3.13 shows the inversion result and its interpretation, which is in accordance with the GPR results (Fig. 3.10). Quaternary sediments of medium resistivity and strong heterogeneity overlie the resistive sandstone, displaced in the middle of the profile. Whereas towards the end of the profile the sandstone is massive and dry, it seems to be weathered at the other side and/or fractured so that water lowers the resistivity. In-between there is a clay layer, clearly visible as highly conductive feature that reaches down to greater depth, where GPR does not have penetration anymore. In summary, both methods support each other and lead to a more reliable interpretation. The GPR reflections could be further included as structural constraints, as demonstrated in the following example.

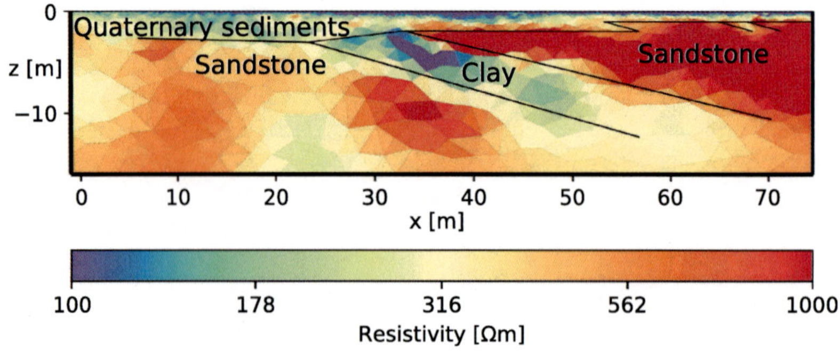

FIG. 3.13 Resistivity image of a shallow fault scarp at Kerry Head, Ireland. The top of the sandstone (resistive body), is covered by Quaternary sediments.

3.4.2 Large-scale fault imaging with structural information

In small scales (profiles of up to 1 km length), ERT measurements can be performed by so-called multi-electrode devices with electrode spacings up to 10–20 m. For larger penetration depths, this is not possibly anymore, due to restrictions in cable length and necessary power. An alternative are large-scale dipole-dipole experiments (e.g., Storz et al., 2000), where large currents (up to 50 A) are injected into a dipole of limited length (50–200 m) and the potentials are continuously recorded by data loggers (Oppermann and Günther, 2018) on receiving dipoles at large distances. The resulting time series need to be extracted, calibrated and the voltages are determined by numerical methods. Oppermann and Günther (2018) develop a new method and compare it to existing ones for a large-scale ERT survey in Thuringia.

We present a survey over the eastern flank of the Leinetal Graben, where Triassic layers are displaced by an offset of more than 300 m. This fault system has already been investigated by seismic reflection methods (Tanner et al., 2015). Several boreholes helped to interpret the seismic in terms of the present layers and to understand the salt-related tectonic processes. For details of the geology and seismic data, refer to Tanner et al. (2015) and references therein. On one of the profiles (profile 2), we performed a large-scale ERT experiment using fifteen 200 m and 400 m long current dipoles and 50 m long voltage dipoles in a total profile length of 2.4 km (Günther et al., 2011). Note that the seismic profile was just 1.8 km long due to inaccessibility of the forest with the seismic source. Nevertheless, these data represent important structural information that can be used in the regularization process. This technique, described in detail by Doetsch et al. (2012) for three-dimensional GPR reflections, deactivates the penalties in the smoothness-constrained inversion so that the usually smooth structures are avoided and structured model can be obtained.

A total number of 230 quadrupole data could be extracted from the time series by a Fourier transform method and used for inversion with a very good data fit of 4%. Fig. 3.14 shows the resulting resistivity distribution, the independent ERT inversion (Fig. 3.14A) already depicts a detailed image. In Fig. 3.14B, we have used the most confident of the seismic reflectors (black lines) from Tanner et al. (2015) as structural prior information and added their geological interpretation. In some cases, the interfaces improve existing contrasts that were otherwise blurred. In other cases, like the near-vertical main fault, constraints do not lead to significant changes, where there is obviously no resistivity contrast across the seismic boundary. Generally, the main features are kept so that the interpretation can also be made using ERT on its own.

In Fig. 3.14B, at the surface, in the west, we see conductive material (20–30 m) that can be attributed to Jurassic (marl), Keuper (claystones), underlain by more resistive Muschelkalk (limestone), underlain by upper Bunter sandstones. At 1450 m profile distance, the near-surface resistivity

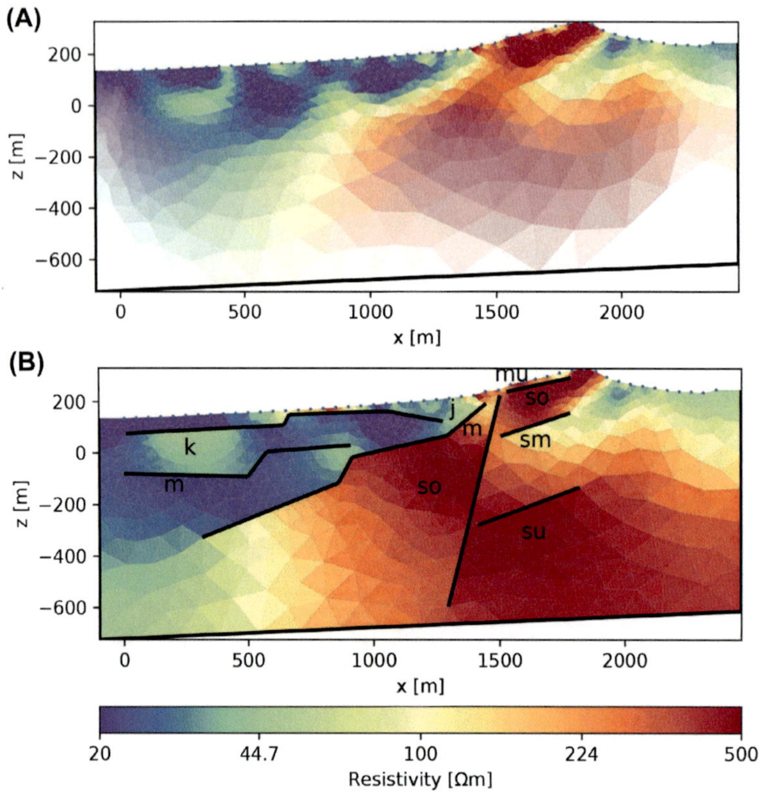

FIG. 3.14 Resistivity distributions inferred from inversion of the large-scale ERT data across the shoulder of the eastern Leinetal Graben. (A) Independent result without any constraints. (B) result using the interpreted seismic reflectors (*thick black lines*) as structural information along with the geological interpretation from Tanner et al. (2015): j-Jurassic, k-Keuper, m-Muschelkalk, so/sm/su - upper/middle/lower Bunter sandstone.

changes abruptly into higher value. All layers have been strongly eroded; the Triassic was eroded such that the very resistive upper Bunter sandstone can be found under a thin Muschelkalk layer. Beyond the graben shoulder, the layer boundaries dip steeply, and below the upper Bunter sandstone the middle Bunter sandstone formations show lower resistivity due to a higher clay content. Below, again higher resistivity from the lower Bunter sandstone appears, however it is not outcropping at the end of the profile, but is covered by rather conductive Triassic material.

In total, ERT adds some important information that could possibly improve the interpretation, e.g., that right next to the main fault, the upper Bunter sandstone reaches closer to the surface than expected. However, due to the strong contrast to the resistive blocks, the overall conductive Triassic material cannot be clearly distinguished. Further small-scale ERT could provide more details.

ERT is a cost-efficient and widespread geophysical technique that can be successfully applied to image faults. This requires significant resistivity contrasts between the contacting geological units. Mostly the layers have different clay content, porosity or fluid content and are displaced wide enough in case of normal or reverse faults. In other cases (like strike-slip faults), the active zones are filled by clay or graphite mineralization and show therefore as conductive anomalies. In all cases, we recommend a combination with other techniques, first to raise the reliability of the interpretation, but also for improving the ERT image by structural constraints (from GPR or reflection seismics).

3.5 Gravimetry and magnetics

Gravimetry and magnetics belong to the group of potential field methods (e.g., Blakely, 1995). With regard to the exploration of sub-surface structures, they make use of small spatial variations in the Earth's gravity and magnetic fields that are caused by lateral contrasts in the physical properties of different lithologies. Hence, these methods do not require active sources, like seismics or electrical resistivity tomography (see Sections 3.2 & 3.4). Because the observed quantities represent the sum effect of anomalous sources in the subsurface, both methods are theoretically able to deliver structural information over a large range of scales. However, the interpretation of the acquired data is subjected to the principle of equivalence, i.e., the interpretation is ambiguous and therefore requires external knowledge to constrain it.

3.5.1 Gravity and magnetic anomalies – definition and instruments for measurement

Gravimetry or magnetics are ideal methods to detect faults, if these faults cause sufficient large lateral contrasts in density or magnetization. The application of both methods is based on the geological interpretation of anomalies in the Earth's gravity or magnetic field. Gravity and magnetic anomalies are defined as the deviation of the observed quantities, i.e., gravitational acceleration (or simply gravity) and magnetic flux density, respectively, from the expected value of a reference Earth. Although the magnetic flux density is a vector field, magnetic exploration is often based on the interpretation of the anomalies of the Earth's total magnetic field (e.g., Nabighian et al., 2005; Lowrie, 2007; Hinze et al., 2013). The magnetic flux density (**B**, in tesla, or kilograms per second squared per ampere) is defined as:

$$\mathbf{B} = \mu_0(\mathbf{H} + \mathbf{M}),$$

where μ_0 is the magnetic constant or the permeability of free space (in henries per metre or newtons per ampere squared), **H** is the magnetic field strength (in ampere per metre) and **M** is the magnetisation of the rock in question (also in ampere per metre).

The most-frequently used anomaly in gravimetric exploration is the Bouguer anomaly (Δg_{BA}):

$$\Delta g_{BA} = g_{obs} - g_0 - \delta g_h - \delta g_{bpl} + \delta g_{terr}$$

In some applications, the free-air anomaly (Δg_{FA}) is also used:

$$\Delta g_{FA} = g_{obs} - g_0 - \delta g_h$$

where g_{obs} is the observed gravity, g_0 is the normal gravity, δg_h is the free air reduction, δg_{bpl} is the Bouguer plate reduction, and δg_{terr} is the terrain reduction (e.g., Li and Götze, 2001; Hackney and Featherstone, 2003; Lowrie, 2007; Hinze et al., 2013). Gravity anomalies can be given in metres per second squared. However, geophysicists still widely use the cgs unit, Gal ($1\ \text{Gal} = 10^{-2}\ \text{m s}^{-2}$) in gravity exploration.

Potential field anomalies are caused by lateral variations of the relevant rock physical parameters, i.e., density and magnetization (or just susceptibility, if no remanent magnetization exists). Horizontally-layered strata will not generate an interpretable potential field anomaly. Depending on the explored subject and the related physical rock contrast, anomalies caused by geological structures can typically amount up to 10^2 mGal (Bouguer anomalies) and to $10^1 - 10^4$ nT (anomalies of the Earth's total magnetic field).

The observation of gravity and magnetic anomalies requires the application of specific sensors (e.g., Lowrie, 2007). Standard gravimeters are designed to observe the vertical component of the gravitational acceleration vector; beyond land surveys, there are also instruments available for airborne and shipborne surveys. A recent development is the operation of modern gravity gradiometers on moving platforms, i.e., planes and ships, which allow the observation of gravity gradients (e.g., Fairhead et al., 2017). The derived directional information provides a spatial solution that is often superior to that of conventional gravity data (Dransfield and Christensen, 2013).

For the observation of the Earth's magnetic field, magnetometers are used, mostly either devices which allow the observation of the intensity of the Earth's magnetic total field (e.g., precession proton magnetometers, optically-pumped caesium/potassium magnetometers) — which is sufficient for many applications — or instruments which allow the observation of the vector components of the field (e.g., fluxgate magnetometers). The interpretability of magnetic anomalies can further benefit from the use of magnetic gradiometers and tensor magnetometers (Schmidt and Clark, 2000). All kinds of magnetometers can be operated on movable platforms. However, airborne gradiometer surveys are expensive, because only few instruments are as yet available, and high resolution requires low flight velocities.

The standard equipment for terrestrial gravity and magnetic exploration is less expensive than that for other geophysical methods; at least gravity surveys can be conducted even in urban environments. Airborne surveys, however, have the advantage that they can explore remote areas. Disadvantages are the limited spatial resolution of conventional exploration data acquired with scalar gravimeters and the ambiguity of the interpretation of the gravity and magnetic anomalies.

In the recent past, satellite missions have contributed much to an enhanced knowledge of the global Earth's gravity and magnetic fields due to improved coverage and availability of data. The spatial resolution of satellite data is continuously improving; e.g., the spatial resolutions of the gravity missions, in order of oldest to youngest, CHAMP, GRACE, and GOCE were circa 550 km (Reigber et al., 2005), 140 km (Mayer-Guerr, 2007), and 90 km (gradients of the gravity field; Bouman et al., 2015), respectively. From the recent magnetic missions, lithospheric magnetic field models with a resolution of 333 km (CHAMP; Maus et al., 2008) and about 250 km and better (combination of CHAMP and SWARM data; Olsen et al., 2017) were derived. Hence, the spatial resolution of satellite data typically does not come close to that of area-wide terrestrial and airborne data. Satellite data on its own only allows the interpretation of large lithospheric structures. Combined models of terrestrial and satellite data, e.g., the Earth gravitational model (EGM) 2008 (Pavlis et al., 2012), which also includes satellite altimetry, can achieve a higher resolution.

3.5.2 Gravity and magnetic anomalies - interpretation

With the potential field methods, gravimetry and magnetics, vertical or dipping lithological boundaries can be determined. Thus, they provide important complementary information to for instance, seismic methods, which prefer planar, only shallowly-dipping layers. If the lateral contrast in the rock physical parameters is sufficient, faults can cause steep horizontal gradients in the observed anomalies, i.e., closely spaced, elongated isolines in anomaly maps (Figs 3.15 and 3.16). Prominent examples are, e.g., the gravity anomalies that are caused by the boundary faults of the Upper Rhine Graben, a Tertiary rift structure in Central Europe (e.g., Rotstein et al., 2006). The strong contrast between the low-density graben sediments and the high-density crystalline rocks of the adjacent highlands results in pronounced gravity gradients, which extend over a distance of more than 200 km (Fig. 3.15). In contrast, faults in the basement of the Upper Rhine Graben beneath the thick sedimentary overburden result in less pronounced gravity anomalies, due to the smaller density contrasts.

In general, different types of faults can cause differently-shaped lateral density or magnetization contrasts. The relevant factor for faults is the displacement vector relative to the fault geometry. The shape and amplitude of the related gravity or magnetic anomaly depend on position, length, and

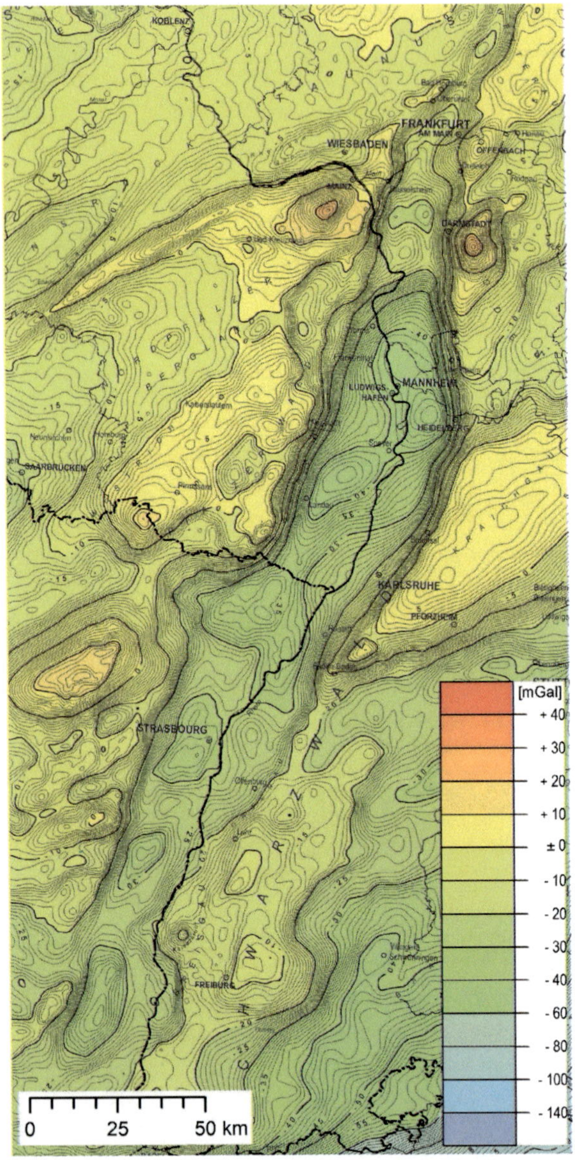

FIG. 3.15 Bouguer anomaly map (Leibniz Institute for Applied Geophysics, 2010) of the Upper Rhine, located in the German-French border area. The contrast between the low-density graben sediments and the high-density crystalline rocks of the adjacent highlands results in pronounced gravity gradients along the major graben faults. Topography from DTK1000© GeoBasis-DE/BKG 2010.

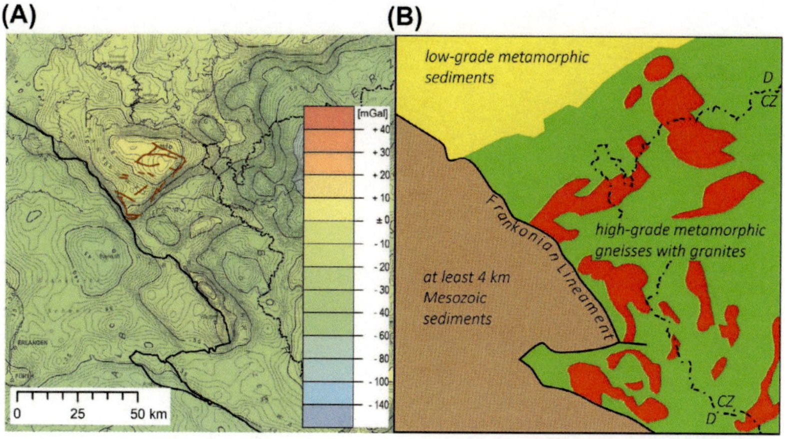

FIG. 3.16 (A) Bouguer anomaly map (Leibniz Institute for Applied Geophysics, 2010) of the western margin of the Bohemian Massif in NE Bavaria (Germany). The *thick black line* indicates the Franconian Lineament and adjacent faults; it separates Carboniferous-Mesozoic sediments of the South German Basin to the east from the Variscan basement, including gneisses and granites, to the west. The resulting large density contrast is clearly visible in the anomaly pattern. The thick brown lines are faults within the Münchberg Gneiss Massif. Only small variations in density occur here, thus the contours lines do not reflect well the fault traces. Topography from DTK1000© GeoBasis-DE/BKG 2010. (B) Generalized geological map of the same region of NE Bavaria. The *black line* is the Franconian Lineament, red bodies are granites.

orientation of the fault, and the rock physical parameters of the hanging- and footwall lithologies, i.e., the resulting contrast.

Apart from layers of differing magnetic or density properties that are truncated and juxtaposed against each other at the fault, sometimes the fault core itself can be interpreted from potential field anomalies, especially magnetic anomalies, if it is associated with mineralisation or magnetic susceptibility changes due, for instance, to hydrothermal alteration (e.g., Wemegah et al., 2015; Airo, 2015).

As mentioned previously, interpretation of gravity and magnetic anomalies is ambiguous. The same anomaly can be caused by similar sources at different depths in the subsurface, which all have the same density or magnetisation contrast against adjacent lithologies, but different geometries. In general, the maximum depth to a gravity or magnetic source can be estimated from the half-width of the anomaly; the definition of the half-width depends on the model used. Moreover, observed anomalies are often superimposed effects of various sources in the subsurface. Whereas 'density' is a natural property (scalar) of each geological body, the ability of a geological body to cause magnetic anomalies depends on its magnetic susceptibility (a tensor in the case

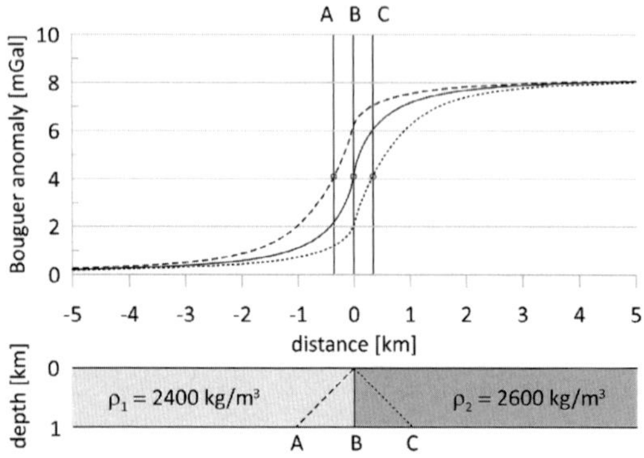

FIG. 3.17 Gravity anomaly profiles across a fault that dips at different angles between two geological bodies of different densities. For a vertical fault, the point of half-amplitude falls directly over the fault trace (case B). For inclined faults (cases A and C), the offset of this point from the surface trace of the fault indicates the direction in which the fault dips.

of anisotropic rock properties) or rather on its total magnetization (induced and remanent magnetization, both vectors), and, hence, on the amount and type of magnetic minerals.

From a gravity anomaly that is caused by a single body, the mass of the body can be derived. The maximum or minimum of the anomaly is located vertically above the centre of mass of the body. For a gravity profile perpendicular to the strike of a vertical fault that can be approximated to an infinitely long vertical contact, the inflection point of the symmetric anomaly coincides with the position of the fault trace. For a dipping fault, the inflection point is offset against the fault trace in the direction of the fault dip (Fig. 3.17). The amplitude of a gravity anomaly decreases inversely to the square of the distance from the source body.

In contrast to a gravity anomaly, the shape of a magnetic anomaly is generally more complex, because the overall magnetization of a rock is a vector property. It consists of the induced magnetization that a rock acquires in the present Earth's magnetic field due to its susceptibility, and the remanent magnetization that it has acquired in the geological past. Even in the existence of only induced magnetization, the shape of the related anomaly does not only depend on the geometry and susceptibility of the source body but also on its geographical latitude, i.e., the orientation of the source body relative to the inclination of the Earth's magnetic field (Fig. 3.18). For this reason, magnetic anomalies generally exhibit a dipole character, i.e., they become asymmetric and the maximum of a magnetic anomaly is shifted with respect to the centre of the source (e.g., Breiner, 1999; Hinze et al., 2013). The amplitude of a magnetic anomaly decreases inversely either to the third power or the square of the distance to the source body, mainly depending on its depth extent (e.g., Breiner, 1999).

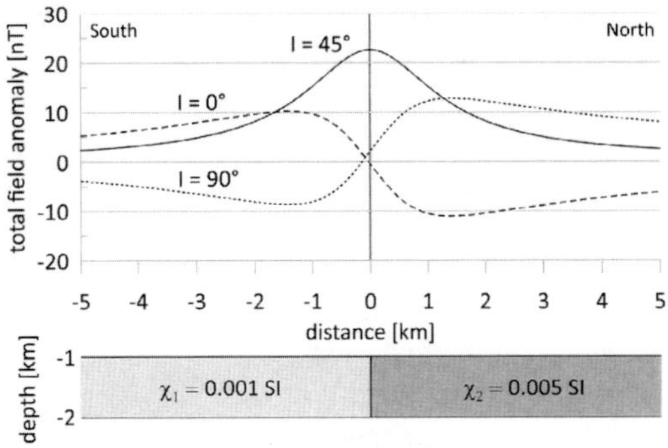

FIG. 3.18 Total magnetic intensity anomaly profile from south (left) to north (right) over a vertical contact (interface) with different geomagnetic inclinations. Note the varying asymmetry and amplitude of the anomaly caused by the dipole character of the geomagnetic field.

Only at the magnetic poles (vertical inclination of the Earth's magnetic field) and the magnetic equator (zero inclination), are the maxima and minima of the anomaly, respectively, located vertically above a body of high susceptibility. Therefore, magnetic interpretation benefits much from mathematical transformation — mostly reduction to the pole (RTP), sometimes reduction to the equator (RTE) — that converts the observed anomalies to a symmetric shape as would be observed at the magnetic poles or the magnetic equator (e.g., Baranov and Naudy, 1964; Swain, 2000). One important consequence of the strong dependency of magnetic anomalies on the inclination of the Earth's magnetic field is that exploration profiles should be therefore preferably orientated north-south, at least in areas of low to medium magnetic latitudes, otherwise the information about an anomaly, which is required for interpretation, remains incomplete. Hence, N-S striking faults are difficult to explore. Moreover, in nature, the interpretation of magnetic anomalies can become even more complicate, if the relevant source body carries also a remanent magnetization that was acquired in the geological past, in addition to the magnetization induced by the recent Earth's magnetic field. In this case, the intensity and the direction of the resulting total magnetization is the addition of both magnetization vectors. As the remanent magnetization is difficult to determine, the interpretation of magnetic anomalies is often limited to the assumption of only induced magnetization.

For a reliable interpretation of an observed anomaly with respect to the location, shape, and physical properties of the source body, it might be necessary to extract the single anomaly caused by a body from the overall observed potential field anomalies, as these represent the superimposed effects of all relevant sources in the subsurface. This can be achieved by separating anomalies of different wavelengths. Apart from graphical methods, the most convenient approaches are analytic methods that operate on equi-dimensional grids in the spatial domain (e.g., Griffin, 1949) and wavelength filters (e.g., Nabighian et al., 2005) that operate in the spectral domain. In addition, the upward continuation of potential field anomalies also allows the separation of regional (long-wavelength) and residual (short-wavelength) components of the observed anomaly fields. Concerning faults, the wavelength of an anomaly depends on the depth of the fault. However, the extent of the survey area and the distance of adjacent measuring points limit the maximum and minimum wavelength of the anomalies that can be resolved by a survey.

Beyond these filters, which isolate specific components of the anomaly fields, other filter techniques are available to enhance the short- and long-wavelength components of the anomalies. Most common is the application of derivative filters. For instance, the calculation of the horizontal gradients, the first vertical derivative, and the second vertical derivative allows a better localization of the edges of a source body, e.g., the position of the dipping contact caused by a fault that is related to horizontal contrasts in density or magnetization. Thus, such filters can help to map the trace of a fault more precisely – examples of the gravity and magnetic anomalies of the Upper Rhine Graben are presented by Rotstein et al. (2006) and Edel et al. (2007). With increasing order of the derivative, these methods become more sensitive to local anomalies (anomalies of short wavelength, shallow sources) and, thus, can be increasingly affected by noisy data. To overcome this amplification of noise, other enhancement filters, such as analytic signal (Roest et al., 1992), tilt angle or tilt-angle derivatives (Salem et al., 2008) were developed.

For dipping faults, a multi-edge analysis of the observed potential field anomalies allows a first estimation of the dip. Multiscale edge mapping is an automatic process of picking edges in potential field data by wavelet analysis at a variety of different scales, i.e., in anomalies that are computed by upward continuation to different heights (Hornby et al., 1999; Archibald et al., 1999; Holden et al., 2000). The multiscale edges are the points of maxima of the horizontal gravity gradient at different heights, which are related to the sharp discontinuities of underground sources (Wang et al., 2009). With increasing height, the edges move in the down-dip direction, moreover the degree of movement of the edges depends on the dip angle of the fault, which allows delineation of relative dip angles (Archibald et al., 1999).

In many cases, simple 2-D models can be used to calculate the gravity or magnetic anomaly related to a fault. A vertical or dipping thin sheet can approximate an isolated fault core zone, whereas a vertical or dipping contact or - depending on its location at depth - a semi-infinite horizontal thin layer or a combination of several of such thin layers can approximate the contrast between the hanging-wall and the footwall (e.g., Grauch and Hudson, 2011; Hinze et al., 2013). For such elementary bodies, simple equations exist that can also be used to plan a gravimetric or magnetic survey across a fault. Moreover, the approximation of a fault by such simple, idealized sources, allows a first estimation of some relevant parameters of the fault, e.g., the maximum depth of the semi-infinite thin layer by using the half-width of the observed anomaly (Sheriff, 2002), or the fault displacement based on the amplitude of the observed anomalies. However, this might require the assumption of additional parameters, like the contrast of the physical parameters or the depth to the top of the thin layer.

For the quantitative interpretation of a more complex geology, sophisticated algorithms and modelling software are available, which allow the calculation of the gravity and magnetic anomalies of a combination of arbitrarily-shaped bodies. Modelling can be carried out either in 2-D, 2.5-D, 2.75-D, or in 3-D, depending on the complexity of the structure (Talwani, 1959; Götze and Lahmeyer, 1988). As a rule of thumb, 2-D approaches are appropriate if the length of a structure is more than five times its width. Hence, for fault modelling, 2-D approximations are, in most cases, feasible. Forward modelling is based on assumptions about the geometry and the physical parameters of geological bodies. The anomaly calculated from a model can be compared to the observed anomaly, and the model must be iteratively improved until a satisfying fit is achieved. Due to the ambiguity in the interpretation, any constraining information should be considered for gravity and magnetic modelling, i.e., information from other sources, such as seismic reflectors or well information as geometrical constraints, and plausible densities or magnetic susceptibilities/magnetizations as rock physical constraints. Moreover, models should be kept as simple as possible. Hence, gravity and magnetic forward modelling is an appropriate tool to check geological hypotheses for plausibility.

Regarding the qualitative and quantitative interpretation of gravity and magnetic anomalies, airborne gravity gradiometry offers new perspectives, as the gradients provide a better resolution than the conventional scalar fields, i.e., the observed gradients provide even more insight into the trace and morphology of a fault surface.

3.6 Seismology

Earthquakes causes elastic deformation of the surrounding rock, a small part of which is converted into seismic energy. These waves travel through the Earth and can be observed by seismic recording stations or arrays on the Earth's surface. Thus, seismology is able to observe fault activity, at any depth or location on this or even other planets, as long as the event has enough energy to be detected. Furthermore, the hypocentres of earthquakes are able to identify fault zones that were previously unknown. The reasons why faults slip and produce earthquakes are covered below and in Chapters 2, 4 and 6. Here we concentrate on the methods to extract the position of hypocentre and kinematics of the fault movement from seismic energy.

3.6.1 Detecting and illuminating faults by earthquake hypocentre distribution

3.6.1.1 Localization of earthquakes

Earthquakes are caused by sudden slip on a portion of a fault surface (co-seismic slip, Chapter 2). A small part of the elastic deformation energy is converted into seismic energy. Seismic waves (compressional (P) and shear (S) waves) are released by the co-seismic process and travel through the Earth at different velocities, depending on the material they pass through and the type of wave.

The recording and analysis of these waves at the Earth's surface can be used to deduce information about the rupture process along the fault. The degree of detail that can be achieved depends on many factors, like, e.g., the number of recording stations, their distribution, or the quality of the recordings. A precise hypocentre location, which is the point of origin of the earthquake, is crucial to associate an earthquake with the location of a particular fault.

Ground motion, caused by seismic waves arriving at the surface, can be measured with seismometers. They convert the ground motion into electric power which can be converted into digital values and stored at a recording device or transmitted to seismological observatories for further investigation. Seismometers must be combined with a timing device. Precise timing is crucial for the analysis of seismograms (time series of ground motion recording) at various locations. An example of such a registration can be seen in Fig. 3.19.

Seismological measurements usually use instruments for the translational movement, with a vertical component and two horizontal components (usually north-south and east-west). Some measurements require only the vertical component, which is less noisy, and they are important for the arrival of P-waves. The horizontal components are more important for the later-arriving S-waves.

The design of a station network with number of stations, the network aperture and inter-station distances, depend on the specific tasks that need to be solved. Other site-specific parameters, like the coherency of the relevant waves or the required detection threshold have to be taken into account as well. The use of seismic arrays (Harjes, 1990; Rost and Thomas, 2002; Schweitzer et al., 2012) requires uniform sensors and improved timing. Seismic arrays are an ensemble of stations in an appropriate configuration that works like an antenna. They allow the calculation of apparent velocity and azimuth of the crossing seismic waves, in addition to the arrival times. The summation of the recordings at the individual array elements improve the signal-to-noise ratio.

Since a couple of years, highly sensitive and portable, rotational motion sensors have been made available. These sensors measure the three rotational components of the wavefield. The observation of the combined translational

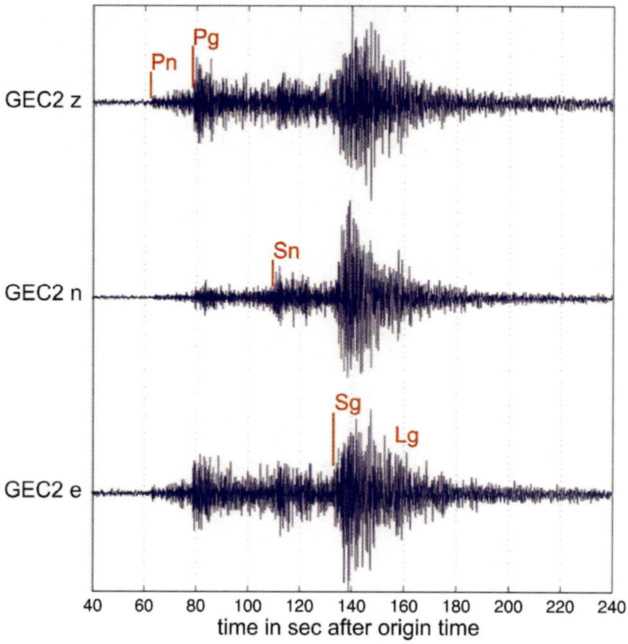

FIG. 3.19 Recording of an earthquake south of Basel, Switzerland (47.76°N, 7.54°E) at a depth of about 20 km on the fourth of May 2018, with the vertical component (z), and the two horizontal components, north-south (n) and east-west (e). The recording station GEC2 is a key element of the GERESS Array, north of Passau (Germany). There was a distance of about 475 km between station and epicentre. The arrival time of the main crustal phases Pn, Pg, Sn, and Sg are marked. The Lg wavetrain, which follows the arrival of the Sg phase, has normally the highest amplitudes (ground motion). The recording is bandpass-filtered for the frequency range of 0.8–6.0 Hz.

and rotational motion enables processing of single-station recordings that would usually require a network or array of conventional three-component measurements. The measurement of the rotational component of the seismic wavefield enables direct identification of the S-wave constituents, for example, Schmelzbach et al. (2018).

For the hypocentre determination, the source of an earthquake is approximated to a point source. The dynamic source process with an extended rupture surface - a complex time-dependant rupture process and the source mechanism - is neglected. The calculated epicentre coordinates represent the start point (nucleus) of the rupture. The origin time is the start of the fault rupture.

The earthquake location process is an optimization task for the four unknowns; X_e, Y_e, Z_e, and T_0 of the rupture process, where X_e, Y_e, Z_e, are the Cartesian coordinates of the hypocentre (earthquake location) at the depth of Z_e, and T_0 is the origin time.

The standard location procedures are based on kinematic principles, with the use of the travel times of various P- and S-waves between source and different seismological stations. The earthquake location process is non-linear in the sense that there is no linear relationship between the observed arrival times and the desired spatial and temporal parameters of the source, which could be solved analytically. An additional problem is caused by the fact that background noise and other site effects at the recording stations influence the determined arrival times.

The first method to solve the location problem was developed by Ludwig Geiger (Geiger, 1910, 1912). His method is based on an iterative search process to minimize the least-square error between the observed and predicted arrival times at the stations. A minimum of three arrival times from three different seismic stations or data from a single seismic array is required.

Modern locating routines are based on nonlinear inversion techniques (e.g., Lomax et al., 2000; Schweitzer, 2001). They minimize the differences between the parameters of the observed seismic waves (e.g., arrival times, time differences and slowness vectors) and the theoretical parameter calculated for the velocity depth model. The accuracy of the calculated earthquake hypocentre strongly depends on how close the used model is to the real geological situation. Velocity models can vary from a simple half space with constant velocities, a laterally homogeneous-layered model or a complex three-dimensional model, which requires detailed knowledge about the geological structures and rock properties along the travel path of the seismic waves. A simple model for the Earth's crust is seen in Fig. 3.20. The crust is modelled as a layer over a half space with a constant P and S velocities. The theoretical travel times of the various P and S-phases are displayed together with seismograms at two arbitrary distances and their corresponding ray paths in the layer.

In case of a single event, a dense network of seismic stations or arrays, and a realistic velocity depth model is required, for both P- and S-waves, to achieve a high-quality hypocentre. For an ensemble of earthquakes, the application of relative location methods minimizes the hypocentre uncertainty between different earthquakes. The double-difference method (Waldhauser and Ellsworth, 2000) minimize the residuals between observed and theoretical travel-time differences (double-differences) for pairs of earthquakes at each station, while linking together all observed event-station pairs. The relative coordinates are then transferred into absolute coordinates with the reference coordinates of the earthquake.

Usually, the reference earthquake is the one with the smallest hypocentre uncertainty. This method is particularly suitable for small events or micro events, which are measured only at a limited number of seismic stations, close to the hypocentre.

The hypocentre is usually calculated as the point at which the rupture process starts. The structure, orientation and size of the rupture zone may vary over a wide range. Its size can be estimated from the magnitude of the earthquake which is a measure of the radiated energy. How to use recordings

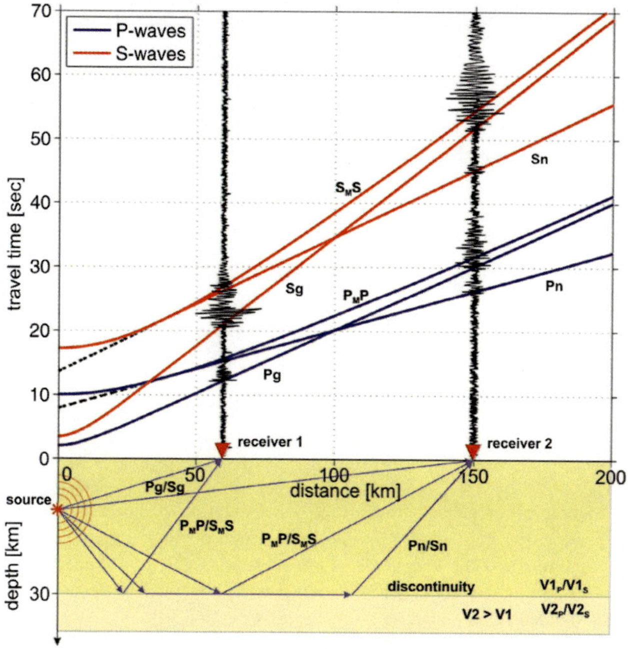

FIG. 3.20 Theoretical travel-time curves for P and S-waves and corresponding seismogram sketches at the distance 60 and 150 km (upper part) for a simple crust model with a layer over half space (lower part). Ray paths of the waves are marked by *blue lines*. The Pg/Sg phases are the direct wave in the layer (crust). P_MP/S_MS are the reflected phases at the discontinuity between layer and half space. Pn/Sn travel along the discontinuity with the velocity of the half space.

of earthquakes for the determination of the orientation of the rupture zone is explained in more detail in Section 3.6.2.

3.6.1.2 *What can be learnt from earthquakes?*

There are three principal ways to associate an earthquake or earthquake sequence with the location and orientation of a fault:

- Spatial and temporal distribution of earthquakes
- The location of a single earthquake in combination with the dynamic properties of the source, such as a fault-plane solution
- Spatial distribution of aftershocks that follow a main earthquake

3.6.1.2.1 Spatial and temporal distribution of earthquakes

The observation of the seismicity over a long period provides one possibility to get an image of the fault zone structure from the spatial distribution of earthquake hypocentres. In regions with a low seismic activity (seismicity rate), this might need a long time. To decrease the observation time, one must lower the detection threshold for the observed region. The Gutenberg-Richter

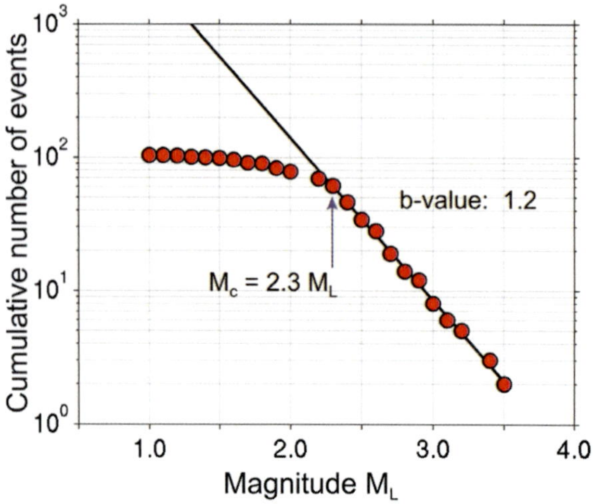

FIG. 3.21 Example of a frequency-magnitude distribution of seismic events within a defined region with a b-value of 1.2. The *black line* is derived from the linear regression of the observed events for magnitudes above the magnitude of completeness (M_c). It is expected that all seismic events with a magnitude above M_c were detected by the seismological network. The gradient of the *black line* is the b-value of the Gutenberg-Richter relationship.

law describes the magnitude-frequency distribution of earthquakes measured for any specific region. It is defined as

$$\log_{10}N = a - bM$$

where N is the number of events with a magnitude $\geq$M within a specific time rate and a and b are constants (Gutenberg and Richter, 1956). The b-value describes the specific relationship between the magnitude and total number of earthquakes and it is commonly close to 1.0 in seismically-active regions. A b-value of 1.0 means that for a given number of magnitude 4.0 and larger earthquakes there will be 10 times as many magnitude 3.0 and larger earthquakes and 100 times as many magnitude 2.0 and larger earthquakes, etc. An example of a magnitude-frequency distribution is seen in Fig. 3.21 with a b-value of 1.2.

There are various methods to improve the detection capability of an investigated area. To increase the density of the network is a well-tried way, but this increases the amount of maintenance. Improvements can be achieved with adapted network designs, the use of arrays instead of stations, and the selection of stations with better site conditions and lower background noise. For some regions with thick sediment coverage, the use of seismic stations in boreholes is necessary to achieve the reduction of noise.

The existence of a single earthquake clearly indicates that an active fault exists at the calculated hypocentre. The analysis of the dynamic properties of the source provides information about the orientation of the fault plane solution of the rupture process, which in turn allows to estimate the orientation of the fault. See Section 3.6.2 for more details.

Aftershock sequences are well-investigated phenomena and have been observed for many earthquakes. They steadily decrease in frequency and magnitude over time, according to Omori's law (Omori, 1894), which is an empirical relation. The absolute number of events and their magnitudes depend on many factors, such as geological properties and the existence of other faults. The parameter for a specific earthquake (decay parameter and productivity of aftershock sequence) are related to the monitoring station network and its detection threshold.

Aftershocks are assumed to be caused by induced stress changes produced by the main shock. They can occur along the fault plane which was activated by the main shock or along fault branches in the environment. The temporal and spatial distribution of aftershocks therefore provides important information about the rupture process and the geometric shape of fault plane of the main shock (e.g., Kilb and Rubin, 2002).

3.6.2 Describing faults by interpretation of source mechanisms

3.6.2.1 The mechanics of earthquakes

More than 90% of all earthquakes are of pure tectonic origin; the remainder are of anthropogenic origin. They are a direct expression of an abrupt stress release in the Earth's crust or mantle. The geomechanical process behind tectonic earthquakes is as follows: Tectonic forces generated by plate motion lead to stress accumulation in the brittle part of the crust. When the strength of rupture or the static friction in the rock is reached, an abrupt dislocation between two distinct, closely-spaced rocks occurs. This process, called earthquake, mostly takes place at the weakest part of the stressed rock unit, usually on an already-existing fault. The accumulated energy is released in the form of deformation, heat and elastic waves. The larger the area of dislocation is, the more energy is released and the higher the earthquake magnitude.

The seismic waves radiate away from the earthquake source and they are measured with seismometers at the Earth's surface (see Fig. 3.19). They can be used to invert for the kinematics of the earthquake process. In the simplest approach, observed polarities (first motions) of P- and S-waves are inverted for a pure shear dislocation, the so-called 'double couple'. Elaborated methods use the full seismic waveforms and invert them for the so-called moment tensor, which comprises not only the shearing part of the source, but all other possible source components, like, e.g., tension or explosion. In the following, the determination of a double-couple focal mechanism will be explained. At the end of the chapter, the concept of the moment tensor is addressed in a short quantitatively way.

3.6.2.2 The concept of the double couple

It is commonly assumed that tectonic earthquakes are mainly caused by shear faulting, i.e., slip on a fault which can be modelled as displacement on an internal surface in an elastic media (e.g., Shearer, 2009). The equivalent force system for such a dislocation is a pair of force couples with no net torque (Fig. 3.22A). This is termed a 'double couple' (DC). The energy radiation of a double couple is highly non-uniform, both for P- and S-waves

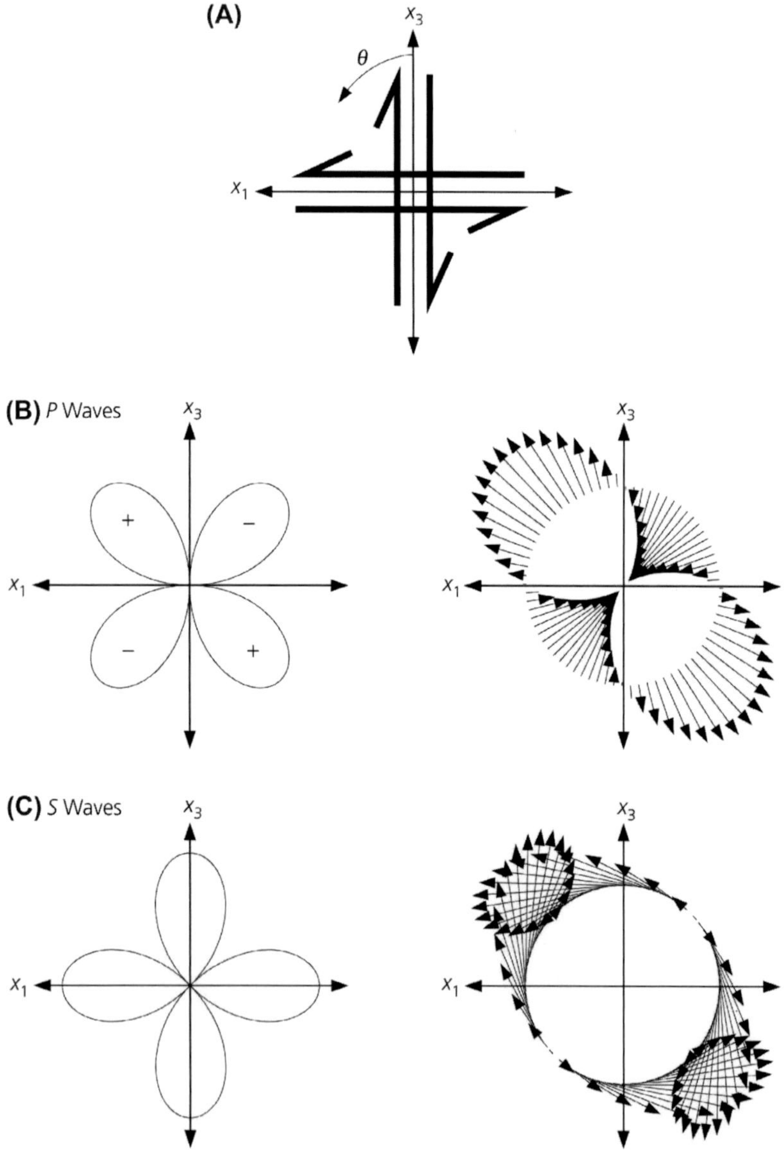

FIG. 3.22 Representation of the double couple (DC) in the x1-x3 plane (A) and corresponding radiation pattern of P waves (B) and S waves (C) in the same plane (Stein and Wysession, 2003). + and − indicate compressional and dilatational onsets, respectively. *Arrows* symbolize the polarization of the wave.

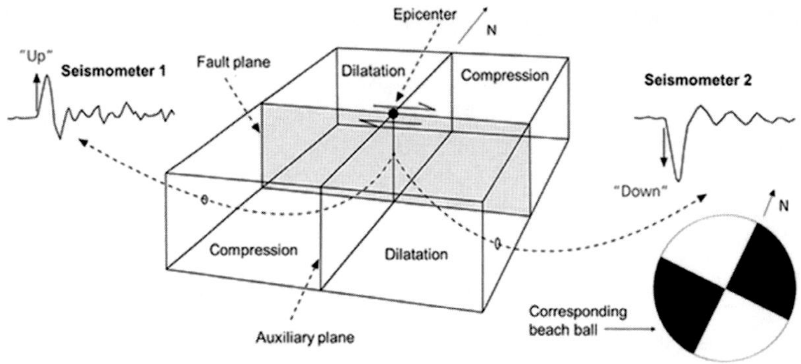

FIG. 3.23 Fault and epicentre together with the radiation of the double couple into the different quadrants of compressional and dilatational motion. Exemplarily, ray paths of the P wave to two seismometers are shown that exhibit opposing first motion of the P wave (up and down). The 'beach ball' that belongs to the shear failure shown is shown in the lower right-hand corner. It represents a right-lateral strike-slip fault on an E-W oriented fault plane. *Modified after Stein and Wysession (2003).*

(Fig. 3.22B and C). Here, it is explained in a quantitatively way. For further information and mathematical formulations, see Aki and Richards (1980).

The P-wave radiation projected on the focal sphere can be divided into four quadrants, which are separated by so-called 'nodal planes' where the displacement is zero (Fig. 3.23). Opposite quadrants have the same polarity, which may have positive or negative initial motion and neighbouring quadrants have different polarities. The positive quadrant is associated with the tensional axis (T axis) leading to upward or compressional motion of the P-wave at the surface receiver (seismometer) and vice versa, i.e., the dilatational quadrant has the compressional axis (P axis) and thus downward motion at the receiver (Fig. 3.23). The S-wave radiation pattern describes a vector displacement that does not have nodal planes, but is perpendicular to the P-wave nodal planes. S-wave motion converges toward the T axis, diverges away from the P axis, and is zero at the so-called nodal points, where both nodal planes intersect (Fig. 3.22C).

One of the two nodal planes that separate compressional from dilatational quadrants represents the "real" fault plane (rupture plane). The other nodal plane is the so-called auxiliary plane. For the fault plane solution alone, there is no possibility to determine which nodal plane is the real fault plane and which is the auxiliary plane. To solve this ambiguity, the distribution of aftershocks (see Chapter 3) or other geological information have to be considered.

3.6.2.3 Determination of focal mechanisms

The complexity of the source radiation of a tectonic earthquake, as described above, allows the inversion and determination of the fault geometry and the relative movement of the fault blocks. It puts strong constraints on the kinematics of the faulting.

A prerequisite for the seismological investigations of the source process are measurements of polarities, amplitudes or waveforms from seismic stations well distributed around the seismic event. The distinct polarities for the stations and the take-off angles of their corresponding rays are plotted in a stereographic projection of the focal sphere. Then, the fault-plane solution is determined by finding those orthogonal planes (nodal planes) that separate the individual observations into compressional and dilatational quadrants, or in other words, splitting the regions of upward and downward first motion. Today, a simple grid-search algorithm is usually performed to find those pairs of nodal planes that are consistent with the observed polarities.

An example for this procedure is shown in Fig. 3.24, for the strongest event of the 2008 earthquake swarm in Vogtland/NW-Bohemia. Fig. 3.24A shows the distribution of the observed 24 P-wave polarities in stereographic projection of the lower hemisphere, in which octagons and triangles mark compressional and dilatational onsets, respectively. They serve as the basic input parameters for the inversion of the focal mechanism, as well as amplitude ratios of P- to S-waves (marked by crosses). The best-fitting pair of nodal planes that were found in the inversion is shown in Fig. 3.24A; they separate the compressional and dilatational onsets. A more simplified or sketched representation of the fault plane solution is the so-called 'beach ball' (because of its appearance), which omits the individual observations and differentiates only between compressional and dilatational quadrants using different colours (usually black for compression and white for dilatation). It allows a quick and

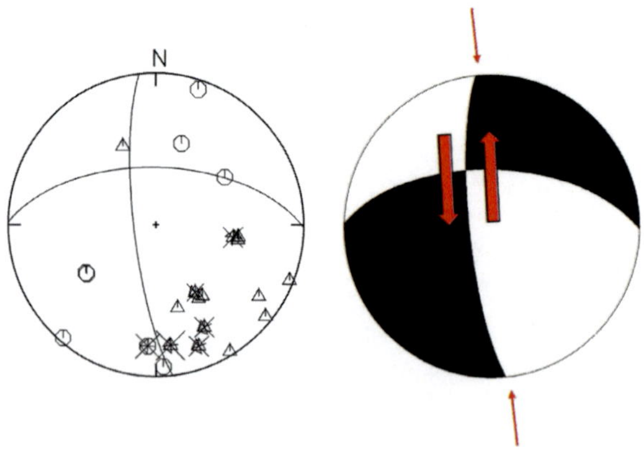

FIG. 3.24 Example of focal mechanism determination from observations of P polarities and P/S amplitude ratios. The mechanism is determined for the strongest event of the October–November 2008 earthquake swarm in Vogtland/NW-Bohemia (Plenefisch, 2009). Left: The input parameters (octagon = compression, triangle = dilatation, cross = additional amplitude ratio) as well as the best-fitting nodal planes calculated with the program FOCMEC (e.g., Snoke, 2003). Right: The corresponding "beach ball" with the preferred fault plane (for further information, see text).

more recognisable assessment of the fault mechanism (Fig. 3.24B). In this case, the beach ball represents a strike-slip fault mechanism with a slight normal fault component (for explanation of different mechanisms, see below).

3.6.2.4 Styles of faulting

In seismology, the focal mechanism is usually described by three parameters: strike, dip and rake (Fig. 3.25, cf. Chapter 1, Fig. 1.3). The strike φ is measured clockwise from north ($0° \leq \varphi \leq 360°$), the dip δ from the horizontal ($0° \leq \delta \leq 90°$), and the so-called rake λ ($-180° \leq \lambda \leq 180°$). The rake is the angle between the fault strike and the slip vector, which defines the movement of the hanging-wall with respect to the footwall. If the rake is positive, there is an upward movement of the hanging-wall and it is termed a reverse faulting, whereas a downward movement (negative rake) is called a normal faulting (Fig. 3.26). For strike-slip motion one can further distinguish between right-lateral (or dextral) and left-Iateral (or sinistral) strike-slip. If an observer is standing on one block and sees the other block move to the right, it is termed right-lateral strike-slip motion and vice versa for left-lateral strike-slip motion.

3.6.2.5 The concept of the moment tensor

Often the source mechanism of an earthquake is not only a double couple, but also more complex. In order to describe the source processes more completely, the concept of the moment tensor has been developed (e.g., Aki and Richards, 1980; Jost and Hermann, 1989). The moment tensor is a 3 × 3 matrix; it is

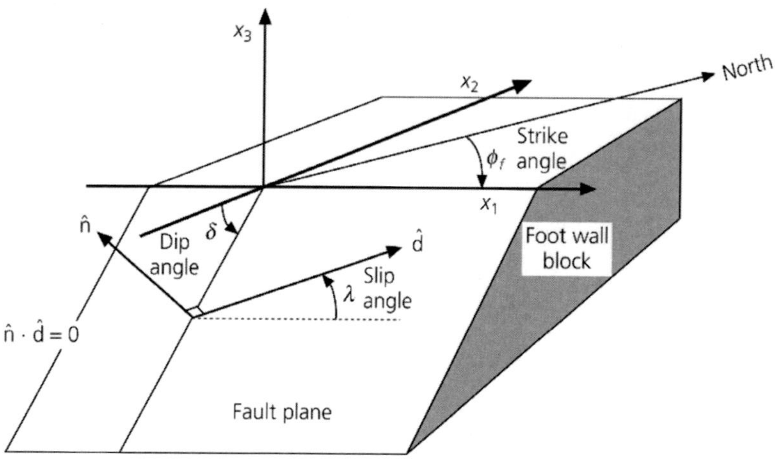

FIG. 3.25 Definition of a planar fault by strike, dip and the direction of the slip vector (Stein and Wysession, 2003). By determining the focal mechanism of an earthquake, the basic geometrical attitude of an active geological fault can be resolved. Moreover, the style of faulting (normal, reverse or strike-slip) can be specified and even the direction of the absolute motion between the associated blocks can be defined.

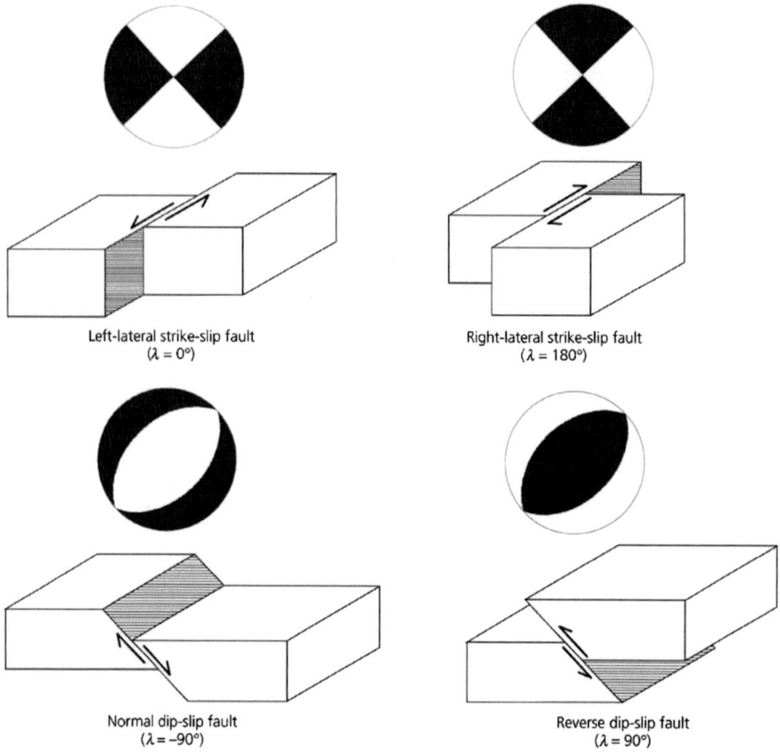

FIG. 3.26 Basic types of fault geometries and styles of faulting together with the corresponding beach ball representation, plotted on a stereographic projection of the lower hemisphere. Black and white colours denote the compressional and dilatational quadrants, respectively. *Modified after Stein and Wysession (2003).*

symmetrical and consists of nine single force couples. The concept of the moment tensor is quite general and comprises not only earthquake sources, but also all seismic sources, such as explosions, implosions, landslides, etc. The full moment tensor consists of an isotropic part and so-called deviatoric moment tensor.

Decomposition of the moment tensor is often needed to find the most appropriate components that represent geological and physical source processes and support interpretations (e.g., Dahm and Krüger, 2014). Different ways exist to decompose the moment tensor and the deviatoric component into different elementary sources (e.g., Jost and Herrmann, 1989). One kind of decomposition is e.g., into a double couple and a so-called 'compensated linear vector dipole' (CLVD). The exact mathematical derivation of the moment tensor and the different decompositions of the moment tensor are beyond the scope of this book and we thus refer to Jost and Herrmann (1989) and Dahm and Krüger (2014).

Several seismological agencies (e.g., USGS, GFZ) routinely provide moment tensors, partly based on automatic analysis; mostly for larger earthquakes and at teleseismic distances (more than 1000 km epicentral distance). In general, the complete seismograms - consisting of body and surface waves and filtered for lower frequencies - are inverted to reveal the moment tensor. It has turned out that the shear faulting component (double couple) of the moment tensor dominates mostly with a ratio of more than 90%. Other components are usually low and of minor importance.

For lower magnitude earthquakes, it is more difficult to invert for the full moment tensor and to resolve non-double-couple components, because higher frequencies have to be considered. Also, more detailed 3D-velocity models are required, which are often not available. Therefore, for small magnitude earthquakes, the double-couple approach, with the classical consideration of phase polarities, is still an adequate and often-used method.

Thanks to increasing computer power, inversions for extended source models and full moment tensor inversions for smaller earthquake magnitudes are advancing fast. The newest development in the field of source determination is the measure and use of rotational ground motion recordings in addition to the classical translational ground motion. This enables the determination of the moment tensor, even in the case of sparse seismic networks (e.g., Donner et al., 2018).

3.6.3 Examples of detecting faults using hypocentre distributions and focal mechanisms

3.6.3.1 Vogtland/NW-Bohemia swarm earthquake area

Earthquake swarms are sequences of many earthquakes that occur in a relatively short period without a specific main shock. Vogtland/NW-Bohemia, a region located on the border between Germany and Czech Republic, is known as one of the most interesting earthquake swarm regions in Europe, where in the course of a swarm thousands of small and intermediate magnitude earthquakes occur within some months (e.g., Klinge et al., 2003; Fischer et al., 2014).

Interestingly, the earthquake swarms of the last two decades in this area took place at more or less the same location in the Czech Republic, near the small village of Nový Kostel. The thousands of hypocentres of each swarm do not build a diffuse cloud, but instead form planar, fault-like structures. Even if the source of the swarms is most likely connected with ascending fluids, the distribution of the hypocentres seems to be located at zones of pre-existing weakness, which are most probably faults.

The epicentres for the swarms since 1997 are shown in Fig. 3.27. The distribution of the epicentres clearly depicts an elongated NNW-SSE oriented zone of around 10 km length and a relatively small width of around 2 km.

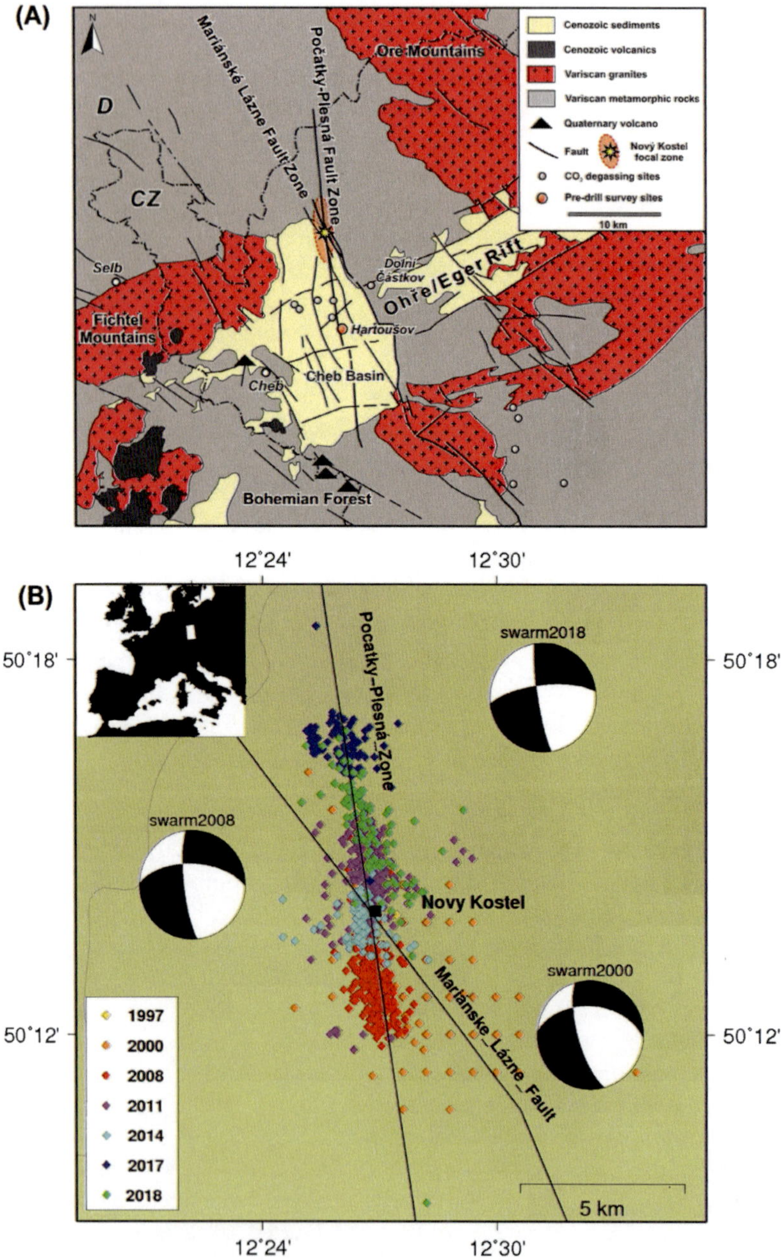

FIG. 3.27 (A) Geological units, major faults, Quaternary volcanoes, Nový Kostel earthquake focal zone and CO_2-degassing sites in the Vogtland/NW-Bohemia region (Bussert et al., 2017), (B) epicentres for the Vogtland/NW-Bohemia major earthquake swarms between 1997 and 2019 (BGR earthquake catalogue), together with a sketch of major faults and representative focal mechanisms for the swarm 2000 (Plenefisch and Klinge, 2003) and 2008 (Plenefisch, 2009).

Thus, the Nový Kostel seismic zone nicely illuminates a distinct zone of weakness. The NNW-SSE oriented seismic zone is in relatively good accordance with the strike of the nearby Mariánské Lázně Fault and other parallel-oriented, smaller faults, like the Počátky-Plesná zone (Bankwitz et al., 2003). The focal depths of the swarm earthquakes (not shown here), range from 6 to 10 km depth and the distribution of the hypocentres indicate a west-dipping fault plane of 70–80°.

The majority of focal mechanisms from the earthquake swarms exhibits strike-slip movements with a slight normal faulting component. A great number of them have an almost N-S oriented nodal plane. According to the hypocentre distribution and the orientation of the surrounding faults, this should be the fault plane. Summing up the findings from seismological investigations - the hypocentre distribution and the focal mechanisms - there is strong evidence for an active, almost N-S oriented and steeply dipping fault in that region, which ruptures mainly as a left-lateral strike-slip fault.

3.6.3.2 Central Apennines, Italy

Another example where the distribution of earthquake hypocentres and their focal mechanisms contributes to the characterization of faults is the region of the central Apennines in Italy. In comparison to the preceding example of the Vogtland/NW-Bohemia, which is situated in an intra-continental area, with a comparatively small fault zone, the faults in the Central Apennines are larger and the tectonic situation is predominantly determined by the surrounding large-scale plate tectonics. It is the general distribution of the seismicity, recorded over decades, that illuminates the faults in the region and in particular the earthquake sequence of 2016 between Amatrice and Assisi, with 3 earthquakes above magnitude Mw 6 (6.2, 6.1 and 6.5). The aftershock series clearly depict the regional faults in the area (Fig. 3.28).

The mountain range of the Apennines runs from the Gulf of Taranto in the south to the southern edge of the Po basin in northern Italy. It formed because of the E-W directed subduction of the Adria microplate beneath Italy. The contemporary tectonics is very complex, because it is additionally influenced by the opening of the Tyrrhenian Basin to the west, and the convergence of the African against the European plate. Thus, the central part of the Apennine mountain range is dominated by extensional tectonics, often manifested by the occurrence of normal faulting earthquakes.

This also holds for the earthquake sequence from August to October 2016, which started close to the small village of Amatrice, with a main shock of moment magnitude Mw 6.2. The sequence continued in October with a magnitude 6.1 and another magnitude 6.5 event, which both took place NNW of the first. The earthquakes killed approximately 300 people and caused severe damage in the region.

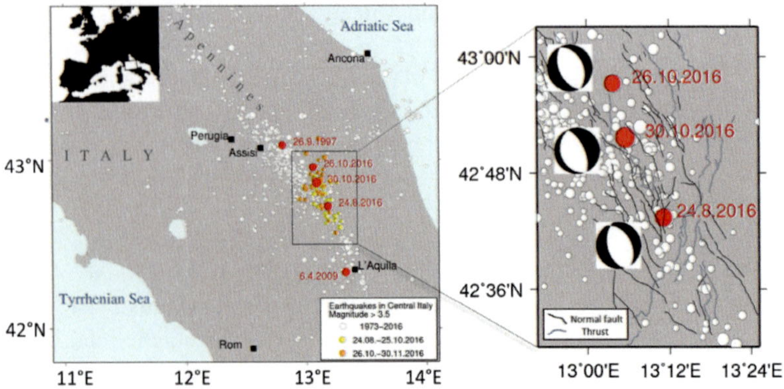

FIG. 3.28 Epicentre distribution of earthquakes in Central Italy between 1973 and 2016 (epicentres taken from USGS). Indicated by *red circles* are the strongest and most-damaging events of the last 20 years. The epicentres of the aftershock sequences of the Amatrice (24.8.2016) and of the Norcia earthquakes (26/30.10.2016) are marked by *yellow* and *orange circles*, respectively. The map on the right shows the epicentres of the three strongest earthquakes from the 2016 sequence (*red circles*), the corresponding focal mechanisms (source USGS) and fault systems (adopted from Villani et al., 2018). Quaternary normal faults are shown as black lines, Miocene-Pliocene thrusts shown as *blue lines*.

Fig. 3.28 shows the epicentres of the three main shocks, as well as those of the corresponding aftershocks. The distribution of the epicentres marks a more or less NNW-SSE elongated zone of around 50 km length. In addition, Fig. 3.28 also indicates the background seismicity of more than 40 years as well as the positions of the two preceding disastrous earthquakes in the region, namely the earthquake sequence close to Assisi in 1997 and that at L'Aquila in 2009. It seems that the earthquake sequence in 2016 filled a gap between the sequences in 1997 and 2009. Together with the earlier aftershock sequences, the epicentre distribution illuminates a wide NWW-SSE oriented fault system in the Central Apennines. The distribution of the earthquake hypocentres can be associated to the so-called Mt. Garzano Fault System and the Mt. Bove-Mt. Vettore Fault System (e.g., Chiaraluce et al., 2017; Centamore et al., 2009; Lavecchia, 1985), which are nearby and of more or less the same strike direction.

The focal mechanisms for the main shocks of the 2016 sequences represent NNW-SSE striking, normal faults. They are thus in accordance with the faulting style of the majority of the known faults (Fig. 3.10) and consistent with an extensional regime with a minimum stress axis oriented perpendicular to the strike of the mountain chain (see e.g., INGV Working Group, 2016).

Integrating earthquake observations and analysis, together with knowledge of the regional fault pattern, can give a relatively consistent picture of the geometry and kinematics of an active fault system.

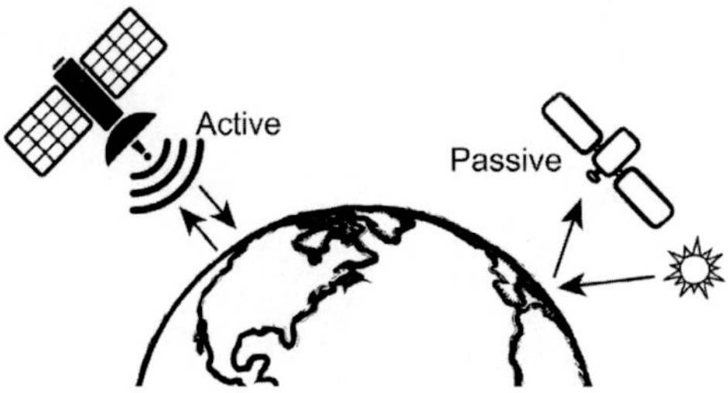

FIG. 3.29 Active sensors illuminate the Earth, and passive sensors use the sunlight and thermal radiation. For active systems, the time delay between emission and return is measured (absolute or relative). Some active systems (such as synthetic aperture radar, SAR) can even operate through cloud cover and at night. Note that this figure only illustrates satellites, but different carriers (aircraft and unmanned aerial vehicles) can also transport active and passive sensors.

3.7 Remote sensing

Remote sensing of faults is achieved by collecting images, points, anomalies and trends from a distance, using vehicles such as satellites, aircrafts, or drones. Past decades have demonstrated the great value of modern acquisition technologies, image analysis methods, and data science for exploring the surface expression and deformation associated with fault dynamics. Remote sensing allows the mapping of lineaments associated with geological structures, such as fractures, joints, dykes, and straight rivers often associated with fault trends, and determining lithology and topography offsets.

Remote sensing of faults also allows the complex analysis from two-dimensional image stacks to time-dependent multi-sensor approaches, to assess fault propagation, interactions, slip, creep, and erosion. Active and passive sensing technologies are used (Fig. 3.29), where the integration of data and interpretation on the physics and source processes at depth are derived through modelling, allowing a robust interpretation of structural features from the surface to depth. This section provides first a short historical overview, and then aims at summarizing the applicable and most promising tools that the portfolio of satellite remote sensing has to offer. Examples are used to illustrate the remote sensing of faults from selected locations at different geodynamic settings around the world.

3.7.1 History and background of remote sensing

Remote sensing has a long history; for more than a century and for many decades have airborne and spaceborn systems been used (Fig. 3.30), respectively. The earliest airborne systems were photo cameras transported by balloons in the 1840s, which were soon replaced by fixed-wing airplanes to realize controlled photographic perspectives of the ground. Miniaturizing cameras and automatic releases by a timer allowed especially lightweight camera systems even to be carried by homing pigeons after 1905. The main

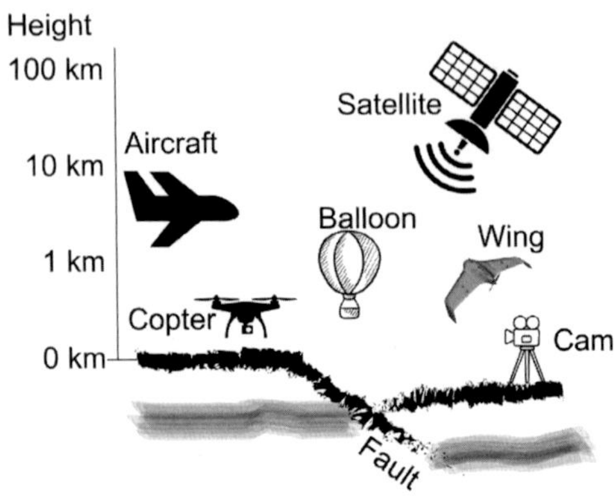

FIG. 3.30 Faults can be remotely sensed from various distances, using different remote sensing carriers, such as satellites, aircrafts, copters drones, balloons, wings or ground-based imaging (camera) systems. As a rule of thumb, close-range systems provide a higher resolution, whereas larger distance systems provide large overviews.

problem was that the flight of these early "drones" was rather difficult to control. Cameras mounted on rockets launched in the late 1940s opened a new era of photogrammetric remote sensing from high altitudes and with wide perspectives. The first cosmonauts and astronauts collected extensive photographic records, which are still available and used today extensively in geoscience. Many more vehicles and satellites were developed for military use, mainly as photogrammetric surveillance tools, the data of which has become partly declassified and is now available for tectonic studies.

Operational remote sensing of the Earth's surface began with the Landsat program. Launched in 1972, it was the first satellite dedicated to the mapping and monitoring of natural and cultural resources. Early studies concentrated, amongst others, on structural geology and geomorphological investigations (Sabins, 1996; Gupta, 2018). Therefore, thanks to the long history of data acquisition, decades of records are available, allowing even seeing faults move and rift zones widen (Hollingsworth et al., 2013).

Satellite systems deliver data from faulted regions worldwide (Lillesand et al., 2015). Commonly-used satellite remote-sensing techniques include (a) the high-resolution optical-imaging Satellite Pour l'Observation de la Terre (SPOT), Landsat multispectral scanner (MSS) and Landsat thematic mapper (TM), the Advanced Spaceborne Thermal Emission and Reflection Radiometer (ASTER), the Shuttle Radar Topographic Mission (SRTM) and other synthetic aperture radar systems. Important for the use of these sensors in tectonic studies and fault dynamics is both data continuity and resolution. Landsat is the longest-running remote-sensing program and provides a precious time-series of land and ocean changes. Spectral imagery in particular, also allows mineral exploration (Sabins, 1996; van der Meer et al., 2012), the detection of faults, and even to quantify the amount of fault slip. With Landsat, the type of sensor therefore dramatically changed from optical to spectral observation, became privatized and commercialized in the early 1980s, and then, more recently, many archives have opened again (through the USGS). After other sensors and satellites were launched in the 1990s and 2000s, widespread scientific use of such data was based on high temporal and spatial resolution and quality, and the low cost (or even free) access. More recently, a number of countries and space agencies have realized new remote sensing satellite platforms, especially through the European Space Agency's (ESA) earth observation program, Copernicus, with an uncomplicated and completely free data policy, and a mounting database that allows to map and monitor faults from desktops around the globe. The method of data analysis has also changed in the past years, from mapping of fault traces and geological interpretation to detailed deformation mapping, as well as kinematic and physics-based studies associated with seismo-tectonic and volcano-tectonic faulting processes.

In parallel with the early operational optical satellites, other bands of the electromagnetic spectrum have also rapidly gained importance (Fig. 3.31).

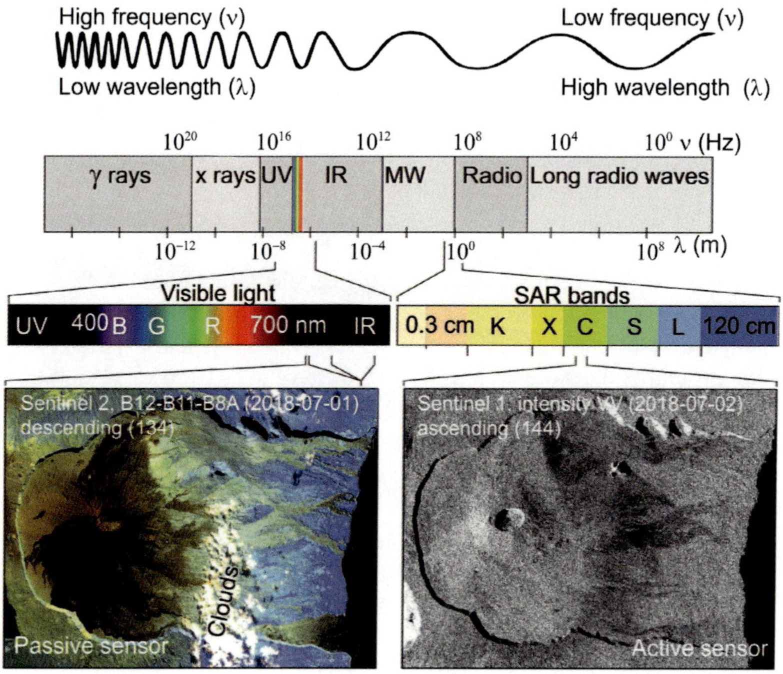

FIG. 3.31 Electromagnetic spectrum and its use in remote sensing. Commonly-used high frequency systems are passive systems, recording in the ultraviolet (UV), the visible light, and the infrared (IR) spectrum. Low-frequency systems are often active systems, recording in medium-wave (MW) and radio spectrum. SAR is a radar system, i.e., radio detection and ranging. Note that cloud cover affects the passive systems, whereas the active SAR systems are able to penetrate clouds. Map views show a passive sensor's view (left side, Sentinel 2 satellite) and an active sensor's view (right side, Sentinel 1 satellite) of the Piton de la Fournaise volcano and its craters.

Specifically, the radar remote-sensing technology, first airborne (in early 1970s) and shortly after that satellite-based radar sensors became non-military aboard the Space Shuttle (SIR-A in 1982). The European Space Agency's (ESA) earth observation program also realized two radar satellites (Sentinel 1A and 1B), the data of which are free to use and push the application and user groups to new levels.

3.7.2 Instruments and data

Faults can be well imaged on data acquired by satellites or aircrafts. While the satellite remote sensing provide systematic and large-scale (background) acquisition plans, aircraft remote sensing is available on-demand and requires dedicated fieldwork and data acquisition planning. Therefore, the satellites typically provide more regular acquisitions over the same target area and over longer periods, but are less flexible for scientific targest and problems. Both of the carriers, satellite or aircraft (Fig. 3.30), may transport similar instruments and thus the data can be similarly used to extract geological knowledge. This, together with advances in computer sciences and numerical techniques (including modelling) allows the identification of faults, the fault movement with strike, dip, slip and rake components (see Figs. 1.4 and 3.25), as well as the recognition of subtle patterns associated with the faulting process. The most common sensor technologies can be categorized into active and passive sensors.

3.7.2.1 Active and passive sensor technologies

The active sensors are located on a carrier (e.g., the satellite) and actively emit a signal or pulse and record either the signal's reflection from an object (two-way signal travel), or the signal's characteristics at a recording station located on the ground (single-way signal travel). Passive sensors in turn, exploit the reflection of sunlight or (thermal) emission from a target, which is detected by a sensor on the carrier. The most commonly-used active sensors are LiDAR and Synthetic Aperture Radar (SAR) systems. For both of these systems, the time delay between emission and return can be measured, as well as the characteristics of the echoed signal. This allows the inference of the location, displacement and slip direction of an active fault. While LiDAR is commonly used to obtain a precise but static image and point cloud of the ground, the SAR techniques (InSAR and pixel offset) enables to quantify dynamic deformation associated with faulting processes.

Common passive sensors that exploit the reflected sunlight or another source of radiation can be recorded by optical and infrared cameras or radiometers. Passive sensors can be used in many ways, such as for classic mapping purposes, but thanks to long time series and huge data catalogues, they are also used to detect landform changes and deformation. The passive remote sensing of geological structures is often based on visible-near infrared

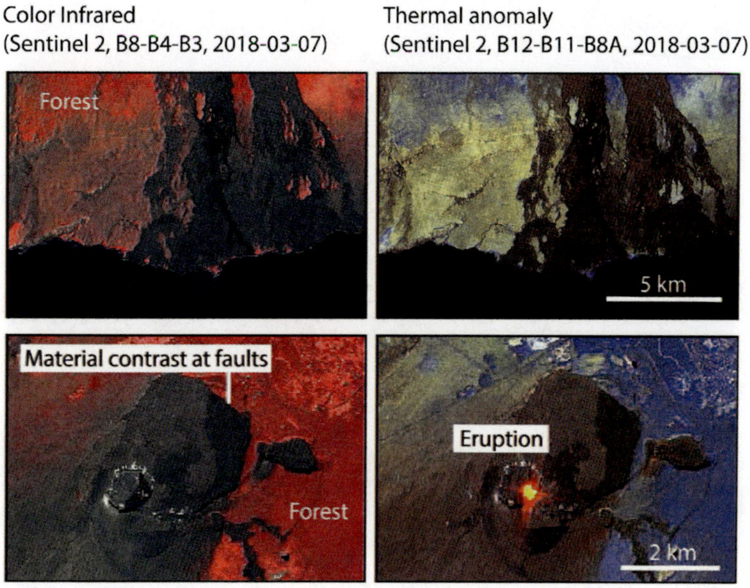

FIG. 3.32 Example of passive remote sensing: Optical satellite identify new structures, vegetation, and thermal anomalies, such as volcano eruption. Faults may be identified by offsets of geological units, different sedimentation and material accumulation on either side of related structural or morphologic barriers. Band selection determines information content. An example from Kilauea volcano.

(NVIR), shortwave infrared (SWIR), mid infrared (MIR and thermal infrared (TIR) bands of the spectrum (Hunt, 1977). The passive sensors with the longest time series (and liberal data policy) are the eight missions of the Landsat program. This data has been extensively used to map faults based on identified linear features, commonly based on gradient data filtering and edge detection (Girard and Girard, 2003). Besides mapping of fault traces of the geological archive, the Landsat data has also been used to identify and investigate new faults and fissures associated with geological hazards (Fig. 3.32). Other systems such as the SPOT satellite (Satellite Pour l'Observation de la Terre) allowed a very high-resolution view and stereo capability to determine fault offsets (Kavak, 2005).

Active and passive sensors may be considered complementary, allowing analysis of the same fault from different perspectives, at different temporal and spatial resolution, and different qualities. Therefore, especially the combined use of the SAR and the optical satellite techniques provide a detailed view of faults and their surface displacements. As LiDAR data has remained very costly, and requires clear skies, a suitable aircraft, and deep experience during acquisition and processing the data, the method is only available to selected and small user group. In contrast, not only the data, but also the tools and software used to study the SAR and satellite optical data are improving and

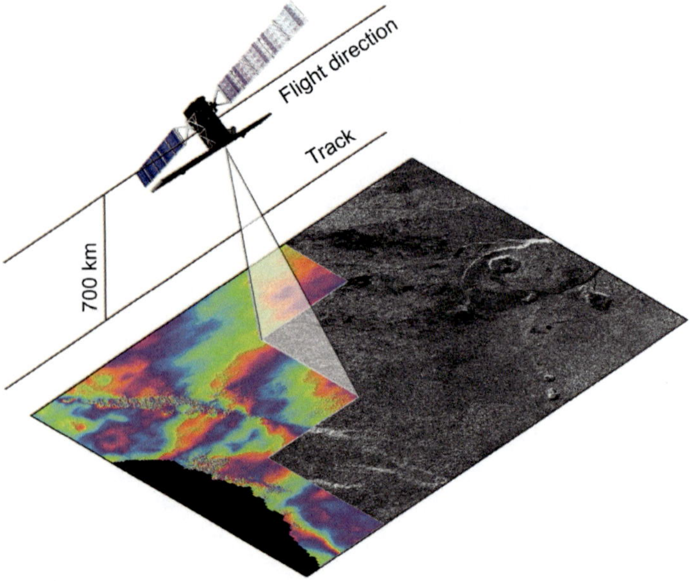

FIG. 3.33 Active remote sensing: Interferometric synthetic aperture radar (InSAR) exploits the phase difference between two or more SAR acquisitions, and performs well, independent of weather and sunlight. This radar interferometric method allows detection of faults by their topography, material contrasts, but also by deformation gradients. The colour changes and number of coloured fringes depict the amount of surface deformation in the satellite's line of sight.

have become freely available, which is why the latter methods are gaining relevance in geoscience (Fig. 3.33).

Remote sensing of fault detection means the acquisition of information and data about a fault or an associated phenomenon without making physical contact with the site. As we have seen above, these are classically either satellite- or aircraft-based, active or passive sensor technologies, but similar tools are now used by drones and on the ground. While there are many studies using different forms of remote sensing for fault detection, there is not a generally applicable workflow. This arises from the complexity, structural and geomorphological particularity, and the vegetation cover and surface water flow, all of which may affect and obscure a fault. In the following, we shall distinguish fault mapping from fault kinematics analysis, each of which have sub-categories. Depending on the needs of the structural geologist, remote sensing data is analyzed in the 2-D image domain, on 3-D geomorphology, on 4-D deformation time series, or a combination thereof.

3.7.3 Fault mapping and kinematics

Faults can be detected using multiple characteristics, ranging from morphological expression, offset of geological units and displacement measurements, to secondary effects such as stream-path changes and erosion that can be mapped and quantified.

3.7.3.1 Fault mapping

One important aim of remote sensing is related to mapping of fault traces. These can be characterized by different signatures, such as lineaments, morphology and displacement discontinuity (Fig. 3.34). But also different geological units on either side of faults can be well mapped using remote sensing (van der Meer et al., 2012).

Lineaments are geological features that are identified as significant lines in a landscape caused by joints and faults, associated with the hidden architecture of the rock basement (Hobbs, 1904). Lineaments are commonly described as mappable features, which are often linear and presumably reflect the subsurface phenomena (O'Leary et al., 1976). Because of the linear expression, spatial domain filtering technologies applied to remote sensing data helped augment and map the lineaments. Traditionally, lineaments are associated with long-term processes, but there is increasing evidence that even short time spans may produce or activate new lineament systems, such as those that occurred after earthquakes (Arellano-Baeza et al., 2006). Lineaments that are expressed as topographic lows are commonly associated with joints, faults and shear zones underground, whereas the lineaments that are expressed as

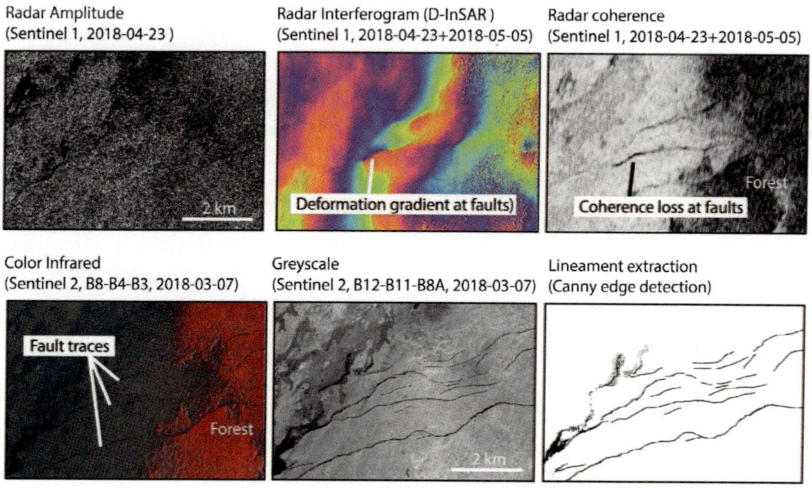

FIG. 3.34 Comparison of different remote sensing data of the same fault structure. Upper row (active remote sensing): Radar amplitude (left), radar interferometry (centre) and radar coherence (right) clearly depict location of deformation gradient and coherence change associated with active faulting. Radar data does not see all faults present in the morphology, however. Lower row (passive remote sensing): Optical data clearly show the location of the faults and lineaments, but do not indicate subtle deformation. Therefore method combination is successful. Example from Kilauea volcano.

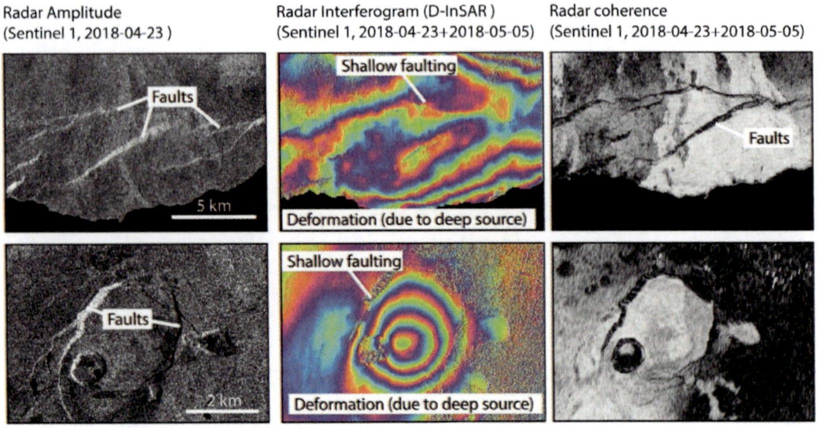

FIG. 3.35 Comparison of different radar results for a faulted coast (upper row) and volcano crater (lower row). Radar amplitude may be used to map scarps associated with faults. InSAR may allow identification of deformation associated with a deep fault, but also active shallow faulting. Radar coherence illustrates prominent changes in the surface.

topographic highs are interpreted as dikes and dike swarms (Ramli et al., 2010), although exceptions from this rule are widespread. Lineaments are commonly extracted from a remote-sensing database by manual interpretation, where the result is strongly biased by the eye and the experience of the observer, or by utilizing computer algorithms. The manual interpretation and tracing of faults may provide very high quality and even allow identifying disconnected traces to lineaments. The algorithms used for automatic extraction, in turn, are less subjective, and initiate with image enhancement and (spatial) filtering, followed by techniques such edge detection, satellite band combination, data fusion, principal component analysis and artificial intelligence (Fig. 3.34). The study of lineaments allows inferring stress fields associated with or even driving the formation of the structural features (Walter and Motagh, 2014). Because of the correlation between lineaments and structures in the field, lineaments may represent fracture networks (Morelli and Piana, 2006). The mapped lineaments may result also from other geomorphological and geological processes, many of them only indirectly associated with joints and faults (Fig. 3.35). They are distinguished by the positive (ridges) and negative lineaments (valleys), and can be best characterized by comparison with an independent remote sensing dataset, such as a digital elevation model.

3.7.3.2 Topography

Active and passive remote-sensing data acquired from satellites and aircrafts may allow imaging from different viewpoints. These are realized from active sensors such as tandem radar (for instance realized by the SRTM mission or the TanDEM-X satellites), or from passive sensors exploiting

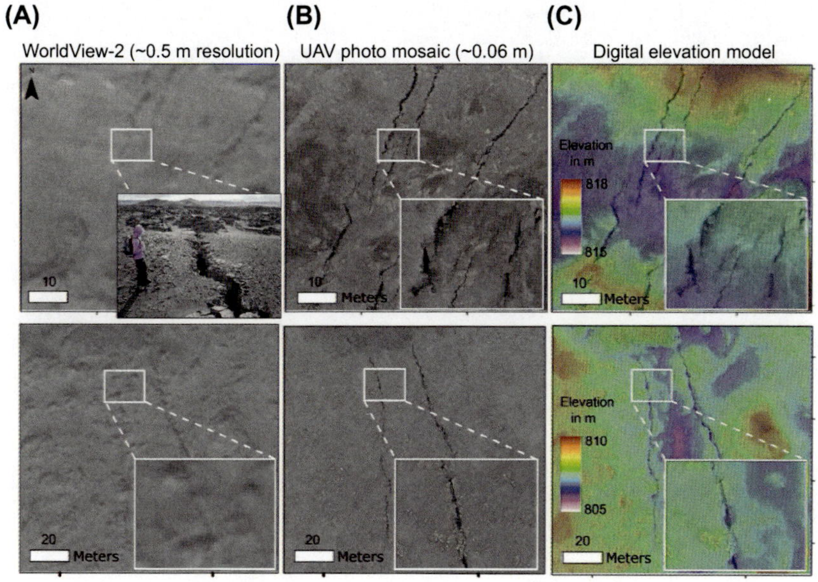

FIG. 3.36 Comparison of different digital elevation models for mapping a fault. Satellite imagery allows large-scale analysis (left), and identification of fault strike. Close-range photogrammetric survey (using drones or aircraft) improve resolution but require dedicated field investigation (centre). The resulting topography and photomosaic helps to identify faults and their complexity (right).

stereo-photogrammetry (realized by ASTER or Pleiades satellites, or structure-from-motion workflow for drone data). Such data allows the generation of three-dimensional point clouds and in a gridded form to derive digital elevation models (DEMs) that are widely used for fault detection (Oguchi et al., 2003). DEMs can be used in multiple ways and analysis; to determine the throw associated with faults, the lateral offset and kink line of erosion channels at faults, and to identify the temporal variation of these features. Graphically, many software programmes allow viewing, gridding and interpreting DEMs, such as the geographic information system applications QGis, GRASS or ArcGIS, or other mapping toolboxes that both support viewing, editing, and analysis of geospatial data. By generating analytical hill shading, slope mapping, azimuth mapping and more, the user is enhancing fault-related lineaments and their geometry in space. Enhancement includes the application of a bandpass filter (often combining low- and high-pass filters) for denoising and certain frequency enhancements. DEMs are ideally used as an additional dataset, supplementing lineament studies (Fig. 3.36), as the 3D aspect enhances the understanding of the fault traces and the related topography and may allow reducing subjectivity, hence distinguishing real from false fault traces/lineaments.

3.7.3.3 Fault kinematics analysis

Fault traces observed by lineaments and morphology provide important clues about the regional or local structure and tectonics, but are challenging to be interpreted in terms of kinematics. This is because even the same fault trace may assemble a number of slip heterogeneities, stiffer and softer materials, and asperities, which are not necessarily depicted by the fault trace alone. To identify the kinematics and therewith investigate the physics of faulting, analysis of displacement, strain and stress is particularly useful.

Fault kinematics can be investigated by remote-sensing deformation analysis. The common strategy for all available sensors is to consider data before, during and after a deformation episode or event. The first item on this list, i.e., data before an episode or event, requires either clairvoyant satellite planning, or has to rely on sensors with a constant and ideally large-scale data-recording plan (a background acquisition plan). Investigation of deformation analysis with such data allows identifying the amount and pattern of deformation. For instance, a vertical fault with strike-slip motion produces a characteristic butterfly pattern of deformation, whereas a dipping fault with a thrust or normal component produces a two-lobe pattern of deformation (Fig. 3.37). The expression of such deformation lobes, whether uplift or subsidence, and the degree of horizontal deformation, allow inferring the fault mechanism, such as distinguishing a dextral from a sinistral fault, or a thrust from a normal fault. These interpretations are commonly based on deformation models.

Deformation mapping of faults is based on two common strategies, first the recording of the distance between the sensor (the satellite) and the ground, and second the recording of the offset of regions and pixels of the ground. The former is more sensitive to vertical movements, whereas the latter is more sensitive to lateral movements.

Distance measurements between the sensor and the ground may be realized by both active and passive sensing. The commonly-used method nowadays is interferometric synthetic aperture radar (InSAR), where the active radar signal is echoed off the ground and recorded again at the satellite. The idea is to combine at least two such recordings from the same fault trace, from a very similar viewing direction. The two SAR scenes contain information about the backscattered intensity, and the phase of the backscattered radar signal. The intensity strongly depends on the reflection characteristics on the ground, i.e., bare rock and an exposed fault trace may lead to a high intensity whereas soft, wet, or vegetated ground often leads to low intensity (Fig. 3.35). These intensities are used to overlay and stack the two images, so that exactly corresponding pixels overlay each other, and then the phase angle can be subtracted. The result is an interferogram that contains information about the topography and deformation. Both these kinds of information are used for fault investigation. The kinematics of faults is commonly investigated by removing the topographic phase from the interferogram, resulting in a differential

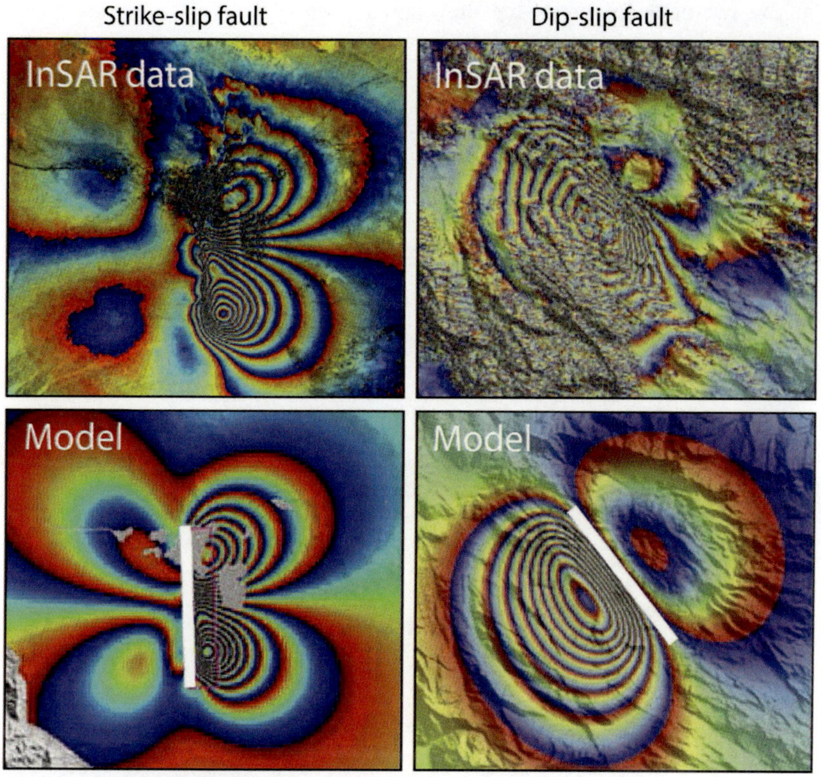

FIG. 3.37 Deformation maps derived from radar interferograms (InSAR, top figures) and a simple analytical models to explain most of the observed displacement (bottom figures). Kinematics of a strike-slip fault (left) and a dip-slip normal fault (right). The strike-slip fault is the 2003 Bam earthquake, Iran; dip-slip is the 2009 L'Aquila earthquake, Italy. Deformation data at these sites allowed identifying unknown faults and/or the interaction between faults and secondary effects at the surface.

interferogram (D-InSAR), which ideally contains only deformation. Careful analysis is needed to ensure the deformation is real and not affected by artifacts arising from atmosphere or other data recording and processing flaws. Deformation results from the D-InSAR analysis allow locating those fault segments that were activated during an earthquake, or creeping and sliding in the time period covered by the two radar images. More sophisticated analysis considers then a large number of data acquisitions, to derive a time series of deformation.

Other methodology to measure the distance between the ground and the sensor are based on visible frequencies of the spectrum. LiDAR and stereo-photography both allow identifying topography, and reproject not only the location of a certain point on the ground but also the position of the imaging sensor. LiDAR and stereo-photography alone do not allow kinematic studies, but by repeat acquisitions the elevation changes and displacements of the ground can be assessed well.

Pixel offset technologies have been applied to both active and passive sensors. The idea comes from computer vision and was developed in the 1980s, where by cross-correlation of a two-dimensional array of information, i.e., the pixels, a systematic shift of a group of pixels can be identified. At least two images of the same target area are compared in this method. To this aim, they first need to be aligned, so that each pixel in image 1 finds its appropriate position in image 2. Then a moving subwindow is defined, for instance 128×128 pixels in size, and its correlation peak on the second image is found. By moving the subwindow over the image domain, one obtains information of systematic shifts of localized regions. The resolution is stepwise increased by shrinking the subwindow dimension to, e.g., 64×64 and 32×32 and 8×8 pixels, after each reduction of the subwindow size the iteration of digital image correlation is repeated. Results allow finally subpixel deformation detections associated with faults (Fig. 3.38). The pixel offset

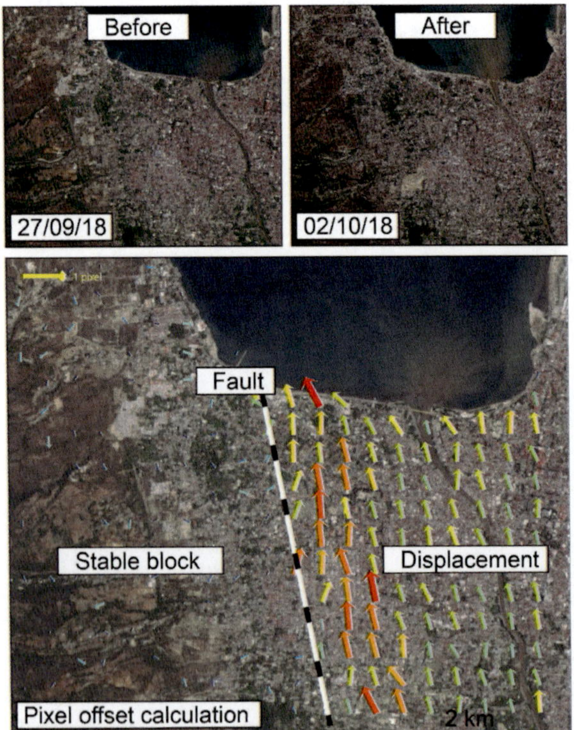

FIG. 3.38 Pixel offsets calculated for a shallow 7.5 magnitude earthquake (Palu, Indonesia). The satellite images before and after the earthquake were acquired by Sentinel-2, but also other images acquired by Landsat or by radar satellites may provide comparable results. Lower image shows the up to 4 m northward displacement of the area east of the identified fault trace (*dashed line*). The fault is a sinistral strike-slip fault with near-vertical dip.

technologies can be applied to radar and optical imagery, and therefore offer a large potential to extend the above-mentioned methods, and allow also extending the time series to very early airborne photogrammetric survey data that were recorded many decades ago. In this way, deformation associated with faulting and extension could be retrieved for a >5 decade long observation period in some tectonically active regions (e.g., Iceland). For both data types, one needs to be aware of artifacts and geometric complexities arising from the sensor technology.

3.7.4 Summary and outlook

Remote sensing has been revolutionized by modern and freely-available satellite mission data. These allow browsing of faulted regions of the Earth, display geological lithologies that are offset on either side of the fault, and allow determination of changes in erosion, water and streamline paths. Remote sensing for fault detection further allows investigating morphological features, such as fault opening and throw, identify spatial and temporal variations of the geometry and displacements associated with faults and the source processes beneath. The number of available tools and data are exponentially increasing, and become ever-widely applicable, which is why also automatic detection technologies are gaining relevance. More developments are expected soon, after the launch of future high-resolution optical and radar satellites. Higher resolution means both, spatial resolution and temporal resolution. New radar satellites are planned to operate in other bands (for coherent analysis in vegetated areas and for infrastructure monitoring). New optical remote sensing satellites may include mini satellites that significantly increase temporal resolution, but will also improve optical capacity. In addition, data science, using tools such as artificial neural networks, will allow to identify and recognize patterns in data quantities that were impossible before.

References

Aki, K., Richards, P.G., 1980. Quantitative Seismology: Theory and Methods. W.H. Freeman, San Francisco.

Geophysical signatures of mineral deposit types in Finland. In: Airo, M.-L. (Ed.), 2015. Geological Survey of Finland, Special Paper, vol. 58, pp. 9–70.

Alford, R.M., 1986. Shear data in the presence of azimuthal anisotropy. In: 56th SEG Annual International Meeting & Exhibition, Expanded Abstracts, pp. 476–479.

Annan, A.P., 2005. Ground-penetrating radar. In: Butler, D.K. (Ed.), Near-Surface Geophysics, SEG Investigations in Geophysics Series, No. 13. Society of Exploration Geophysicists, Tulsa.

Archibald, N.J., Gow, P., Boschetti, F., 1999. Multiscale edge analysis of potential field data. Explor. Geophys. 30, 38–44.

Arellano-Baeza, A.A., Zverev, A.T., Malinnikov, V.A., 2006. Study of changes in the lineament structure caused by earthquakes in South America by applying the lineament analysis to the Aster (Terra) satellite data. Adv. Space Res. 37, 690–697.

Bankwitz, P., Schneider, G., Kämpf, H., Bankwitz, E., 2003. Structural characteristics of epicentral areas in central Europe: study case Cheb basin (Czech Republic). J. Geodyn. 35, 5–32. https://doi.org/10.1016/S0264-3707(02)00051-0.

Baranov, V., Naudy, H., 1964. Numerical calculation of the formula of reduction to the magnetic pole. Geophysics 29, 67–79.

Baysal, E., Kosloff, D., Sherwood, J.W.C., 1983. Reverse time migration. Geophysics 48, 1514–1524.

Blakely, R.J., 1995. Potential Theory in Gravity and Magnetic Applications. Cambridge University Press, 441 p.

Brandes, C., Igel, J., Loewer, M., Tanner, D., Lang, J., Müller, K., Winsemann, J., 2018. Visualisation and analysis of shear-deformation bands in unconsolidated Pleistocene sands using ground-penetrating radar. Sediment. Geol. 367, 135–145.

Breiner, S., 1999. Applications Manual for Portable Magnetometers. Geometrics Co., 58 p.

Bréthaut, D., Hartstra, B., Jaya, M., Moser, T., van Noort, G., Pelissier, M., 2018. Improved prospect evaluation and drilling results in the Dutch North Sea with diffraction imaging. First Break 36 (9), 33–39.

Bouman, J., Ebbing, J., Meekes, S., Abdul Fattah, R., Fuchs, M., Gradmann, S., Haagmans, R., Lieb, V., Schmidt, M., Dettmering, D., Bosch, W., 2015. GOCE gravity gradient data for lithospheric modeling. Int. J. Appl. Earth Obs. Geoinf. 35, 16–30.

Bristow, C.S., Jol, M., 2003. Ground Penetrating Radar in Sediments. Geological Society, London, p. 211. Special Publications.

Buske, S., Gutjahr, S., Sick, C., 2009. Fresnel volume migration of single-component seismic data. Geophysics 74 (6), WCA47–WCA55.

Bussert, R., Kämpf, H., Flechsig, C., Hesse, K., Nickschick, T., Liu, Q., Umlauft, J., Vylita, T., Wagner, D., Wonik, T., Flores, H.E., Alawi, M., 2017. Drilling into an active mofette: pilot-hole study of the impact of CO_2-richmantle-derived fluids on the geo–bio interaction in the western Eger Rift (Czech Republic). Sci. Drill. 23, 13–27. https://doi.org/10.5194/sd-23-13-2017.

Centamore, E., Rossi, D., Tavarnelli, E., 2009. Geometry and kinematics of Triassic-to-recent structures in the northern-central Apennines: a review and an original working hypothesis. Boll. Soc. Geol. Ital. 128 (2), 419–432.

Chalupa, V., Pánek, T., Tábořík, P., Klimeš, J., Hartvich, F., Grygar, R., 2018. Deep-seated gravitational slope deformations controlled by the structure of flysch nappe outliers: insights from large-scale electrical resistivity tomography survey and LiDAR mapping. Geomorphology 321, 174–187.

Chiaraluce, L., Di Stefano, R., Tinti, E., Scognamiglio, L., Michele, M., Casarotti, E., Cattaneo, M., De Gori, P., Chiarabba, C., Monachesi, G., Lombardi, A., Valoroso, L., Latorre, D., Marzorati, S., 2017. The 2016 central Italy seismic sequence: a first look at the mainshocks, aftershocks, and source models. Seismol Res. Lett. 88 (3), 757–771. https://doi.org/10.1785/0220160221.

Chopra, S., Marfurt, K., 2007. Seismic Attributes for Prospect Identification and Reservoir Characterization. SEG Geophysical Developments No. 11, Tulsa, OK, USA.

Christie, M., Tsoflias, G.P., Stockli, D.F., Black, R., 2009. Assessing fault displacement and off-fault deformation in an extensional tectonic setting using 3-D ground-penetrating radar imaging. J. Appl. Geophys. 68, 9–16.

Couples, G., Ma, J., Lewis, H., Olden, P., Quijano, J., Fasae, T., Maguire, R., 2007. Geomechanics of faults: impacts on seismic imaging. First Break 25, 83–90.

Crampin, S., 1985. Evaluation of anisotropy by shear wave splitting. Geophysics 50, 142–152.

Dahm, T., Krüger, F., 2014. Moment tensor inversion and moment tensor interpretation. In: Bormann, P. (Ed.), New Manual of Seismological Observatory Practice 2 (NMSOP-2), Potsdam: Deutsches GeoForschungsZentrum GFZ, pp. 1–37. https://doi.org/10.2312/GFZ.NMSOP-2_IS_3.9.

Demirci, I., Erdogan, E., Candansayar, E., 2012. Two-dimensional inversion of direct current resistivity data incorporating topography by using finite difference techniques with triangle cells: investigation of Kera fault zone in western Crete. Geophysics 77 (1), E67–E75.

Doetsch, J., Linde, N., Pessognelli, M., Green, A.G., Günther, T., 2012. Constraining 3-D electrical resistance tomography with GPR data for improved aquifer characterization. J. Appl. Geophys. 78, 68–76. https://doi.org/10.1016/j.jappgeo.2011.04.008.

Donner, S., Igel, H., Hadziioannou, C., The Romy Group, 2018. Retrieval of the Seismic Moment Tensor from Joint Measurements of Translational and Rotational Ground Motions: Sparse Networks and Single Stations. https://doi.org/10.1007/978-3-319-77359-9_12.

Drahor, M.G., Berge, M.A., 2017. Integrated geophysical investigations in a fault zone located on southwestern part of İzmir city, Western Anatolia, Turkey. J. Appl. Geophys. 136, 114–133.

Dransfield, M.H., Christensen, A.N., 2013. Performance of airborne gravity gradiometers. Lead. Edge 32 (8), 908–922.

Edel, J.B., Schulmann, K., Rotstein, Y., 2007. The Variscan tectonic inheritance of the Upper Rhine Graben: evidence of reactivations in the Lias, Late Eocene–Oligocene up to the recent. Int. J. Earth Sci. 96 (2), 305–325.

Fairhead, J.D., Cooper, G.R.J., Sander, S., 2017. Advances in airborne gravity and magnetics. In: Tschirhart, V., Thomas, M.D. (Eds.), Proceedings of Exploration 17: Sixth Decennial International Conference on Mineral Exploration, pp. 113–127.

Fischer, T., Horálek, J., Hrubcová, P., Vavryčuk, V., Bräuer, K., Kämpf, H., 2014. Intra-continental earthquake swarms in West-Bohemia and Vogtland: a review. Tectonophysics 611, 1–27.

Geiger, L., 1910. Herdbestimmung bei Erdbeben aus den Ankunftszeiten. Nachrichten von der Gesellschaft der Wissenschaften zu Göttingen, Mathematisch-Physikalische Klasse 1910, pp. 331–349. http://eudml.org/doc/58769.

Geiger, L., 1912. Probability method for the determination of earthquake epicenters from the arrival time only. (translated from Geiger's 1910 German article: Herdbestimmung bei Erdbeben aus den Ankunftszeiten. Gesellschaft der Wissenschaften zu Göttingen, Mathematisch-Physikalische Klasse, 331–349). Bull. St. Louis Univ. 8 (1), 56–71.

Geldart, L.P., Sheriff, R.E., 1995. Exploration Seismology. Cambridge University Press, Cambridge, UK.

Ghose, R., Carvalho, J., Loureiro, A., 2013. Signature of fault zone deformation in near-surface soil visible in shear wave seismic reflections. Geophys. Res. Lett. 40, 1074–1078. https://doi.org/10.1002/grl.50241.

Girard, M.-C., Girard, C.M., 2003. Processing of Remote Sensing Data. Taylor & Francis Group, A.A. Balkema, Lisse. ISBN-13: 978-9058092328.

Götze, H.J., Lahmeyer, B., 1988. Application of three-dimensional interactive modeling in gravity and magnetics. Geophysics 53 (8), 1096–1108.

Grasmueck, M., 1996. 3-D ground-penetrating radar applied to fracture imaging in gneiss. Geophysics 61 (4), 1050–1064.

Grauch, V.J.S., Hudson, M.R., 2011. Aeromagnetic anomalies over faulted strata. Lead. Edge 30 (11), 1242–1252.

Griffin, W.R., 1949. Residual gravity in theory and practice. Geophysics 14 (1), 39–56.

Günther, T., Rücker, C., Spitzer, K., 2006. 3-D modeling and inversion of DC resistivity data incorporating topography – part II: inversion. Geophys. J. Int. 166, 506–517. https://doi.org/10.1111/j.1365-246X.2006.03011.x.

Günther, T., Musmann, P., Schaumann, G., Grinat, M., 2011. Imaging of a fault zone by a large-scale DC resistivity experiment and seismic structural information. In: Ext. Abstr., 17th EAGE Near Surface, 12.-14.09.2010; Leicester (UK).

Gupta, R.P., 2018. Remote Sensing Geology. Springer-Verlag Berlin, Heidelberg, Germany. ISBN 978-3-662-55876-8.

Gutenberg, B., Richter, C.F., 1956. Magnitude and energy of earthquakes. Ann. Geofisc. 9, 1–15.

Hackney, R.I., Featherstone, W.E., 2003. Geodetic versus geophysical perspectives of the 'gravity anomaly'. Geophysical Journal International 154, 35–43.

Harjes, H.-P., 1990. Design and siting of a new regional array in Central Europe. Bull. Seismol. Soc. Am. 80 (6), 1801–1817.

Hinze, W.J., von Frese, R.R.B., Saad, A.H., 2013. Gravity and Magnetic Exporation: Principles, Practices, and Application. Cambridge University Press, 512 p.

Hobbs, W.H., 1904. Lineaments of the Atlantic border region. Bull. Geol. Soc. Am. 15 (1), 483–506.

Holden, D.J., Archibald, N.J., Boschetti, F., Jessell, M.W., 2000. Inferring geological structures using wavelet-based multiscale edge analysis and forward models. Explor. Geophys. 31 (6), 617–621.

Hollingsworth, J., Leprince, S., Ayoub, F., Avouac, J.-P., 2013. New constraints on dike injection and fault slip during the 1975–1984 Krafla rift crisis, NE Iceland. J. Geophys. Res. Solid Earth 118, 3707–3727. https://doi.org/10.1002/jgrb.50223.

Hornby, P., Boschetti, F., Horowitz, F., 1999. Analysis of potential field data in the wavelet domain. Geophys. J. Int. 137, 175–196.

Hunt, G.R., 1977. Spectral signatures of particulate minerals in the visible and near infrared. Geophysics 42, 501–513.

Inazaki, T., 2004. High resolution reflection surveying at paved areas using S-wave type land streamer. Explor. Geophys. 35, 1–6.

INGV Working Group on the Amatrice Earthquake, 2016. Second Summary Report on the M6.0 Amatrice Earthquake of August 24, 2016 (Central Italy). https://doi.org/10.5281/zenodo.16624.

Jol, H.M. (Ed.), 2009. Ground Penetrating Radar: Theory and Applications. Elsevier, Amsterdam, 525 p. ISBN: 978-0-444-53348-7.

Jones, I.F., Bloor, R.I., Biondi, B., Etgen, J.T., 2008. Prestack depth migration and velocity building. SEG Geophys. Repr. 25.

Jost, M.L., Herrmann, R.B., 1989. A Student's Guide to and Review of Moment Tensors. Seismol. Res. Lett. 60 (No. 2), 37–57.

Kavak, K.S., 2005. Determination of palaeotectonic and neotectonic features around the Menderes Massif and the Gediz Graben (western Turkey) using Landsat TM image. Int. J. Remote Sens. 26, 59–78.

Kilb, D., Rubin, A.M., 2002. Implications of diverse fault orientations imaged in relocated aftershocks of the Mount Lewis, ML 5.7, California, earthquake. J. Geophys. Res. 107 (B11), 2294. https://doi.org/10.1029/2001JB000149.

Klinge, K., Plenefisch, T., Stammler, K., 2003. The earthquake swarm 2000 in the region Vogtland/NW-Bohemia—earthquake recording at German stations and temporal distribution of events. J. Geodyn. 35, 83–96. https://doi.org/10.1016/S0264-3707(02)00055-8.

Krawczyk, C., Polom, U., Beilecke, T., 2013. Shear-wave reflection seismics as valuable tool for near-surface urban applications. Lead. Edge 32 (3), 256–263.

Lavecchia, G., 1985. II sovrascorrimento dei Monti Sibillini: analisi cinematica e strutturale. Boll. Soc. Geol. Ital. 104, 161–194.

Leibniz Institute for Applied Geophysics, 2010. Bouguer anomaly map of Germany 1: 1,000,000. LIAG, Hannover.

Li, X., Götze, H.J., 2001. Ellipsoid, geoid, gravity, geodesy, and geophysics. Geophysics 66 (6), 1660–1668.

Li, Q., Vasudevan, K., Cook, F., 1997. Seismic skeletonization: a new approach to interpretation of seismic reflection data. J. Geophys. Res. (Solid Earth) 102 (B4), 8427–8445.

Lillesand, T., Kiefer, R.W., Chipman, J., 2015. Remote Sensing and Image Interpretation, seventh ed. Wiley, p. 736. ISBN: 978-1-118-34328-9.

Liu, E., Martinez, A., 2012. Seismic Fracture Characterization – Concepts and Practical Applications. EAGE Publications, 3990 DB Houten, The Netherlands.

Lomax, A., Virieux, J., Volant, P., Berge, C., 2000. Probabilistic earthquake location in 3D and layered models: introduction of a metropolis -Gibbs method and comparison with linear locations. In: Thurber, C.H., Rabinowitz, N. (Eds.), Advances in Seismic Event Location. Kluwer, Amsterdam, pp. 101–134. https://doi.org/10.1007/978-94-015-9536-0_5.

Lowrie, W., 2007. Fundamentals of Geophysics, second ed. Cambridge University Press. 381 p.

Luo, Y., Al-Dossary, S., Alfaraj, M., 2002. Edge preserving smoothing and application. Lead. Edge 21, 136–158. https://doi.org/10.1190/1.1452603.

Lüschen, E., Görne, S., von Hartmann, H., Thomas, R., Schulz, R., 2015. 3D seismic survey for geothermal exploration in crystalline rocks in Saxony, Germany. Geophys. Prospect. 63, 975–989.

Marfurt, K., 2018. Seismic Attributes as the Framework for Data Integration Throughout the Oilfield Life Cycle. SEG Distinguished Instructor Series, No. 21, Tulsa, OK, USA.

Maus, S., Yin, F., Lühr, H., Manoj, C., Rother, M., Rauberg, J., Michaelis, I., Stolle, C., Müller, R.D., 2008. Resolution of direction of oceanic magnetic lineations by the sixth-generation lithospheric magnetic field model from CHAMP satellite magnetic measurements. Geochem. Geophys. Geosyst. 9, Q07021.

Mayer-Guerr, T., 2007. ITG-Grace03s: the latest GRACE gravity field solution computed in Bonn. In: Joint International GSTM and DFG SPP Symposium (Potsdam, 15–17 October 2007).

McClymont, A.F., Green, A.G., Streich, R., Horstmeyer, H., Tronicke, J., Nobes, D.C., Pettinga, J., Campbell, J., Langridge, R., 2008. Visualization of active faults using geometric attributes of 3D GPR data: an example from the Alpine Fault Zone, New Zealand. Geophysics 73 (2), B11–B23.

McClymont, A.F., Villamor, P., Green, A.G., 2009. Fault displacement accumulation and slip rate variability within the Taupo Rift (New Zealand) based on trench and 3-D ground penetrating radar data. Tectonics 28, TC4005.

Miller, R.D., Pullan, S.E., Steeples, J.A., 1994. Field comparison of shallow P-wave seismic sources near Houston, Texas. Geophysics 59, 1713–1728.

Morelli, M., Piana, F., 2006. Comparison between remote sensed lineaments and geological structures in intensively cultivated hills (Monferrato and Langhe domains, NW Italy). Int. J. Remote Sens. 27 (20), 4471–4493.

Moser, T.J., Howard, C.B., 2008. Diffraction imaging in depth. Geophysical Prospecting 56 (5), 627–641.

Musmann, P., Buness, H., 2010. High-resolution seismic imaging of near-surface fault structures within the Upper Rhine Graben, Germany. In: Miller, R.D., Bradford, J.H., Holliger, K. (Eds.), Advances in Near-Surface Seismology and Ground-Penetrating Radar, Chapter 17: SEG Geophysical Developments Series No. 15, pp. 281–296. https://doi.org/10.1190/1.9781560802259.ch17.

Nabighian, M.N., Grauch, V.J.S., Hansen, R.O., LaFehr, T.R., Li, Y., Peirce, J.W., Phillips, J.D., Ruder, M.E., 2005. The historical development of the magnetic method in exploration. Geophysics 70, 33ND–61ND.

Oguchi, T., Aoki, T., Matsuta, N., 2003. Identification of an active fault in the Japanese Alps from DEM-based hill shading. Comput. Geosci. 29, 885–891.

O'Leary, D.W., Friedman, J.D., John, H.A., 1976. Lineament, linear, lineation: some proposed new standards for old terms. Geol. Soc. Am. Bull. 87 (10), 1463–1469.

Olsen, N., Ravat, D., Finlay, C.C., Kother, L.K., 2017. LCS-1: a high-resolution global model of the lithospheric magnetic field derived from CHAMP and Swarm satellite observations. Geophys. J. Int. 211 (3), 1461–1477.

Omori, F., 1894. On after-shocks of earthquakes. J. Coll. Sci. Imp. Univ. Tokyo 7, 111–200.

Oppermann, F., Günther, T., 2018. A remote-control datalogger for large-scale resistivity surveys and robust processing of its signals using a software lock-in approach. Geosci. Instrum. Methods Data Syst. 7, 55–66. https://doi.org/10.5194/gi-7-55-2018.

Pavlis, N.K., Holmes, S.A., Kenyon, S.C., Factor, J.K., 2012. The development and evaluation of the earth gravitational model 2008 (EGM2008). J. Geophys. Res. Solid Earth 117, B04406.

Plenefisch, T., Klinge, K., 2003. Temporal variations of focal mechanisms in the Novy Kostel focal zone (Vogtland/NW-Bohemia)—comparison of the swarms of 1994, 1997 and 2000. J. Geodyn. 35, 145–156. https://doi.org/10.1016/S0264-3707(02)00059-5.

Plenefisch, T., 2009. The October-November 2008 earthquake swarm in Vogtland/NW-Bohemia: first results of focal mechanism determinations. Geophys. Res. Abstr. 11, EGU2009–5262.

Pucci, S., Civico, R., Villani, F., Ricci, T., Delcher, E., Finizola, A., Sapia, V., Marco De Martini, P., Pantosti, D., Barde-Cabusson, S., Brothelande, E., Gusset, R., Mezon, C., Orefice, S., Peltier, A., Poret, M., Torres, L., Suski, B., 2016. Deep electrical resistivity tomography along the tectonically active Middle Aterno Valley (2009 L'Aquila earthquake area, central Italy). Geophys. J. Int. 207, 967–982. https://doi.org/10.1093/gji/ggw308.

Pugin, A.J.-M., Pullan, S.E., Hunter, J.A., Oldenborger, G.A., 2009. Hydrogeological prospecting using P- and S-wave landstreamer seismic reflection methods. Near Surf. Geophys. 7, 315–327.

Randen, T., Pedersen, S.I., Sønneland, L., 2001. Automatic extraction of fault surfaces from three-dimensional seismic data. In: SEG Technical Program Expanded Abstracts 2001, pp. 551–554. https://doi.org/10.1190/1.1816675.

Rappin, D., Schnitzler, J., 2007. How faults express themselves on 3D seismic data: an update using modern processing techniques. SEG Tech. Progr. Expand. Abstr. 26 (1), 2807–2811. https://doi.org/10.1190/1.2793050.

Ramli, M.F., Yusof, N., Yusoff, M.K., Juahir, H., Shafri, H.Z.M., 2010. Lineament mapping and its application in landslide hazard assessment: a review. Bull. Eng. Geol. Environ. 69, 215–233.

Reigber, C., Schmidt, R., Flechtner, F., König, R., Meyer, U., Neumayer, K.-H., Schwintzer, P., Zhu, S.Y., 2005. An Earth gravity field model complete to degree and order 150 from GRACE: EIGEN-GRACE02S. J. Geodyn. 39, 1–10.

Roest, W.R., Verhoef, J., Pilkington, M., 1992. Magnetic interpretation using the 3-D analytic signal. Geophysics 57 (1), 116–125.

Rost, S., Thomas, C., 2002. Array seismology: methods and applications. Rev. Geophys. 40 (3), 1008.

Rotstein, Y., Edel, J.-B., Gabriel, G., Boulanger, D., Schaming, M., Munschy, M., 2006. Insight into the structure of the Upper Rhine Graben and its basement from a new compilation of Bouguer gravity. Tectonophysics 425, 55–70.

Sabins, F.F., 1996. Remote Sensing: Principles and Interpretation. Worth Publishers, New York.

Salem, A., Williams, S., Fairhead, D., Smith, R., Ravat, D., 2008. Interpretation of magnetic data using tilt-angle derivatives. Geophysics 73 (1), L1–L10.

Schmelzbach, C., Donner, S., Igel, H., Sollberger, D., Taufiqurrahman, T., Bernauer, F., Häusler, M., Van Renterghem, C., Wassermann, J., Robertsson, J., 2018. Advances in 6C seismology: applications of combined translational and rotational motion measurements in global and exploration seismology. Geophysics 83 (3). https://doi.org/10.1190/geo2017-0492.1.

Schmidt, P.W., Clark, D.A., 2000. Advantages of measuring the magnetic gradient tensor. Preview 85, 26–30.

Schweitzer, J., Fyen, J., Mykkeltveit, S., Gibbons, S.J., Pirli, M., Kühn, D., Kværna, T., 2012. Seismic arrays. In: Bormann, P. (Ed.), New Manual of Seismological Observatory Practice 2 (NMSOP-2), Potsdam: Deutsches GeoForschungsZentrum GFZ, pp. 1–80. https://doi.org/10.2312/GFZ.NMSOP-2_ch9.

Schweitzer, J., 2001. HYPOSAT – an enhanced routine to locate seismic events. Pure Appl. Geophys. 158, 277–289.

Shearer, P.M., 2009. Introduction to Seismology. Cambridge University Press, Cambridge.

Sheriff, R.E., 2002. Encyclopedic Dictionary of Applied Geophysics, fourth ed. Society of Exploration Geophysicists. 442 p.

Snoke, J.A., 2003. FOCMEC: FOCal MEChanism determinations. International Handbook of Earthquake and Engineering Seismology, pp. 1629–1630.

Stein, S., Wysession, M., 2003. An Introduction to Seismology, Earthquakes, and Earth Structure. Blackwell Pub, Malden, MA.

Štěpančíková, P., Dohnal, J., Panek, T., Joj, M., Smolkova, V., Silhan, K., 2011. The application of electrical resistivity tomography and gravimetric survey as useful tools in an active tectonics study of the Sudetic Marginal Fault (Bohemian Massif, central Europe). J. Appl. Geophys. 74, 69–80.

Storz, H., Storz, W., Jacobs, F., 2000. Electrical resistivity tomography to investigate geological structures of the earth's upper crust. Geophys. Prospect. 48, 455–471.

Swain, C.J., 2000. Reduction to the pole of regional magnetic data with variable field direction and its stabilisation at the low inclinations. Explor. Geophys. 31, 78–83.

Talwani, M., 1959. Rapid gravity computations for two-dimensional bodies with application to the Mendocino submarine fracture zone. Geophysics 64 (1), 49–59.

Tanner, D.C., Musmann, P., Wawerzinek, B., Buness, H., Krawczyk, C.M., Thomas, R., 2015. Salt tectonics of the eastern border of the Leinetal Graben, Lower Saxony, Germany, as deduced from seismic reflection data. Interpretation 3 (3), T169–T181. https://doi.org/10.1190/INT-2014-0221.1.

van der Meer, F.D., van der Werff, H.M.A., van Ruitenbeek, F.J.A., Hecker, C.A., Bakker, W.H., Noomen, M.F., van der Meijde, M., Carranza, E.J.M., Boudewijn de Smeth, J., Woldai, T., 2012. Multi- and hyperspectral geologic remote sensing: a review. Int. J. Appl. Earth Obs. Geoinf. 14 (1), 112–128. https://doi.org/10.1016/j.jag.2011.08.002.

van der Veen, M., Buness, H.A., Bücker, F., Green, A.G., 2000. Field comparison of high-frequency seismic sources for imaging shallow (10-250 m) structures. J. Environ. Eng. Geophys. 5 (2), 39–56. https://doi.org/10.4133/JEEG5.2.39.

Villani, F., Civico, R., Pucci, S., Pizzimenti, L., Nappi, R., De Martini, P.M., The OPEN Emergeo Working Group, 2018. A database of the coseismic effects following the 30 October 2016 Norcia earthquake in Central Italy. Sci. Data 5, 180049. https://doi.org/10.1038/sdata.2018.49.

Wang, H., Luo, Z., Ning, J., 2009. Multiscale edge analysis of gravity data and its application. Geo Spat. Inf. Sci. 12 (3), 230–234.

Waldhauser, F., Ellsworth, W.E., 2000. A double-difference earthquake location algorithm: method and application to the Northern Hayward Fault, California. Bull. Seismol. Soc. Am. 90 (6), 1353–1368.

Walter, T.R., Motagh, M., 2014. Deflation and inflation of a large magma body beneath Uturuncu volcano, Bolivia? Insights from InSAR data, surface lineaments and stress modelling. Geophys. J. Int. 198 (1), 462–473. http://doi.org/10.1093/gji/ggu080.

Wemegah, D.D., Preko, K., Noye, R.M., Boadi, B., Menyeh, A., Danuor, S.K., Amenyoh, T., 2015. Geophysical interpretation of possible gold mineralization zones in Kyerano, south-western Ghana using aeromagnetic and radiometric datasets. J. Geosci. Environ. Prot. 3, 67–82.

Chapter 4

Numerical modelling of faults

Andreas Henk
Institute of Applied Geosciences, Technical University Darmstadt, Darmstadt, Germany

Chapter outline

4.1 Introduction

Numerical methods have been applied successfully to problems in various geoscience disciplines, including geodynamics, structural geology and rock mechanics (e.g., Gerya, 2009; Ramsay and Lisle, 2000; Jing and Hudson, 2002). Such numerical simulations improve not only the quantitative understanding of the underlying physical processes, but also allow for scenario testing and forecasting. With respect to faults, numerical models presented hitherto have addressed aspects of fault integrity, fault fluid flow and earthquake risk assessment, among others. For example, numerical techniques have been used to assess quantitatively the reactivation potential of faults in different tectonic regimes (e.g., McLellan et al., 2004; Buchmann and Connolly, 2009). Fluid flow issues along fault zones have been studied by Cappa (2009), Cappa and Rutqvist (2011) and Schuite et al. (2017), among others. This approach has been expanded also to address earthquake processes (e.g., Rice et al., 2009; Cappa, 2011). More recently, modelling the effect of man-made pore pressure changes on fault stability due to fluid injection (pore pressure increase) and fluid production (pore pressure decrease) has

Understanding Faults. https://doi.org/10.1016/B978-0-12-815985-9.00004-7

attracted a lot of attention (Rutqvist et al. 2016; Zhang et al., 2016; Haug et al., 2018).

Any comprehensive numerical analysis of a fault requires at least a coupled hydromechanical (HM) modelling approach. Special applications may even ask for incorporation of thermal and chemical processes leading to thermo-hydromechanical (THM) and thermohydromechanical-chemical (THM-C) simulations, respectively. Different numerical techniques are available to account for the coupling and the mutual dependence of pore pressure, effective stress, volumetric strain, porosity and permeability (e.g., Minkoff et al., 2003; Rutqvist and Stephansson, 2003). Each technique has its pros and cons, but common to all numerical approaches is the challenge how to describe adequately the complexities of natural fault zones with respect to architecture and petrophysical property distribution. Depending on host rock lithology, deformation history and a possible later hydrothermal overprint, the dimensions and internal structure of damage zone and fault core as well as their hydromechanical characteristics can be quite variable (e.g., Evans et al., 1997; Wibberley et al., 2008; Childs et al., 2009; Faulkner et al., 2010). In particular, fracturing in the damage zone and connectivity of the fractures will determine whether the fault zone acts as barrier or conduit for fluid flow (Chapter 7).

In the following subchapters, an overview of the different numerical techniques commonly used for fault modelling is given. Thereby the focus is on modelling of existing faults rather than formation of new faults, i.e., fracturing and fault propagation (see Chapters 2 and 6). The numerical models have to be populated with specific hydromechanical parameters for each fault zone component. Hence, an overview of the typical parameter range describing host rock, damage zone and fault core is given. Finally, a generic numerical model is presented to illustrate the workflow and potential of fault zone modelling. Parameter studies allow to systematically vary model features like, fault zone architecture, hydromechanical fault rock properties, depth, fluid pressure and strain rate in order to study their impact on fault zone processes like, for example, stress perturbation, strain localisation, fault reactivation and fluid transfer. Such numerical simulations do not only help to improve our quantitative understanding of faults from a scientific point of view, but also serve as an important tool for a safe utilization of the subsurface.

4.2 Numerical methods for hydromechanical fault zone modelling

Several numerical techniques have been used for hydromechanical modelling of fault zones and incorporation of faults into larger models, respectively. The various methods can be grouped in three main categories, i.e., continuum- and discontinuum-based numerical techniques as well as hybrid (continuum/discontinuum) methods. Each approach has its pros and cons, which will be discussed in the following paragraphs.

At first glance, continuum methods – by the name – appear to contradict modelling of a discrete discontinuity like a fault, but discretization schemes and special element types like contact or interface elements offer possibilities to represent localized deformation and even differential slip, respectively. Among these continuum approaches, the **Finite Difference Method (FDM)** is a well established numerical method for solving the partial differential equations (PDE's) describing hydromechanical processes, i.e., fluid flow and deformation, in rocks (e.g., Smith, 1986; Ramsay and Lisle, 2000; Anderson et al., 2015). Thereby, the differential equations are replaced by difference equations over a certain interval in space and – for time-dependent solutions – also over a certain time interval. This approach requires to subdivide the domain of interest into a grid of calculation points, so-called nodes, which – at least for the standard FDM – have to be distributed regularly (Fig. 4.1A). This leads to an important drawback of this method as it implies that irregular subsurface geometries and/or lithological distributions cannot be properly represented. Advancements like the general FDM and special numerical solution schemes as well as further developments like the **Finite Volume Method (FVM)** can overcome these shortcomings and also allow for irregular node distributions (e.g., Perrone and Kao, 1975; Leveque, 2002). However, a limitation remains as the continuity requirement adhered to FDM does not allow for modelling of new discrete fracture formation and fault propagation, respectively. Worked examples of fault zone modelling with FDM have been presented by Cappa (2009) and Zhang et al. (2016), for example. An established software combination for hydromechanical simulations based on FDM/FVM – but by no means the only one – is the coupling of codes TOUGH

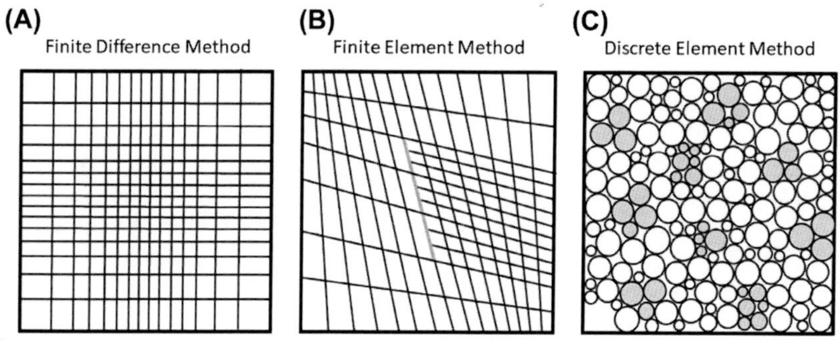

FIG. 4.1 Sketches showing characteristics of the various numerical methods with respect to the discretization of the model domain. (A) Application of the standard finite difference method is based on a rectangular grid geometry. (B) The finite element method also offers the use of irregular grids. In addition, contact elements (in grey) can be defined at discrete discontinuities, which allows for differential sliding between independently meshed parts of the model. (C) A variant of the discrete element method uses individual elements with regular shape (e.g., circles in 2D, spheres in 3D) to describe the model geometry. Optionally, these elements can be bonded together initially (in grey) but can break apart during the course of the numerical simulation.

(developed at Lawrence Berkeley National Laboratory, USA) for modelling of multiphase flow and of FLAC (Itasca, USA) for geomechanical calculations (Rutqvist, 2011; Jeanne et al., 2014). Among others, it offers the incorporation of several faulting-related features like, strain-softening Coulomb failure and slip-weakening as well as fault permeability changes (Rutqvist, 2011).

The other most commonly-used continuum technique for hydromechanical modelling is the **Finite Element Method (FEM)**. Its strengths are the ability to describe complex model geometries and heterogeneous material distributions as well as non-linear material behavior (e.g., Zienkiewicz et al., 2013; Simpson, 2017). The principle is to divide or to discretize the entire model domain into numerous elements (Fig. 4.1B). Joint corner points – and additional mid-side points in case of higher-order elements – ensure the continuity requirement. For each element, so-called basis or trial functions are formulated that represent a local approximation of the underlying PDE's. Subsequently, these equations are combined in to a global set of equations comprising the entire model domain. Boundary conditions are applied and the entire set of equations is solved based on the principle of minimum potential energy (Zienkiewicz et al., 2013). The result is an approximate solution for the entire model. The quality of the approximation depends – beside the convergence criteria selected – on the basis function used for the element equations (linear vs. higher-order) and the element size and model resolution, respectively. As a continuum method, FEM has similar restrictions regarding modeling of fracturing processes as FDM. However, an interesting extension of the classical FEM, also relevant for fault modeling, is the use of so-called interface or contact elements to represent existing discontinuities, e.g., fractures and/or faults (depending on model scale). These elements can transmit shear and normal stresses according to the frictional properties (cohesion and angle of friction) assigned to them. Their use allows for differential displacement between independently-discretized parts of the model. However, contact elements have to be incorporated into the initial model geometry at sites of existing discontinuities or where fracturing and relative displacements are expected to occur in the course of the modeling (Fig. 4.1B). Thus, they cannot account for arbitrarily oriented fracture formation. Recent developments like FEM with embedded discontinuities (ED-FE; e.g., Ibrahimbegovic and Melnyk, 2007) and extended FEM (XFEM; e.g., Fries and Belytschko, 2010; Prevost and Sukumar, 2015) aim to overcome these limitations. Some recent examples of fault zone modelling studies based on FEM can be found in Pereira et al. (2014) and Schuite et al. (2017).

A discontinuum approach that is suitable for fault modelling is the **Discrete Element Method (DEM)**. Initially designed for mechanical problems, it has been expanded to also cover hydromechanical simulations. The basic principle is to divide the model geometry into numerous individual elements with regular (e.g., circles in 2D, spheres in 3D) or irregular (e.g., polygons in 2D, polyhedra in 3D) shapes that interact with each other on the basis of contact laws and equations of motion (Fig. 4.1C; e.g., Cundall, 1971; Lisjak and Grasselli, 2014). The strength of the approach is that the elements can move relative to each other and that even the contact of elements initially bonded together can break during the course of the simulation. Thus, this numerical technique is well suited for modelling of fractured rock masses as well as of new fracture formation and fault propagation. However, as the discrete elements cannot break internally, fracturing is still controlled by the geometry of the element edges. In addition, the properties derived from standard rock mechanical tests cannot be used directly but have to be transferred to specific parameters for use in the DEM model, which depend on element size and bonding. Examples of hydromechanically-coupled fault modelling based on DEM are provided by Yoon et al. (2015, 2016).

The **Combined Finite Discrete Element Method (FEM/DEM)** is a rather new development and intends to make use of the advantages of both numerical techniques (e.g., Munjiza, 2004; Lisjak and Grasselli, 2014). The approach allows to insert cracks in a formerly continuous model domain based on a failure criterion and the local stress conditions. Thus, it is ideal for modelling of fracture formation and propagation independent of any pre-defined contacts or element shapes. In combination with flow simulations the approach has been used for modelling of hydraulic fracturing processes (e.g., Profit et al., 2016) and reactivation of faults due to pore pressure reduction (Ferguson et al., 2016), among others.

4.3 Material parameters of fault zone rocks required for modelling

No matter which numerical method is used, hydromechanical properties have to be assigned to the fault zone and its various subunits. Depending on the particular fault zone to be modelled, anisotropy effects have to be considered, i.e., material parameters differing in directions parallel and perpendicular to the fault zone. In general, hydraulic properties required for modelling comprise porosity and permeability, whereas the basic mechanical properties necessary to simulate stresses and deformation in the elastic domain are Young's modulus and Poisson's ratio (Chapter 2). The frictional limit and onset of plastic deformation can be described by the Mohr-Coulomb Law, which requires the knowledge of cohesion, angle of internal friction as well as the corresponding residual values for material behaviour in the post-failure domain (Chapter 2). If necessary, even more complex material laws such as, for example, Hoek-Brown, CamClay or temperature- and/or strain rate - dependent creep laws can be considered (e.g., Jaeger et al., 2007).

In general, the fault zone components, i.e., fault core and damage zone, exhibit petrophysical properties that are quite different from the surrounding host rock (Fig. 4.2; Caine et al., 1996; Wibberley et al., 2008; Faulkner et al., 2010).

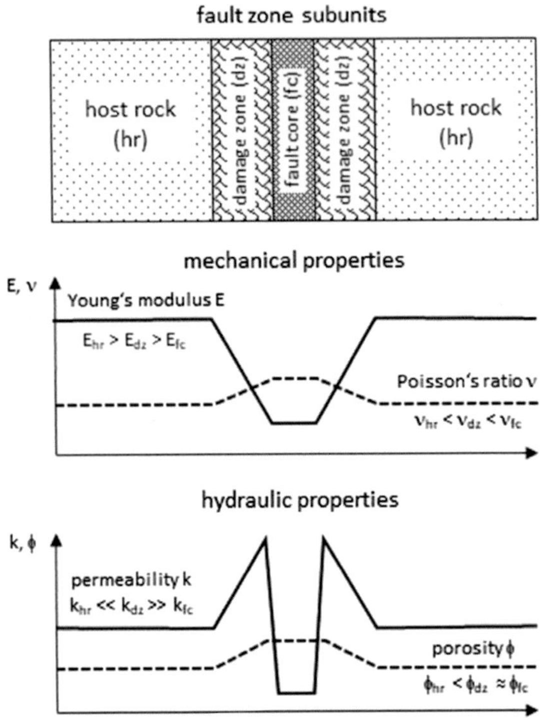

FIG. 4.2 Some general trends regarding the spatial distribution of hydromechanical properties across a fault zone. No single value or horizontal scale can honour the wide range of petrophysical rock properties as well as the damage zone and fault core widths observed in nature. This pattern is further complicated in cases of fault zones with multiple fault cores (Faulkner et al., 2010), discrete fractures with high aperture that act as high permeability pathways and/or fault zones which experienced a later hydrothermal overprint.

Due to fracturing, permeabilities in the damage zone can be several orders of magnitude higher than in the protolith, while Young's modulus values are somewhat reduced. For various host rock lithologies, Cappa (2009 and references therein) and Faulkner et al. (2010) provide a parameter range of 10^{-14} to 10^{-16} m^2 for permeability and 10 and 50 GPa for Young's modulus of the damage zone. If a weak fault gouge is developed in the fault core, Young's modulus values can be even lower, i.e., 1–10 GPa. Permeabilities in the central part of the fault zone are frequently significantly reduced to 10^{-17} to 10^{-21} m^2 (Cappa, 2009 and references therein). The frictional properties of the fault core can also be quite different from those of the host rock and damage zone, especially, if a weak fault gouge exists. The friction coefficient (= tangent of the friction angle) is usually between 0.6 and 0.85 (Byerlee, 1978), but can be reduced to 0.1–0.2, depending on the type of clay minerals present and the slip rate (Di Toro et al., 2011).

This basic pattern in the distribution of hydraulic and mechanical parameters across a fault zone (Fig. 4.2) is modified substantially in cases of multiple fault cores (Faulkner et al., 2010) and/or discrete fractures with higher aperture. Even further complexities may arise from a later hydrothermal overprint of the fault zone rocks. While hydrothermal alteration will further weaken the fault and reduce its bulk permeability, precipitation of minerals like quartz and barite can also substantially increase the strength of the damage zone and fault core, potentially leading to even higher Young's modulus than the original country rock. Such an overprint can change the basic characteristic of the fault with time, i.e., from a permeable and mechanically-weak feature to an impermeable and strong one.

The various petrophysical properties outlined above provide a first-order approximation, which may already be sufficient for generic fault zone modelling, However, if not a generic but a real fault, with specific properties is to be investigated, a detailed rock mechanical testing program has to accompany the numerical study. Various routine hydraulic and rock mechanical lab testing procedures are available to determine the parameters required for modelling. With respect to fault zone rocks, however, the collection of appropriate samples may be a challenge. Exploration and production wells are usually not designed to core fault zones and may even try to avoid drilling through faults because of potential borehole stability problems. However, even if a fault is cored, fracturing in the damage zone and the weak fault gouge frequently lead to poor core recovery. Similar problems are already encountered in faults exposed in surface quarries, making it frequently rather difficult to gather representative and sufficiently large samples for hydraulic and rock mechanical lab testing.

Another aspect regarding the population of a numerical model with petrophysical properties is the discrepancy in scale between the geometrical details of the fault zone and the sample size for rock testing on one side and the spatial resolution of the simulation on the other: the grid size used for numerical modelling is usually at least one, frequently several orders of magnitude larger. Hence, upscaling techniques are required for both hydraulic and mechanical parameters to account for material inhomogeneity and the combined effect of rock and fractures, respectively. A possible option is the **Discrete Fracture Network (DFN)** approach, which uses, among others, a stochastic fracture description to determine spatial variations in permeability and Young's modulus tensors of fractured rock masses, which then can be used by continuum and discontinuum models (e.g., Dershowitz et al., 2000; Lei et al., 2017). This upscaling issue also holds for another application of fault modelling, i.e., the incorporation of faults into larger hydromechanical reservoir and basin models, e.g., models with lateral dimensions of tens to hundreds of kilometres. For such applications, further modelling concepts have been developed to honour the importance of faults for localized fluid flow and stress field perturbations. For example, the use of fault transmissibility multipliers to account for the bulk permeability of faults is common practice in hydraulic modelling (e.g., Walsh et al., 1998; Manzocchi et al., 1999). The mechanical aspect is addressed by fault facies models that account for the different properties of the fault zone subunits (e.g., Fredman et al., 2008; Braathen et al., 2009) and adaptive meshes with refined grids in the vicinity of the faults (e.g., Qu et al., 2015; Fachri et al., 2016).

However, the basic problem remains: as the discrepancy between the spatial resolution of such field- to basin-size models and the inhomogeneity of a fault zone becomes larger, they lose the detail of models that focus only on a single fault.

4.4 An example of numerical modelling

A model based on FEM and using a highly simplified fault geometry is presented to illustrate the basic principles, typical workflow and output of a numerical fault zone simulation. The results of some parameter studies highlight the potential of the approach with respect to quantitative insights into fluid flow, stress perturbations and deformation in relation to the fault. The modelling utilizes the FEM software ANSYS (Ansys Inc., USA) for pre-processing (mesh generation, assignment of material properties and boundary conditions), the solution, as well as post-processing and visualization of the calculation results.

4.4.1 Modelling concept and parameters

The model further outlined below is a 2D plane strain representation perpendicular to the strike of a single fault zone that dips 60 degrees (Fig. 4.3). Incorporation of any along-strike variations or representation of a real fault would require a 3D modelling approach, but basic principles of fault zone processes can still be derived from such 2D models.

Hydromechanical fault zone modelling in this generic example is based on poroelastoplastic material behaviour and fluid flow through a porous medium (see, for example, Wang (2000), Shapiro (2015) and Cheng (2016) for more comprehensive treatment of these subjects). For the mechanical calculations, effective stresses are of particular importance, as stresses applied to a saturated porous medium are distributed between the solid rock matrix, i.e., the minerals, and the pore fluid. Using the geological notation that compressive stresses and pressure both have positive signs, this relationship can be expressed as;

$$\sigma'_{ij} = \sigma_{ij} - \alpha p \delta_{ij} \tag{4.1}$$

In this formula, σ'_{ij} and σ_{ij} are the components of the effective and total stress tensor, p is pore pressure and δ_{ij} is Kronecker's delta, defined as $\delta_{ij} = 1$ for $i = j$ and $\delta_{ij} = 0$ for $i \neq j$ (e.g., Wang, 2000). α is the Biot coefficient, which ranges between zero and one. A value of one is usually considered for

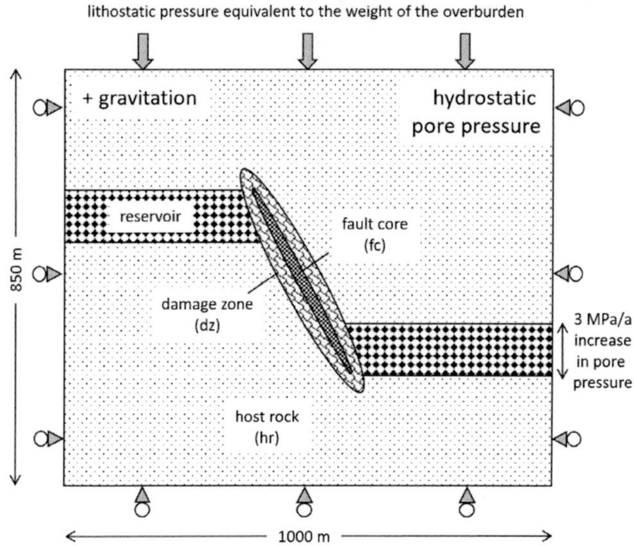

FIG. 4.3 Sketch of modelling concept (not to scale) showing model dimensions, fault zone geometry and the boundary conditions selected for the coupled hydraulic and mechanical simulation.

soils, whereas for hard rocks, values between 0.2 and 0.9 have been reported (e.g., Shapiro, 2015).

Linear poroelastic material behaviour relating strain ε_{ij} and effective stress σ'_{ij} via rock mechanical properties can be expressed according by;

$$\varepsilon_{ij} = \frac{1+\upsilon}{E}\sigma'_{ij} - \frac{\upsilon}{E}\sigma'_{kk}\delta_{ij} \tag{4.2}$$

with σ'_{kk} being the sum of the effective principal stresses and E_{ij} and ν_{ij} being Young's modulus and Poisson's ratio, respectively (e.g., Wang, 2000). The elastic limit and onset of plastic material behaviour is determined by the Mohr Coulomb criterion, using effective normal stress σ'_n according to;

$$\tau_{crit} = c\prime + \sigma'_n \tan\varphi\prime \tag{4.3}$$

τ_{crit} is the critical shear stress at failure, c′ is effective cohesion and is effective angle of internal friction (= arc tan coefficient of internal friction μ′) (e.g., Jaeger et al., 2007; Chapter 2).

For the hydraulic calculations, fluid flow is described by Darcy's law, which in a general 3D form can be stated as

$$\nabla\left[\frac{k}{\eta}(\nabla p - \rho_f g)\right] = S\frac{\partial p}{\partial t} \tag{4.4}$$

In this equation, k denotes the intrinsic permeability of the porous medium, η the fluid viscosity, p the pore pressure, ρ_f the fluid viscosity, S the specific storage (as a function of porosity) and t the time (e.g., Wang, 2000). Hydraulic and mechanical calculations are coupled via changes in effective stress and associated volumetric strain, which affects porosity and permeability. This leads to pore pressure changes, which in turn affect the effective stresses again.

Modelling in the present study involves a number of parameter studies, which differ regarding the hydromechanical properties assigned to the damage zone and fault core relative to those of the host rock (Table 4.2). The first two scenarios describe a weak fault without (model A) and with (model B) a damage zone, situated within an otherwise homogeneous protolith. Models C and D use the same fault properties as A and B, but additionally include a permeable layer offset by the fault. These models are used to study the impact of pore pressure changes in the reservoir layer of the hanging-wall block on fluid flow through and across the fault as well as on fault stability. Model E uses the same geometry as scenario C, but with a permeable fault core, thus, mimicking conditions of a breccia-like fault core, without a low-permeability fault gouge. Finally, model F addresses the effect of a hydrothermal overprint of the fault zone, which precipitated minerals that decreases the permeability and increases the strength of damage zone and fault core.

4.4.2 Model geometry and discretization

Model dimensions are $1000 \times 850\ m^2$, arranged around a fault zone of 230 m length and dipping 60 degrees (Fig. 4.3). Following the scaling relationships given by Johri et al. (2014), a total fault zone thickness of 20 m is assumed. Thereby the thickness of the fault core is 2 m and the width of the damage zones on both sides of the fault is 9 m. About 45,000 eight-node elements with quadratic trial functions and poroelastoplastic capabilities are used to describe the model geometry. Thereby, the edge length of the elements in the fault core is about 0.2 m.

4.4.3 Hydromechanical rock properties

This generic model is populated with different hydraulic and mechanical parameters for host and reservoir rocks as well as the fault zone subunits using a parameter range similar to those outlined above. Some models (models A and B), depict the fault being situated in a homogeneous host rock, whereas others (models C to F) incorporate also a more permeable layer offset by the fault. The specific properties assigned to host rock and reservoir are given in Table 4.1, whereas Table 4.2 lists the variable values of damage zone and fault core used for the six parameter studies.

TABLE 4.1 Hydromechanical properties of host rock and reservoir used for generic modelling.

	Host rock (h)	Reservoir
permeability k (m^2)	10^{-18}	10^{-14}
porosity ϕ (–)	0.1	0.2
Young's modulus E (GPa)	50	40
Poisson's ratio ν (–)	0.25	0.25
solid density ρ (kg m^{-3})	2400	2400
Biot coefficient α (–)	0.7	0.7
cohesion C (MPa)	1	5
friction angle φ (degrees)	40	30

4.4.4 Boundary and initial conditions

Hydraulic as well as mechanical boundary and initial conditions have to be defined for numerical modelling (Fig. 4.3). For the hydraulic calculations, an impermeable lower boundary, is assumed, whereas no flow in horizontal direction is allowed on the sides of the model. The initial state is a hydrostatic pore pressure field throughout the model. The mechanical calculations assume a free upper boundary, whereas the lower boundary is fixed with respect to vertical displacements ('roller boundary condition'). The sidewalls are free to move in a vertical direction and confined in horizontal direction. However, specific horizontal displacements could be assigned to the sidewalls in order to generate any desired horizontal stress magnitudes. A pressure equivalent to the weight of the overburden can be applied to the top of the model, if a deeply-buried fault is to be modelled. In the present model, the lithostatic load acting on top of the model is assumed to be equivalent to 2000 m of overburden, i.e., the centre of the modelled fault is located at a depth of 2425 m. The initial stress field, with its vertical and horizontal components, is the result of loading by its own weight and these boundary conditions. No further tectonic stress components are considered.

4.4.5 Modelling results

The modelling results illustrate exemplarily how a fault affects the stress and flow field in its surroundings. The magnitude and the spatial extent of the changes will depend on the particular material properties selected for host rock, damage zone and fault core, but some general trends can be derived from the six parameter studies outlined in Table 4.2.

TABLE 4.2 Overview of the modelling scenarios and the specific hydromechanical properties assigned to damage zone and fault core in each model. Information provided in bold summarizes the main characteristics of each of the scenarios.

	Damage zone (dz)		Fault core (fc)		
Scenario	Hydraulic	Mechanical	Hydraulic	Mechanical	Fault zone type
A	–	–	$\mathbf{k_{fc} < k_{hr}}$ $k_{fc} = 10^{-19}\ m^2$ $\phi_{fc} = 0.15$ $\alpha_{fc} = 1.0$	$\mathbf{E_{fc} \ll E_{hr}}$ $E_{fc} = 5$ GPa $C_{fc} = 15$ MPa $\varphi_{fc} = 20$ degrees	**weak, low perm fc without dz without reservoir layer**
B	$\mathbf{k_{dz} \gg k_{hr}}$ $k_{dz} = 10^{-15}\ m^2$ $\phi_{dz} = 0.15$ $\alpha_{dz} = 0.9$	$\mathbf{E_{dz} < E_{hr}}$ $E_{dz} = 30$ GPa $C_{dz} = 5$ MPa $\varphi_{dz} = 30$ degrees	$\mathbf{k_{fc} \ll k_{dz} \gg k_{hr}}$ $k_{fc} = 10^{-19}\ m^2$ $\phi_{fc} = 0.15$ $\alpha_{fc} = 1.0$	$\mathbf{E_{fc} < E_{dz} < E_{hr}}$ $E_{fc} = 5$ GPa $C_{fc} = 15$ MPa $\varphi_{fc} = 20$ degrees	**weak, low perm fc with dz without reservoir layer**
C	–	–	$\mathbf{k_{fc} < k_{hr}}$ $k_{fc} = 10^{-19}\ m^2$ $\phi_{fc} = 0.15$ $\alpha_{fc} = 1.0$	$\mathbf{E_{fc} \ll E_{hr}}$ $E_{fc} = 5$ GPa $C_{fc} = 15$ MPa $\varphi_{fc} = 20$ degrees	**weak, low perm fc without dz with reservoir layer**
D	$\mathbf{k_{dz} \gg k_{hr}}$ $k_{dz} = 10^{-15}\ m^2$ $\phi_{dz} = 0.15$ $\alpha_{dz} = 0.9$	$\mathbf{E_{dz} < E_{hr}}$ $E_{dz} = 30$ GPa $C_{dz} = 5$ MPa $\varphi_{dz} = 30$ degrees	$\mathbf{k_{fc} \ll k_{dz} \gg k_{hr}}$ $k_{fc} = 10^{-19}\ m^2$ $\phi_{fc} = 0.15$ $\alpha_{fc} = 1.0$	$\mathbf{E_{fc} < E_{dz} < E_{hr}}$ $E_{fc} = 5$ GPa $C_{fc} = 15$ MPa $\varphi_{fc} = 20$ degrees	**weak, low perm fc with dz with reservoir layer**
E	–	–	$\mathbf{K_{fc} \gg k_{hr}}$ $k_{dz} = 10^{-15}\ m^2$ $\phi_{fc} = 0.15$ $\alpha_{fc} = 0.9$	$\mathbf{E_{fc} < E_{hr}}$ $E_{fc} = 30$ GPa $C_{fc} = 5$ MPa $\varphi_{fc} = 30$ degrees	**strong, high perm fc without dz with reservoir layer**
F	$\mathbf{k_{dz} \gg k_{hr}}$ $k_{dz} = 10^{-17}\ m^2$ $\phi_{dz} = 0.15$ $\alpha_{dz} = 0.9$	$\mathbf{E_{dz} = E_{hr}}$ $E_{dz} = 50$ GPa $C_{dz} = 5$ MPa $\varphi_{dz} = 40$ degrees	$\mathbf{k_{dz} \gg k_{fc} = k_{hr}}$ $k_{fc} = 10^{-19}\ m^2$ $\phi_{fc} = 0.15$ $\alpha_{fc} = 1.0$	$\mathbf{E_{fc} < E_{dz} = E_{hr}}$ $E_{fc} = 30$ GPa $C_{fc} = 15$ MPa $\varphi_{fc} = 30$ degrees	**hydrothermal overprint with dz with reservoir layer**

Models A and B describe a fault embedded in an otherwise homogeneous host rock, whereas models C to F additionally consider a more permeable reservoir layer offset by the fault. The latter models are used to study the effects of the increase in reservoir pore pressure on fault stability and fault fluid flow.

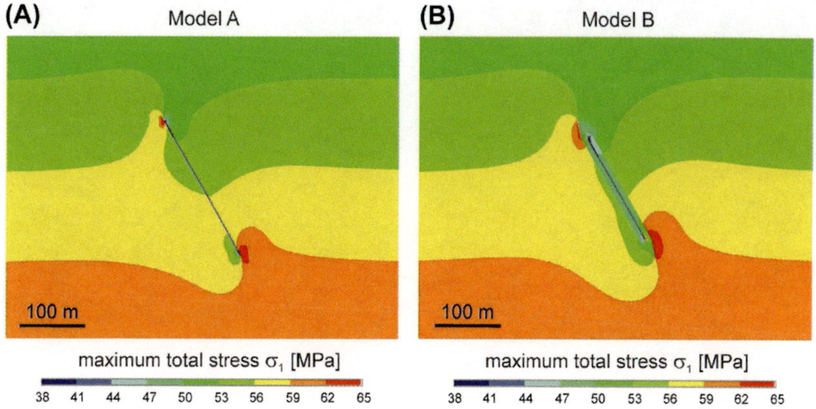

FIG. 4.4 Models A and B: Initial maximum total stress (s_1) for models A (A) and B (B), i.e., the models are subject only to their own weight and consider a hydrostatic pore pressure field. Note the stress perturbation caused by the fault zone that modifies the stress magnitudes, which regularly prevail at these depths up to a distance of 150 m from the fault.

Models A and B, for example, illustrate the stress perturbations caused by a weak fault zone without (model A) and with (model B) a damage zone in an otherwise homogeneous host rock. Elastic deformation in the fault core and damage zone, which is already caused by the weight of the model and the lithostatic load acting on the top model boundary, modifies the stress field in the vicinity of the fault. This differential loading across the fault results in a typical stress pattern, which for the upper part of the fault shows higher stresses in the footwall and lower stresses in the hanging wall (relative to the stress level that would normally occur at this depth) and vice versa for the lower part of the fault (Fig. 4.4A). These stress perturbations increase with the total width of the fault zone and affect a region up to 150 m distance from the fault (Fig. 4.4B).

Models C and D differ from the previous models by incorporation of a more permeable reservoir layer offset by the fault. As in the previous models, mechanical properties of the fault zone are such that no plastic strain occurs under the model's weight, i.e., the fault is initially stable. Starting with an equilibrium hydrostatic pore pressure field, the effect of an increasing pore pressures in the hanging-wall reservoir layer (e.g., due to injection or upward-moving fluids) on fault stability and fault fluid flow is investigated. For this purpose, the pore pressure on the right side of the reservoir layer is progressively increased over 10 years at a rate of 3 MPa/a. In model C, the low permeability of the fault core largely inhibits flow through the fault to the reservoir layer on the footwall side (Fig. 4.5A). However, increasing pore pressure on the hanging-wall side reduces the effective stresses in the fault zone, ultimately leading to failure and plastic deformation of the fault core (Fig. 4.5B). Reactivation of the fault initiates at the intersection with the

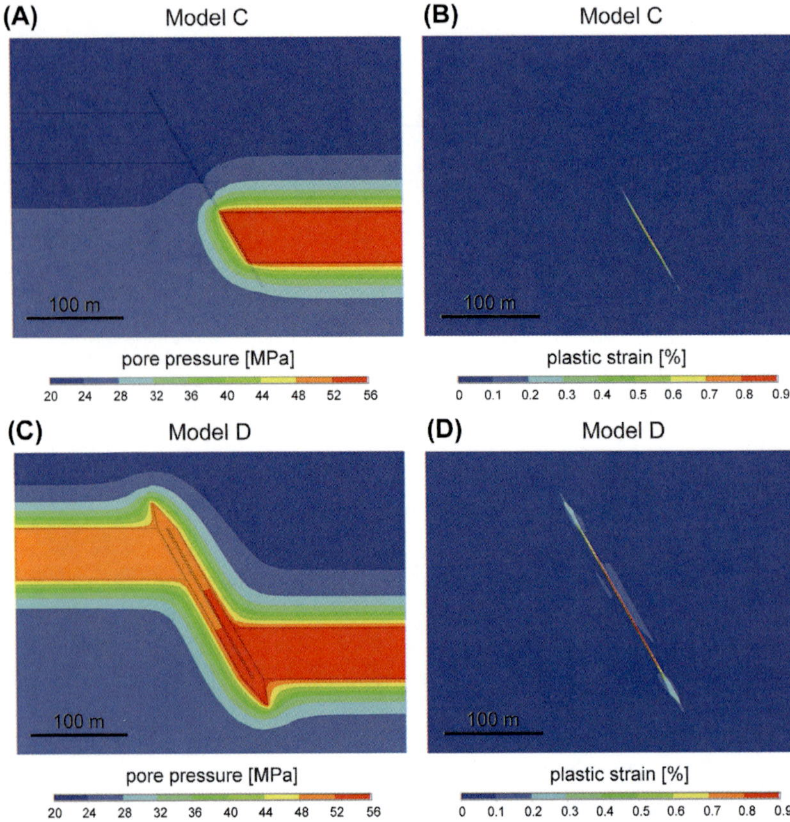

FIG. 4.5 Models C and D: Pore pressure (A and C) and equivalent plastic strain (B and D) after 10 years of pore pressure increase at a rate of 3 MPa a^{-1} applied to the right boundary of the reservoir section. The strain plots illustrate spreading and intensifying of the plastic deformation zone in the fault core and progressive reactivation of the initially stable fault, respectively.

lower reservoir and subsequently spreads up- and down-dip the fault plane. At higher pore pressures, plastic strain also extends into the host rock at the lower tip of the fault, which would imply further propagation of the fault.

In model D, the high permeability of the damage zone results in the fault being a conduit for fluid flow and the pore pressure increase in the hanging-wall block is also transferred to the reservoir on the footwall side (Fig. 4.5C). Due to the particular geometry and permeability distribution chosen for the fault zone, the pore pressure front propagating through the damage zone on the footwall side is somewhat delayed with respect to the hanging-wall side. Again, the pore pressure increase results in plastic strain and reactivation of the fault. In comparison to model C, larger parts along the entire length of the fault core and partly also in the damage zone are affected

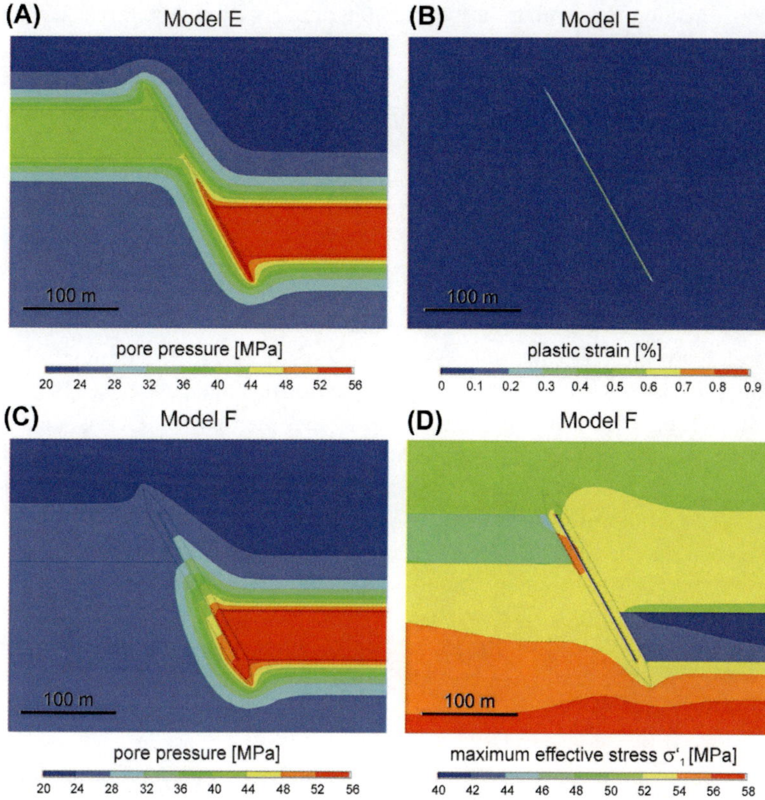

FIG. 4.6 Models E and F: Pore pressure (A and C) and equivalent plastic strain (B) after 10 years of pore pressure increase at a rate of 3 MPa a^{-1} at the right boundary of the reservoir section. For Model F, the assumed hydrothermal overprint and related strengthening of the fault zone suppress any plastic straining and reactivation of the fault, respectively. The corresponding distribution of the maximum effective stress (σ'_1) is shown in (D).

(Fig. 4.5D). In this scenario, high strain zones extend into the host rock at both fault tips.

Model E describes a fault without a weak fault gouge in the centre, but with a highly fractured and permeable fault core. This again allows for pressure communication and flow between the two reservoir sections offset by the fault (Fig. 4.6A). However, the different material properties (in comparison to a weak, low-perm fault core) largely suppress plastic deformation and fault reactivation (Fig. 4.6B).

Finally, model F addresses the effects of a late hydrothermal overprint of the fault zone leading to permeability reduction and strengthening of damage zone and fault core due to precipitation of minerals such as quartz, calcite or barite. For the properties selected, the fault is a semi-permeable baffle, rather than a complete barrier to fluid flow. Thus, the pressure increase in the footwall reservoir section is less than in the previous models (Fig. 4.6C), but the fault core and damage zone do not experience plastic strain, i.e., the fault remains stable, even for pore pressure conditions, which led to substantial fault reactivation in the other scenarios. The pore pressure increase and the different Biot coefficients assigned to the various model units lead to a heterogeneous effective stress pattern (Fig. 4.6D) with minima in the hanging-wall reservoir and the fault core.

4.5 Conclusions

Various numerical techniques are available for modelling of fault zones and related hydromechanically-coupled processes. The most commonly-used methods hitherto are the finite difference (FD), the finite element (FE) and the distinct element (DE) methods. Each has its own pros and cons, and the decision, which is the most appropriate for a particular study, has to honour the processes to be modelled and the geometrical complexity of the fault zone. If modelling of a specific fault zone is intended, a comprehensive testing program both in the lab and in-situ should be considered to determine the appropriate mechanical and hydraulic properties of the various fault zone units and the host rock.

Generic models, with strongly idealized fault geometries and properties taken from the literature, can provide important insights into stress-field perturbations, fault reactivation as well as fluid flow through and across faults. The modelling work presented should be considered as an appetizer for application to more complex fault geometries and case studies of actual faults. It provides a toolbox which can be easily expanded to incorporate more realistic fault zone structures, e.g., with anisotropic material properties varying with distance from the fault core, lenticular shear bodies and/or discrete shear planes. In the present paper, six model scenarios are presented that differ regarding the geometry and hydromechanical properties of a fault zone embedded in a homogeneous host rock and offsetting a more permeable reservoir layer, respectively. The results show, among others, the crucial role of the fault zone geometry and the hydromechanical properties for the stability and hydraulic behaviour of the fault. This also underlines, beside any further advancements in numerical modelling techniques, the need for robust information on hydromechanical fault zone properties. Several other challenges to future work are capturing details of fault zone architecture in the numerical models and upscaling the geometry and properties of the fault zone to the dimensions of the numerical grid. Thus, numerical fault zone modelling offers interesting scientific perspectives with a wide range of applications, ranging from studies of hydrocarbon reservoirs and fault reactivation through man-made pore pressure changes, with the aim for a better understanding of earthquake processes and more accurate earthquake forecasting models.

References

Anderson, M.P., Woessner, W.W., Hunt, R.J., 2015. Applied Groundwater Modeling: Simulation of Flow and Advective Transport, second ed. Academic Press, p. 630.

Buchmann, T.J., Connolly, P.T., 2009. Contemporary kinematics of the Upper Rhine Graben: a 3D finite element approach. Glob. Planet. Chang. 58, 287–309.

Braathen, A., Tveranger, J., Fossen, H., Skar, T., Cardozo, N., Semshaug, S.E., Bastesen, E., Sverdrup, E., 2009. Fault facies and its application to sandstone reservoirs. AAPG Bull. 93/7, 891–917.

Byerlee, J.D., 1978. Friction of rock. Pure Appl. Geophys. 16 (4–5), 615–626.

Caine, J.S., Evans, J.P., Forster, C.B., 1996. Fault zone architecture and permeability structure. Geology 24 (11), 1025–1028.

Cappa, F., 2009. Modelling fluid transfer and slip in a fault zone when integrating heterogeneous hydromechanical characteristics in its internal structure. Geophys. J. Int. 178, 1357–1362.

Cappa, F., 2011. Influence of hydromechanical heterogeneities of fault zones on earthquake ruptures. Geophys. J. Int. 185/2, 1049–1058.

Cappa, F., Rutqvist, J., 2011. Modeling of coupled deformation and permeability evolution during fault reactivation induced by deep underground injection of CO_2. Int. J. Greenh. Gas Con. 5 (2), 336–346.

Cheng, A.H.D., 2016. Poroelasticity. Springer International Publishing, 877 p.

Childs, C., Manzocchi, T., Walsh, J.J., Bonson, C.G., Nicol, A., Schopfer, M.P.J., 2009. A geometric model of fault zone and fault rock thickness variations. J. Struct. Geol. 31/2, 117–127.

Cundall, P.A., 1971. A computer model for simulating progressive large scale movements in blocky rock systems. In: Proceedings Symposium Int. Soc. Rock Mech., Nancy Metz, 1, Paper II–8.

Dershowitz, B., LaPointe, P., Eiben, T., Wei, L., 2000. Integration of discrete feature network methods with conventional simulator approaches. SPE Reservoir Eval. Eng. 3/2, 165–170.

Di Toro, G., Han, R., Hirose, T., De Paola, N., Nielsen, S., Mizoguchi, K., Ferri, F., Cocco, M., Shimamoto, T., 2011. Fault lubrication during earthquakes. Nature 471, 494–498.

Evans, J.P., Forster, C.B., Goddard, J.V., 1997. Permeability of fault-related rocks, and implications for hydraulic structure of fault zones. J. Struct. Geol. 19/11, 1393–1404.

Fachri, M., Tveranger, J., Braathen, A., Røe, P., 2016. Volumetric faults in field-sized reservoir simulation models: a first case study. AAPG Bull. 100/5, 795–817.

Faulkner, D.R., Jackson, C.A.L., Lunn, R.J., Schlische, R.W., Shipton, Z.E., Wibberley, C.A.J., Withjack, M.O., 2010. A review of recent developments concerning the structure, mechanics, and fluid flow properties of fault zones. J. Struct. Geol. 32, 1557–1575.

Ferguson, W., Bere, A., Rodriguez, C., Félix, L., Marsili, M., Medeiros, L., 2016. Modelling of a deepwater Brazilian field to assess fault reactivation and the in situ stresses during production. First Break 34/6, 39–47.

Fredman, N., Tveranger, J., Cardozo, N., Braathen, A., Soleng, H., Røe, P., Skorstad, A., Syversveen, A.R., 2008. Fault facies modeling: technique, and approach for 3D conditioning and modeling of faulted grids. AAPG Bull. 92/9, 1–22.

Fries, T.P., Belytschko, T., 2010. The generalized/extended finite element method: an overview of the method and its applications. Int. J. Numer. Methods Eng. 84, 253–304.

Gerya, T., 2009. Introduction to Numerical Geodynamic Modelling. Cambridge University Press, 358 p.

Haug, C., Nüchter, J.A., Henk, A., 2018. Assessment of geological factors potentially affecting production-induced seismicity in North German gas fields. Geomech. Energy Environ. 16, 15–31.

Ibrahimbegovic, A., Melnyk, S., 2007. Embedded discontinuity finite element method for modeling of localized failure in heterogeneous materials with structured mesh: an alternative to extended finite element method. Comput. Mech. 40/1, 149–155.

Jaeger, J., Cook, N.G., Zimmerman, R., 2007. Fundamentals of Rock Mechanics, fourth ed. Wiley-Blackwell. 488 p.

Jeanne, P., Rutqvist, J., Dobson, P.F., Walters, M., Hartline, C., Garcia, J., 2014. The impacts of mechanical stress transfers caused by hydromechanical and thermal processes on fault stability during hydraulic stimulation in a deep geothermal reservoir. Int. J. Rock Mech. Min. Sci. 72, 149–163.

Jing, L., Hudson, J.A., 2002. Numerical methods in rock mechanics. Int. J. Rock Mech. Min. Sci. 39 (4), 409–427.

Johri, M., Zoback, M., Hennings, P., 2014. A scaling law to characterize fault-damage zones at reservoir depths. AAPG Bull. 98/10, 2057–2079.

Lei, Q., Latham, J.P., Tsang, C.F., 2017. The use of discrete fracture networks for modelling coupled geomechanical and hydrological behaviour of fractured rocks. Comput. Geotech. 85, 151–176.

Leveque, R.J., 2002. Finite Volume Methods for Hyperbolic Problems. Cambridge University Press, 578 p.

Lisjak, A., Grasselli, G., 2014. A review of discrete modeling techniques for fracturing processes in discontinuous rock masses. J. Rock Mech. Geotech. Eng. 6, 301–314.

Manzocchi, T., Walsh, J.J., Nell, P.A.R., Yielding, G., 1999. Fault transmissibility multipliers for flow simulation models. Pet. Geosci. 5, 53–63.

McLellan, J.G., Oliver, N.H.S., Schaubs, P.M., 2004. Fluid flow in extensional environments; numerical modeling with an application to Hamersley iron ores. J. Struct. Geol. 26, 1157–1171.

Minkoff, S.E., Stone, C.M., Bryant, S., Peszynska, M., Wheeler, M.F., 2003. Coupled fluid flow and geomechanical deformation modelling. J. Pet. Sci. Eng. 38, 37–56.

Munjiza, A., 2004. The Combined Finite-Discrete Element Method. John Wiley & Sons Ltd., 352 p.

Pereira, L.C., Guimarães, L.J.N., Horowitz, B., Sánchez, M., 2014. Coupled hydro-mechanical fault reactivation analysis incorporating evidence theory for uncertainty quantification. Comput. Geotech. 56, 202–215.

Perrone, N., Kao, R., 1975. A general finite difference method for arbitrary meshes. Comput. Struct. 5, 45–58.

Prevost, J.-H., Sukumar, N., 2015. Multi-scale X-FEM faults simulations for reservoir-geomechanical models. In: 49th US Rock Mechanics/Geomechanics Symposium. American Rock Mechanics Association.

Profit, M., Dutko, M., Yu, J., Cole, S., Angus, D., Baird, A., 2016. Complementary hydro-mechanical coupled finite/discrete element and microseismic modelling to predict hydraulic fracture propagation in tight shale reservoirs. Comput. Part. Mech. 3/2, 229–248.

Qu, D., Røe, P., Tveranger, J., 2015. A method for generating volumetric fault zone grids for pillar-gridded reservoir models. Comput. Geosci. 81, 28–37.

Ramsay, J.G., Lisle, R.J., 2000. The Techniques of Modern Structural Geology: Volume 3: Applications of Continuum Mechanics in Structural Geology. Academic Press Inc., 1061 p.

Rice, J.R., Dunham, E.M., Noda, H., 2009. Thermo- and hydro-mechanical processes along faults during rapid slip. In: Hatzor, Y., Sulem, J., Vardoulakis, I. (Eds.), Meso-scale Shear Physics in Earthquale and Landslide Mechanics. CRC Press (Taylor & Francis Group), pp. 3–16.

Rutqvist, J., 2011. Status of the TOUGH-FLAC simulator and recent applications related to coupled fluid flow and crustal deformations. Comput. Geosci. 37, 739–750.

Rutqvist, J., Rinaldi, A.P., Cappa, F., Jeanne, P., Mazzoldi, A., Urpi, L., Guglielmi, Y., Vilarrasa, V., 2016. Fault activation and induced seismicity in geologic carbon storage - lessons learned from recent modeling studies. J. Rock Mech. Geotech. Eng. 8, 775–966.

Rutqvist, J., Stephansson, O., 2003. The role of hydromechanical coupling in fractured rock engineering. Hydrogeol. J. 11, 7–40.

Schuite, J., Longuevergne, L., Bour, O., Burbey, T.J., Boudin, F., Lavenant, N., Davy, P., 2017. Understanding the hydromechanical behavior of a fault zone from transient surface tilt and fluid pressure observations at hourly time scales. Water Resour. Res. 53, 10558–10582.

Shapiro, S., 2015. Fundamentals of poroelasticity. In: Fluid-Induced Seismicity. Cambridge University Press, pp. 48–117.

Simpson, G., 2017. Practical Finite Element Modeling in Earth Science Using Matlab. Wiley-Blackwell, 272 p.

Smith, G.D., 1986. Numerical Solution of Partial Differential Equations: Finite Difference Methods, third ed. Oxford University Press. 354 p.

Walsh, J.J., Watterson, J., Heath, A.E., Childs, C., 1998. Representation and scaling of faults in fluid flow models. Pet. Geosci. 4, 241–251.

Wang, H.F., 2000. Theory of Linear Poroelasticity—With Applications to Geomechanics and Hydrogeology. Princeton University Press, 287 p.

Wibberley, C.A.J., Yeilding, G., Di Toro, G., 2008. Recent advances in the understanding of fault zone internal structure: a review. In: Wibberley, C.A.J., et al. (Eds.), The Internal Structure of Fault Zones: Implications for Mechanical and Fluid Flow Properties, vol. 299. Geological Society Special Publications, London, pp. 5–33.

Yoon, J.S., Zimmermann, G., Zang, A., Stephansson, O., 2015. Discrete element modeling of fluid injection induced seismicity and activation of nearby fault. Can. Geotech. J. 52 (10), 1457–1465.

Yoon, J.S., Zang, A., Stephansson, O., Zimmermann, G., 2016. Modelling of fluid-injection-induced fault reactivation using a 2D discrete element based hydro-mechanical coupled dynamic simulator. Energy Procedia 97, 454–461.

Zhang, Y., Clennell, M.B., Plane, C.D., Ahmed, S., Sarout, J., 2016. Numerical modelling of fault reactivation in carbonate rocks under fluid depletion conditions – 2D generic models with a small isolated fault. J. Struct. Geol. 93, 17–28.

Zienkiewicz, O.C., Taylor, R.L., Zhu, J.Z., 2013. The Finite Element Method: Its Basis and Fundamentals, seventh ed. Butterworth-Heinemann. 756 p.

Chapter 5

Faulting in the laboratory

André Niemeijer[a], Åke Fagereng[b], Matt Ikari[c], Stefan Nielsen[d], Ernst Willingshofer[e]

[a]*Department of Geoscience, Utrecht University, Utrecht, The Netherlands;* [b]*School of Earth & Ocean Sciences, Cardiff University, Cardiff, United Kingdom;* [c]*MARUM—Center for Marine Environmental Sciences and Faculty of Geosciences, University of Bremen, Bremen, German;* [d]*Department of Earth Sciences, Durham University, Durham, United Kingdom;* [e]*Department of Earth Sciences, Utrecht University, Utrecht, The Netherlands*

Chapter outline

Understanding Faults. https://doi.org/10.1016/B978-0-12-815985-9.00005-9

5.1 Fault friction in the quasi-static regime

5.1.1 Laboratory measurements of friction

There are several different experimental configurations that can be used to determine fault rock or fault gouge friction under variable loading conditions (Fig. 5.1). One of the earliest setups used, is the so-called saw-cut method (Fig. 5.1A), in which a cylindrical sample of rock, steel or ceramic is cut at an angle of 30 or 45 degrees to the vertical axis creating a simulated fault which can be roughened or filled with fine-grained wear material, called fault gouge. Alternatively, a cylindrical rock sample can be fractured within the experimental apparatus and sliding friction of the resulting fracture can be measured.

The cylinder with or without a simulated fault is jacketed with rubber or metal, depending on the experimental conditions and the type of confining medium. The jacketed sample is then loaded into a pressure vessel, which can be pressurized with oil or gas. Experiments in oil-filled pressure vessels typically are done with a rubber jacket and temperature is limited to ~200 °C. Higher temperatures can be achieved in gas-pressured vessels in which metal jackets are usually used. The type of metal (e.g., copper) is selected such that the jacket strength is negligible at the experimental conditions.

The confined cylinder is loaded uniaxially by displacing a loading piston at a constant rate, thereby creating shear on the simulated fault. Both shear stress and normal stress vary during the loading and are calculated from the principal stresses as:

$$\sigma_n = \left(\frac{\sigma_1 + \sigma_3}{2}\right) - \left(\frac{\sigma_1 - \sigma_3}{2}\right)\cos 2\alpha \tag{5.1}$$

$$\tau = \left(\frac{\sigma_1 - \sigma_3}{2}\right)\sin 2\alpha \tag{5.2}$$

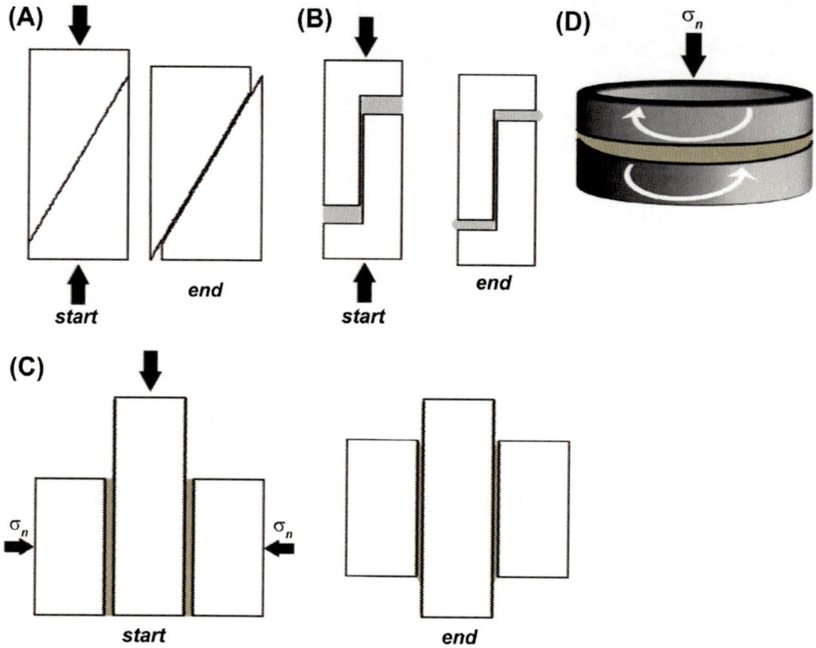

FIG. 5.1 Different possible configurations for measuring friction of simulated faults and fault gouges (not to scale) (A) Saw-cut set-up for use in a triaxial pressure vessel where the confining pressure supplies the fault normal stress (B) Direct-shear set-up for use in a triaxial pressure vessel where the confining pressure supplies the fault normal stress (C) Double-direct shear set-up for use in a biaxial loading frame (D) rotary shear set-up.

Here, σ_3 is the confining pressure and is equal to σ_2, σ_1 is the axial stress and α is the angle of the inclined simulated fault. Axial force is typically measured internally inside the pressure vessel and thus measures the total differential force, which can be easily converted to stress using the contact area of the ellipsoidal sliding interface. A disadvantage of this method is that the contact area continuously changes with accumulated displacement. The corresponding change in normal stress can be corrected by having a servo-control of the confining pressure, lowering it with ongoing displacement. However, this is not feasible when displacement occurs rapidly such as in a failure event. An alternative set-up that can be used in a triaxial pressure vessel is shown in Fig. 5.1B. This assembly consists of two L-shaped semi-cylindrical pistons, which sandwich a thin gouge layer (1–2 mm), and, when assembled, form a cylinder that is jacketed to seal the sample from the confining fluid or gas. Shear displacement along the gouge layer is accommodated by inserting easily deformable plugs. The plugs can be composed of any material that has negligible strength under the experimental conditions of interest and is usually a polymer, such as putty. The confining pressure supplies the normal stress to the simulated fault and, if kept constant, will yield a constant normal stress while shearing. Maximum displacements that can be reached are limited by the size of the plug-filled gaps and are typically on the order of 5–6 mm. Axial force measured internally is easily converted to a shear stress on the sample by dividing over the contact area.

A second experimental configuration used to determine friction is the biaxial double-direct shear configuration (Fig. 5.1C). This configuration consists of a set of equally sized side blocks, between which a center block is displaced and has the advantage that the contact area remains constant during sliding. When a gouge layer is used, the layer thins as the center block is displaced. The maximum displacement is limited by how quickly the gouge layer thins and by the length of the center block. Two hydraulically driven pistons provide the normal and shear stress, controlled independently. Typically, the horizontal axis provides the normal stress that is held constant through a servo-feedback system controlling the force and the vertical axis is driven at a prescribed displacement rate while measuring the shear force. Both bare rock and gouge experiments can be done in a biaxial double-direct shear configuration. Experiments with fluids can be done by submerging the block assembly in a fluid or by performing the experiments inside a pressure vessel. In the latter case, the entire block assembly needs to be sealed from the confining fluid (oil) which requires custom-made rubber jackets. The confining pressure contributes to the normal stress on the gouge layer and thus needs to be taken into account. Fluid access to the gouge layer is provided through holes in the three forcing blocks which are connected to fluid pressure intensifiers via ports in the pressure vessel. To distribute the fluid over the entire area of the gouge sample, porous plates are inserted in the block faces. In this way, fluid pressure on both sides of the gouge layer can be controlled and manipulated, thus also allowing for the measurement of fault-perpendicular permeability. A variation of the biaxial configuration is the single-direct shear configuration, where two forcing blocks induce shear deformation in the sample and the resistance at the outer edges of the forcing blocks is either negligible or accounted for. An example of this system is the "shear box" commonly used to test soils for engineering purposes.

A third experimental configuration, often used to measure friction, is the rotary shear assembly (Fig. 5.1D). Here, cylindrical or ring-shaped samples are pressed together using a servo-controlled driving ram, driven by either hydraulics or an electromotor. In the case of bare rock samples, an aluminum support ring is typically employed to provide extra support for the sample. For gouge samples, several different confinement methods are employed. The simplest method is to use rings composed of a low friction material, typically

PTFE (Teflon), or alternatively metal alloy rings with a low-friction coating (such as molybdenum sulphide). Other methods rely on a combination of steel rings and O-rings to provide both sample containment and sealing of pore fluids. Regardless of the confinement method, gouge loss due to extrusion of material is common and efforts to minimize the gouge loss can result in additional friction between the cylinder or rings and the containment rings. Despite this obvious drawback, the rotary shear configuration is advantageous for studying the evolution of friction over large slip displacements, since in principle this displacement can be infinite. This is the reason that experiments conducted with a controlled seismic slip velocity of 1 m/s and higher are almost exclusively done in a rotary shear configuration (a notable exception is the direct shear configuration in Saber et al., 2016). Incorporating confining fluids and temperature is a challenge in the rotary shear configuration and only a few apparatuses exist that allow experiments to be done at elevated temperature in the presence of fluids (Niemeijer et al., 2008) and results from experiments at a maximum temperature of 700 °C have been reported (Van Diggelen et al., 2010). In a rotary shear configuration, displacement and sliding velocity vary from the center of the cylinder or inner edge of the ring to the outer boundaries. Typically, one of two approaches is used to deal with these variations. In experiments employing ring-shaped samples, displacement, velocity and shear stress are calculated for the average radius of the annulus. Alternatively, for experiments using cylindrical samples, shear stress across the sample surface can be assumed to be constant (i.e., there is little or no dependence of shear stress on velocity), leading to the following equations for shear stress and equivalent velocity (see also Shimamoto and Tsutsumi, 1994):

$$\tau = \frac{3T}{2\pi \cdot \left(R_o^3 - R_i^3\right)} \tag{5.3}$$

$$V_{eq} = \frac{4\pi R \cdot \left(R_o^2 + R_o R_i + R_i^2\right)}{3 \cdot (R_o + R_i)} \tag{5.4}$$

Here, T is the measured torque (N·m), R_o and R_i are the outer and inner radii, respectively (m) and R is the rotation rate (rotations per second). For a solid cylinder, R_i is zero and the equivalent velocity is defined at 2/3 of the cylinder radius.

5.1.2 General observations of steady state friction

It has been recognized since the 1970s that the overall level of friction of most rock-forming minerals is between 0.6 and 1.0, as was summarized for a large set of experiments by Byerlee (1978). In this seminal paper, Byerlee plotted maximum shear stress data as a function of (effective) normal stress for a large number of rock types and found that all data could be fitted with a linear equation:

$$\tau = \mu\sigma_n^{\text{eff}} + C \tag{5.5}$$

At room temperature and in the absence of a pore fluid, most data is fitted well with a maximum friction coefficient of 0.85 at normal stresses below 200 MPa, whereas the data at stresses above 200 MPa gives a slope of 0.6 with an intercept of 50 MPa. This general observation of a friction of 0.85 or 0.6 is typically quoted as Byerlee's law or rule. Notable exceptions to this rule are phyllosilicate and clay minerals that typically have values of friction less than 0.6, particularly when wet. Note that for experiments involving gouges, the maximum friction might not be very representative, because the preparation method, grain size (distribution) and time under load (i.e., the initial porosity) all play a role in regulating maximum friction. From soil mechanics studies it is known that overconsolidated gouge will show a peak in friction due to initial dilation before weakening to a residual friction level.

In general, the addition of aqueous fluids to gouges has little effect on the level of steady state sliding friction, at least in the absence of any time-dependent compaction mechanisms and for non-sheet-like minerals. In contrast, water causes significant weakening in clay- or phyllosilicate gouges. In these gouges, water is thought to be actively involved in shearing, as it is adsorbed on the (001) surfaces of most of these mineral types, with the notable exception of talc which is hydrophobic. The strength of water-saturated clay- and phyllosilicate gouge seems to be mostly controlled by the strength of the bonding between the adsorbed water and the (001) interface (Moore and Lockner, 2004). It should be noted that it is crucial to understand the drainage conditions of a particular experimental set-up. Poor drainage could lead to the development of locally high pore-fluid pressures, which would reduce effective normal stress and lead to a lowering of the apparent friction (see Faulkner et al., 2018 and references therein).

The strength of binary or ternary mixtures of strong and weak minerals depends non-linearly on the content of the weak phase. To make matters worse, different combinations of strong/weak minerals give different effects. With quartz as the strong phase, it seems that the strength of a mixture of quartz and a weak clay or phyllosilicate remains high for weak phase contents of up to 30 wt% (e.g., Logan and Rauenzahn, 1987; Takahashi et al., 2007; Tembe et al., 2010). In contrast, with calcite as the strong phase, weakening is already reported to occur for weak phase contents as low as 20 wt% (Giorgetti et al., 2015). It seems that the grain size of the strong phase plays an important role in disrupting the development of a through-going connected network of

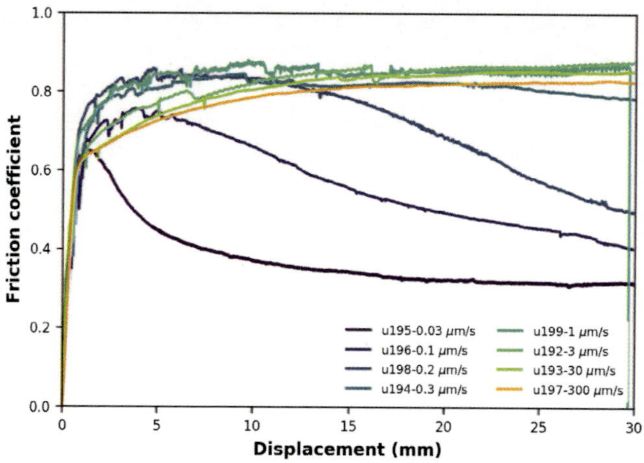

FIG. 5.2 Evolution of friction with displacement for a simulated fault gouge of 80 wt% quartz and 20 wt% muscovite. Strong weakening is only observed when shearing is slow. Experiments were done using a hydrothermal rotary shear apparatus at a temperature of 500 °C, an effective normal stress of 120 MPa and a fluid pressure of 80 MPa. *From Niemeijer (2018).*

weak phase grains, i.e., a foliation. Alternatively, if the strong phase can accommodate shear deformation via solution-transfer processes, weakness can be established once enough displacement has accumulated. This was demonstrated in experiments using halite as a rock-analogue with minor amounts (20 wt%) of muscovite and kaolinite (Bos and Spiers, 2001, 2002; Bos et al., 2000; Niemeijer and Spiers, 2005, 2006) and recently under hydrothermal conditions for quartz-muscovite mixtures (Fig. 5.2, Niemeijer, 2018). Strong weakening with accumulated displacement under hydrothermal conditions and low driving velocity was also observed in experiments using Westerly granite (Blanpied et al., 1995, 1998a,b) and quartz (Chester and Higgs, 1992; Chester, 1994) as fault gouges. In the case of pure quartz, it was inferred that deformation was accommodated by solution-transfer processes, which might have played a role in the experiments with Westerly granite as well. An alternative explanation offered for the weakening is localization of deformation in zones with a higher phyllosilicate content (Blanpied et al., 1995).

5.1.3 Rate-and-state friction

Experiments performed in the broad velocity range of 1 nm/s to 1 mm/s address the nucleation phase of earthquake slip or the broad range of aseismic slip phenomena (from creep at plate tectonic loading rates of tens of mm/year or ~1 nm/s to episodic slow slip and tremor at several mm/day or ~0.3 μm/s). In the late 1970s and early 1980s, Dieterich (1978, 1979, 1981) and Ruina (1983) proposed a set of empirical friction or shear stress equations that can capture the details of experimental results, in particular the transient friction response due to a quasi-instantaneous change in load-point velocity. A theoretical example of experimental data is shown in Fig. 5.3. The so-called Rate-and-State Friction (RSF) equations used to fit such experimental data are given by:

$$\mu = \mu_0 + aln\left(\frac{V}{V_0}\right) + bln\left(\frac{V_0\theta}{d_c}\right) \tag{5.6}$$

$$\frac{d\theta}{dt} = 1 - \frac{V\theta}{d_c} \tag{5.7a}$$

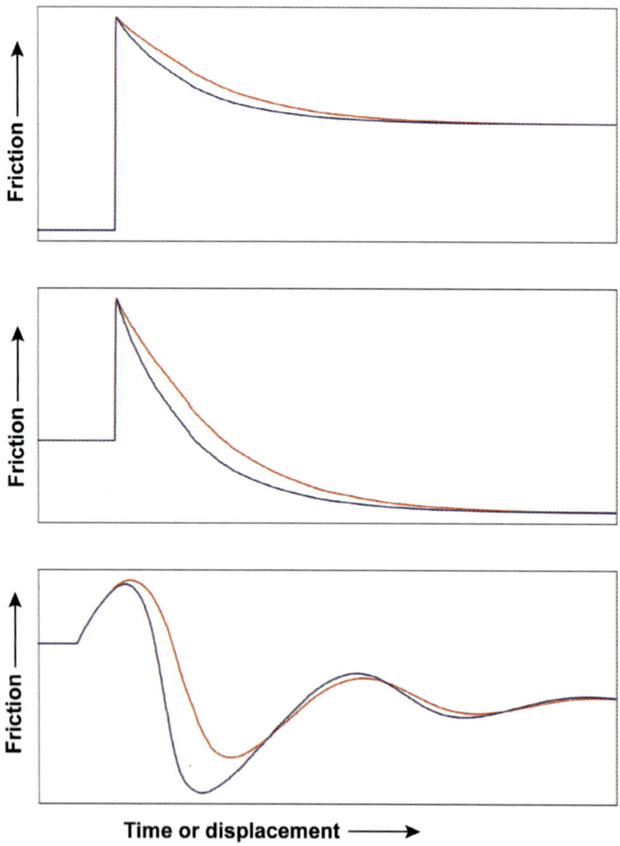

FIG. 5.3 Response of friction to an instantaneous up-step in load-point velocity, as predicted by the two different evolution laws Eq. (5.7a) (Dieterich, red curves) and Eq. (5.7b) (Ruina, blue curves). Panel (A) shows a velocity-strengthening response, panel (B) a velocity-weakening response and panel (C) a velocity-weakening response in a more compliant loading system leading to the onset of oscillations. A lower stiffness, smaller d_c, more negative *(a-b)* or higher effective normal stress (see Eq. 5.9) would eventually result in stick-slip (laboratory earthquakes).

$$\frac{d\theta}{dt} = -\frac{V\theta}{d_c} ln\left(\frac{V\theta}{d_c}\right) \tag{5.7b}$$

Here, μ is the coefficient of friction, defined as shear stress/effective normal stress, ignoring cohesion, μ_0 is the coefficient of friction at a reference velocity V_0, V is the instantaneous slip velocity, a is a parameter that quantifies the direct effect, b is the parameter that describes the evolution effect, d_c is a characteristic or critical slip displacement over which the state variable, θ, evolves. The state variable θ has units of time and is thought to represent the average lifetime of grain-scale asperity contacts. The concept of RSF finds its origin in two observations, namely that the true area of contact of any interface is always smaller than the nominal contact area and that this area of contact changes over time. The two evolution equations embody two views of how a population of contacts evolves during contact. Eq. (5.7a) is sometimes called the slowness or ageing (Dieterich) law, because in this formulation, the frictional contact area continues to evolve in the absence of slip, whereas in Eq. (5.7b), slip is needed for the state variable to evolve. Accordingly, the latter equation is called the slip (Ruina) law.

Other types of state variable evolution equations have been proposed, but these are not typically used in numerical simulations or experiments. Besides the effect of sliding velocity on friction, other parameters, such as effective normal stress and temperature, have effects on transient and steady-state friction, which have been captured in different extended formulations of the evolution of the state variable. Similarly, these extended formulations are rarely used to fit laboratory data or in numerical simulations.

In a typical rock friction experiment, simulated faults are sheared at a pre-described load-point velocity until a steady state friction is reached. At this point, the load-point is instantaneously changed an order or half order of magnitude (e.g., 1−10 or 1−3 μm/s) and the frictional response is recorded. In order to obtain the values of the RSF parameters a, b and d_c, Eqs (5.3) and (5.4) are solved together with an equation that describes the elastic response of the system:

$$\frac{d\mu}{dt} = K(V_{lp} - V) \tag{5.8}$$

Here, K is the stiffness of the system, V_{lp} is the load-point velocity and V is the current slip velocity. This set of equations is solved by numerical integration, typically using a fifth order Runge-Kutta method (Reinen and Weeks, 1993), and used to calculate the evolution of friction with time or displacement, which can be compared to the measured response. The values of a, b, and d_c are changed until a satisfactory fit to the experimental data is obtained, which can be done using an iterative least-squares method.

The RSF laws are useful because they are capable of reproducing both experimental data and seismological observations. Linear stability analysis of a single-degree of freedom spring-slider model with a frictional contact governed by RSF shows that the system is stable for infinitesimal small perturbations if

$$K > K_c = -\frac{\sigma_n^{\text{eff}}(a-b)}{d_c} \tag{5.9}$$

In this framework, frictional instabilities, either earthquakes or slow slip and tremor, can occur accordingly because the fault-weakening rate, K_c, exceeds the rate of elastic unloading, leading to a force imbalance. Therefore, instabilities can only nucleate for negative values of $(a-b)$. Many experimental efforts are thus aimed at establishing the conditions under which $(a-b)$ is negative. Note that a small critical slip distance or high effective normal stress would promote the occurrence of instabilities, depending on the value of $(a-b)$. In the case where $K \approx K_c$, complex sliding behavior results, with the occurrence of slow earthquakes (Leeman et al., 2016; Scuderi et al., 2017).

In numerical simulations of earthquake occurrence using RSF equations and spring-slider analogues, the details of the ratio a/b, the value of d_c and the distribution of effective normal stress control the style of rupture and the amount of aseismic pre- and afterslip. During the nucleation of an earthquake, a limited area of the fault slips aseismically until a critical nucleation size is reached. This critical nucleation size can be approximated using RSF parameters as well:

$$h^* = \frac{G}{K_c} = -\frac{Gd_c}{\sigma_n^{\text{eff}}(a-b)} \tag{5.10}$$

where G is the shear modulus of the sample (fault). Other definitions have been proposed (e.g., Rubin and Ampuero, 2005), also based on a linear slip-weakening law (Harbord et al., 2017; Okubo and Dieterich, 1984), but the formulations are similar and give comparable results (McLaskey and Yamashita, 2017).

5.1.4 Observations of variations in velocity dependence of friction at room temperature

For studies applied to crustal faults, the parameter $(a-b)$ holds the most importance because it determines the possibility of a sliding instability that allows earthquakes to nucleate. At room temperature, a fundamental factor controlling $(a-b)$ has been found to be the mineral composition of the simulated fault (gouge). Earlier studies have documented that the occurrence of stick-slip instabilities, which are considered a laboratory analogue for the seismic cycle (Brace and Byerlee, 1966), tend to be suppressed in phyllosilicate-rich materials (e.g., Brace, 1972; Shimamoto and Logan, 1981).

These observations are supported by subsequent studies using the velocity step test, which show that the presence of phyllosilicates tends to result in positive $(a - b)$ or velocity-strengthening frictional behavior, especially in the presence of water (Ikari et al., 2009; Logan and Rauenzahn, 1987; Morrow et al., 1992; Niemeijer and Spiers, 2006; Saffer and Marone, 2003).

A corollary to this observation is that velocity-weakening behavior tends to occur in frictionally strong materials, i.e., those with little to no phyllosilicates. Thus, the mechanism of velocity-weakening friction is thought to be related to the competition between dilatancy and the growth rate of grain-scale contact asperities (e.g., Beeler et al., 2007; Dieterich and Kilgore, 1994). The rate-dependence $(a - b)$ is also influenced to varying degrees by a wide range of other parameters, including slip rate, total shear displacement, normal pressure, water content, structural heterogeneity, and the condition of the material (e.g., intact rock vs powder). The effect that these factors have on friction velocity-dependence commonly differs for phyllosilicate-rich and phyllosilicate-poor materials. For example, water content has a large effect on phyllosilicate-rich materials, but little effect on stronger framework minerals (Morrow et al., 2000), at least at room temperature. Large total shear strain or displacement can cause $(a - b)$ to switch from positive to negative, but this effect is limited in phyllosilicate-rich gouges (Beeler et al., 1996; Ikari et al., 2011; Scruggs and Tullis, 1998). In many materials, including phyllosilicate-rich gouges, higher slip velocities tend to result in velocity-strengthening behavior, which suggests that the opposite, i.e., velocity-weakening, can occur at the very low slip rates typical of plate motions (Ikari and Kopf, 2017; Weeks, 1993). Note that velocity and temperature are often almost directly interchangeable, so that a higher temperature can shift the velocity-weakening interval to higher velocities (e.g., Niemeijer et al., 2016).

5.1.5 Strength recovery (healing)

On natural faults, the ability to heal, or regain strength in the interval following failure, is a key requirement for repeated rupture. Laboratory experiments can measure frictional healing by isolating the rate of time-dependent frictional strengthening using the slide-hold-slide tests. In these tests, sliding at a constant velocity is periodically interrupted by holding the driving rate at zero (Figure 5.4). Following a controlled amount of time, further shearing at the previous velocity is re-initiated. Upon sliding immediately following the hold, a peak in friction above the original sliding level is typically observed. In most cases, the difference between this peak and the background sliding friction depends linearly on the logarithm of the hold time, so that this dependence quantifies the rate of frictional healing (Dieterich, 1972).

Laboratory observations have confirmed that one of the mechanisms of frictional healing of interfaces is the time-dependent growth of contact asperities (Dieterich and Kilgore, 1996), probably via local crystal-plastic creep

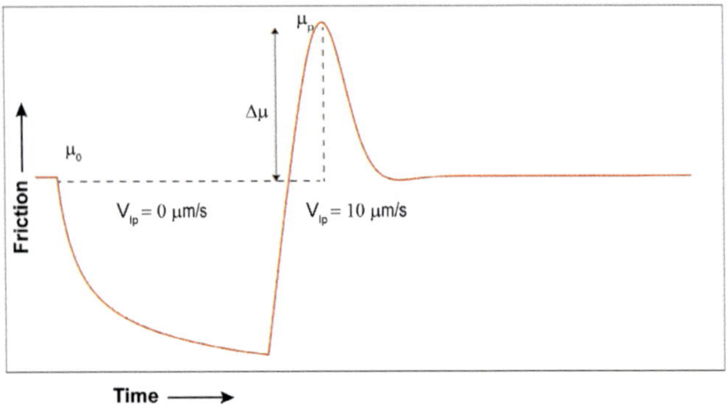

FIG. 5.4 Evolution of friction as a function of time in a slide-hold-slide test. The peak friction upon reloading at the reference load-point velocity typically varies linearly with the logarithm of the hold duration in seconds.

and/or fluid-assisted contact growth. In gouges, porosity reduction during holds acts as an additional healing mechanism, since this porosity needs to be recovered upon a re-slide through dilation against the normal stress. Through the effect of healing on the value of the *b*-parameter, the mechanisms responsible for frictional healing are therefore also responsible for velocity-weakening friction.

Correspondingly, measurements of frictional healing have shown large healing rates in gouges consisting of strong framework minerals and low healing rates in phyllosilicate-rich gouges (Carpenter et al., 2016). Laboratory-measured healing rates have been used to infer the recurrence interval of repeating earthquake sequences (Marone and Saffer, 2015; Marone et al., 1995).

5.1.6 Effect of hydrothermal conditions on velocity dependence of friction

Adding fluids and heat to friction experiments has a profound effect on the rate-and-state frictional parameters, in particular for the framework silicates, such as quartz and feldspars, and less so for the phyllosilicate minerals. Seminal works by Blanpied et al. (1995, 1998a,b) and Chester and Higgs (1992) and Chester (1994) investigated the effect of temperature on friction and the RSF parameters of fault gouges composed of Westerly granite and quartz, respectively. They showed that $(a - b)$ shows systematic changes with temperature; at constant velocity, $(a - b)$ demonstrates a three-regime behavior: a low temperature mild velocity-strengthening (i.e., $(a - b) > 0$) regime, an intermediate temperature velocity-weakening (i.e., $(a - b) < 0$) regime and a high-temperature strong velocity-strengthening regime. The temperature at which the transitions from one regime to the next occur, depends on the sliding velocity: higher sliding velocity results in a higher transition temperature. Similarly, three-regime behavior had already been speculated on by Shimamoto in experiments using rock-salt gouge as the simulated fault (Shimamoto, 1986; Shimamoto and Logan, 1986). In these experiments, a transition from velocity-weakening to velocity-strengthening behavior with increasing confining pressure was observed. Microstructural observations indicated that, at high confining pressures, ductile (i.e., crystal-plastic) flow was active, indicating that the transition from potentially seismic (i.e., velocity-weakening) to aseismic (i.e., velocity-strengthening) was due to a transition from frictional to ductile flow.

Key aspects of frictional behavior include a linear dependence of shear stress on normal stress and a volumetric component to the deformation, through dilation via grain sliding, i.e., granular flow. When time-dependent, thermally-activated creep mechanisms, such as dislocation or diffusion creep, are fast enough, these mechanisms can compete with granular flow, keep porosity low, and produce strong velocity-strengthening behavior. Due to the time- and temperature-dependence of the creep mechanism, the transition from friction-to-flow shifts with sliding velocity and temperature, as observed in the experiments on Westerly granite and quartz mentioned earlier.

On the other end of the spectrum, i.e., at low temperature and/or higher sliding velocity, the second transition from velocity-weakening to velocity-strengthening is postulated to be due to a maximum porosity being reached; or in other words sliding occurs in an incohesive gouge with negligible porosity changes as a function of sliding velocity. In such a gouge, sliding strength would only be controlled by the friction of the grain contacts which, on average, does not change in size or angle. This three-regime behavior has been reported for a variety of simulated fault gouges, including Westerly granite (Blanpied et al. 1995, 1998a,b; Mitchell et al., 2013, 2016), quartz (Chester and Higgs, 1992; Chester, 1994, 1995; Niemeijer et al., 2008), calcite (Verberne et al., 2010, 2015), illite-quartz (den Hartog et al., 2012a,b) and muscovite-quartz mixtures (den Hartog et al., 2013). Other studies have only reported the first transition at low temperature, such as for plagioclase and pyroxene gouge (He et al., 2013), gabbro (He et al., 2006; Mitchell et al., 2015), blueschist (Sawai et al., 2016), and for several natural fault gouges obtained from outcrops or drill holes, such as the Alpine Fault (Boulton et al., 2014; Ikari et al., 2015; Niemeijer et al., 2016), the Zuccale Fault (Niemeijer and Collettini, 2014), and the Costa Rica subduction zone (Ikari et al., 2013; Kurzawski et al., 2016).

All these studies have in common that $(a - b)$ changes with temperature and sliding velocity, which has important implications for earthquake nucleation and propagation. Despite the seemingly large body of experimental work, considerable gaps in knowledge exist with respect to the deformation mechanisms that are responsible for the observed frictional behavior. An important role of solution-transfer processes has been identified (e.g., Bos and Spiers, 2002; Chen and Spiers, 2016; Niemeijer, 2018), but other deformation mechanisms, such as time-dependent crack growth and grain-size reduction, but also dislocation or diffusion creep at very local, highly stressed contacts are expected to play a role as well. In order to extrapolate laboratory results to natural temporal and spatial scales, it is crucial to understand the relative contribution of the different mechanisms to the overall frictional behavior so that other effects, such as, for instance, that of varying pore-fluid pressure might be anticipated and quantified. This can only be achieved by systematic experimental studies coupled with microstructural observations and thermodynamics-based micromechanical models (e.g., Chen and Spiers, 2016). Ultimately, these results need to be verified and linked with micro-structural observation from outcrops of fault rock before being incorporated in large-scale numerical models of the seismic cycle.

5.2 Fault friction in the dynamic regime

5.2.1 Dynamic weakening mechanisms in gouges and solid rocks

As explained in the previous section of this chapter, friction experiments carried out under dynamic, i.e., seismic slip velocity conditions have mostly been done using rotary shear apparatuses. This technique was first applied by Spray using a friction welding apparatus to demonstrate that melting of rocks can occur at high slip rate and normal stress (Spray, 1987) and it was later further developed by Shimamoto et al. (Shimamoto and Tsutsumi, 1994; Tsutsumi and Shimamoto, 1997). Initial experiments were focused on solid rock samples, usually in a cylindrical shape, because of the relative simplicity of the experimental set-up. In recent years, high velocity experiments on gouge have become ubiquitous (e.g., Smith et al., 2013).

The initial series of experimental studies using high velocity rotary shear apparatuses on solid, cohesive rocks demonstrated a rather universal dramatic weakening, regardless of rock type, at velocities higher than some critical velocity, typically in the order of 0.1−1 m/s. The details of the weakening, in particular, the displacement over which weakening occurs, depends on rock type, normal stress, fluid environment and sliding velocity. This is because several different weakening mechanisms might be activated during high velocity sliding, which include flash heating and weakening (e.g., Rice, 1999, 2006), melt lubrication (e.g., Tsutsumi and Shimamoto, 1997; Hirose and Shimamoto, 2005), thermal decomposition and pore pressure production, thermal pressurization (e.g., Sulem and Famin, 2009; Brantut et al., 2010), silica gel lubrication (e.g., Di Toro et al., 2004), the operation of diffusion creep to produce nano-granular gouge (e.g., De Paola et al., 2015) and powder lubrication (e.g., Reches and Lockner, 2010).

Almost all of the weakening mechanisms identified or proposed to have been active during high velocity frictional sliding share one common factor and that is that they are triggered by a temperature rise because of frictional heating. Temperature rise from frictional heating can be estimated using the following equation (e.g., Rice, 2006):

$$\Delta T = \frac{\tau\delta}{\rho Ch} \tag{5.11}$$

Here, τ is the shear stress ($N \cdot m^{-2}$), δ is the amount of slip (m), ρ is the solid density in the reference state ($kg \cdot m^{-3}$), C is the specific heat capacity ($J \cdot kg^{-1}K^{-1}$) and h is the layer thickness (m). This equation gives the temperature rise at the centre of a layer, assuming that all internal energy is converted to heat which is used to raise temperature (adiabatic heating) and is valid as long as the layer thickness is larger than $4\sqrt{\alpha_{th}\delta/V}$, where α_{th} is the thermal diffusivity (m^2s^{-1}) and V is the slip velocity (m/s). The temperature change at the edge of the layer is half that given by Eq. (5.11). If we now assume a typical average slip velocity of 1 m/s, a total slip of 1 m and a thermal diffusivity of 0.7×10^{-6} (Rice, 2006), the minimum layer thickness for which Eq. (5.11) is valid is 3.3 mm. Using a typical value of 2.7×10^6 for ρC (i.e., the specific heat capacity per unit volume (units PaK^{-1}), a slip of 1 m and a constant shear stress of 75 MPa, representative for a friction of 0.6 on a fault at 7 km depth, the temperature rise in the center of a 10 mm thick layer would be 2780 °C, whereas it would be half that at the edge of the layer. It is clear from this simplified calculation that frictional heating during fast slip should give rise to rapid temperature increase, which triggers several of the following weakening mechanisms.

5.2.2 Melt lubrication

Melt lubrication was the first mechanism recognized to be activated during seismic slip and is directly linked to the only unequivocally recognized field evidence for ancient earthquakes - pseudotachylytes (frictional melts, see e.g., Rowe and Griffith, 2015 for a review). The heat generated during sliding can raise the temperature of the sliding interface to such a degree that minerals may melt. The first experiment visibly producing melt were carried out using a solid sample of metadolerite by Spray (1987) who described "a cherry-red glowing interface". Efforts to improve the technique for high velocity friction experiments resulted in a number of studies of the effects of frictional melt on the sliding strength of cohesive, mostly igneous rocks with low quartz content (Tsutsumi and Shimamoto, 1997; Hirose and Shimamoto, 2005; del Gaudio et al., 2009; Niemeijer et al., 2011). These experimental results with visible melting were obtained for a range of conditions, with normal stress between 0.5 and 40 MPa and sliding velocities from 0.1 up to 6 m/s. In all cases, the evolution of friction is characterized by an initial peak, followed by modest weakening which transitions into strengthening and a second peak

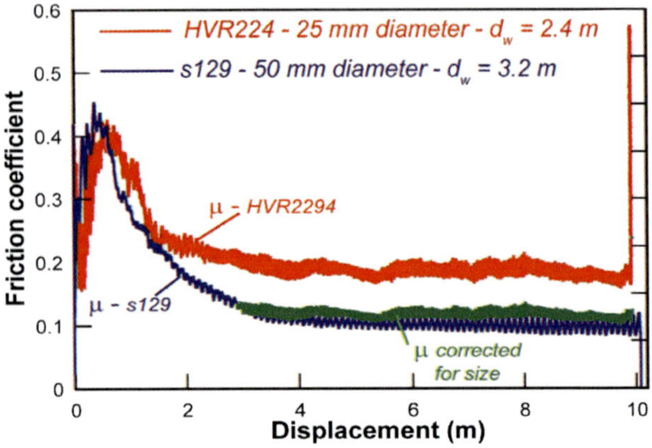

FIG. 5.5 Plots of the evolution of friction of two high velocity experiments, which were performed on solid cylindrical samples of gabbro of identical composition under identical experimental conditions of slip velocity, acceleration and normal stress (1.3 m/s, 3.25 m/s and 10 MPa respectively), but using different sample sizes, 25 mm diameter cylinders and 50 mm diameter. Both the absolute value of friction and the displacement over which friction drops (the slip weakening distance) are different for the different sample sizes as a result of (1) a difference in the distance over which the melt has to travel before being squeezed out of the sample and (2) a larger variation in slip velocity across the slipping surface for solid cylindrical samples. *From Niemeijer et al. (2012).*

after which strong weakening results in a low steady state friction value of 0.1 or below (Fig. 5.5). The first weakening is typically interpreted to be related to flash heating and weakening (Hirose and Shimamoto, 2005), whereas the second weakening has convincingly been shown to be related to the formation and growth of a through-going melt layer, lubricating the fault.

Observations using high speed recording (Niemeijer et al., 2011), as well as microphysical thermodynamic models (Nielsen et al., 2008, 2010a,b), demonstrate that the frictional heat production causes initial melting to occur at highly-stressed contact points and accumulation of melt on the sliding interface occurs with ongoing sliding. The strength of the molten layer is controlled by the thickness of the melt layer and the melt viscosity. Even though the melt viscosity might increase once full melting is achieved (due to the melt becoming increasingly silica-rich as higher melting-temperature minerals melt), the fault strength goes down because of the accumulation of melt, increasing the melt layer thickness and decreasing the shear strain rate. The drop in friction occurs exponentially with displacement and can typically be described by an equation of the form (e.g., Hirose and Shimamoto, 2005, Mizoguchi et al., 2007):

$$\mu = \mu_{ss} + (\mu_p - \mu_{ss}) exp\left[\ln(0.05)\left(\frac{d - d_p}{d_w}\right)\right] \tag{5.12}$$

Here, μ is the coefficient of friction, with μ_{ss} the value at steady state, μ_p that at the (second) peak, d is the displacement, d_p is the displacement at the (second) peak and d_w is a characteristic weakening distance at which 95% of the strength is lost (hence ln(0.05)). This empirical equation is typically used to fit experimental data of high velocity friction experiments, regardless of the weakening mechanism activated. The significance of the characteristic weakening distance, d_w, is that it describes the energy required to accelerate the slip on the fault. The value of d_w is a critical input parameter in numerical models of earthquake nucleation and propagation, particularly those that use a linear slip-weakening law. A slip weakening or breakdown distance can also be obtained from seismological observations (e.g., Tinti et al., 2009) which are often compared to those obtained in experimental high-velocity friction studies. However, the values obtained in seismological studies are highly model-dependent and are fault-averaged values. In contrast, the experimentally obtained value depends on multiple environmental parameters, such as normal stress, presence of a pore fluid but also on sample size, at least for the case of weakening through melt lubrication. Understanding of the operating weakening mechanism is crucial to be able to extrapolate experimentally obtained values to natural faults and earthquakes. For the case of melt lubrication, a microphysical, thermodynamically-based model predicts that the slip weakening distance should be inversely related to effective normal stress, which was confirmed by experimental data. Although this provides more confidence in the extrapolation of the model and experimental data to natural conditions, we still have to consider that natural faults and stresses on them are highly heterogeneous so that spatial variations of the slip weakening distance must exist.

Recently, frictional melting was also achieved in samples with high quartz content (Lee et al., 2017), illustrating that the temperature can rise high enough to melt quartz, in contrast to lower melting point minerals such as biotite. The shape of the friction-displacement curves as well as the steady-state friction are very similar to those of lower quartz-content samples, suggesting that the melt lubrication weakening mechanism is rather universal, with mostly the slip weakening distance being affected by rock type and boundary conditions. Slip weakening distance becomes larger when the rock contains mostly high melting temperature minerals such as quartz and it becomes smaller for higher effective normal stress and sliding velocity.

Bulk melting and associated weakening in in-cohesive rocks, i.e., gouges, has only been achieved in samples of rock salt. Other weakening mechanisms appear to be more efficient in gouges, i.e., these are activated before the temperature rises high enough to achieve melting.

5.2.3 Flash heating and flash weakening

As discussed in the previous section on quasi-static friction, the real area of contact of any sliding interface is much smaller than the nominal contact area.

The contacting asperities are highly stressed because they support all of the applied stress and they thus form the first location where temperature increase due to frictional heating occurs. This asperity heating was first treated theoretically by Rice (1999, 2006) and is based on the notion that the shear strength of asperities degrades continuously with increasing temperature. As a simple model, we can assume that a temperature T_w exists at which the shear strength of the asperity is negligible with respect to the reference strength at ambient temperature. We estimate the time θ_w necessary for an asperity to reach T_w using:

$$\theta_w = \frac{\pi \alpha_{\text{th}}}{V^2} \left(\frac{\rho C (T_w - T)}{\tau_c} \right)^2 \tag{5.13}$$

with the parameters as defined in Eq. (5.11), and τ_c is the reference shear strength of the asperity (Rice, 2006). This solution is based on one-dimensional heat conduction with heat input from both sides of the asperity. The time necessary for the temperature to reach the weakening temperature is only relevant if it is shorter than the lifetime of the asperity, θ, which is simply given by displacement over which an asperity exists, D_a, divided by velocity, V. We can now define a weakening velocity, V_w, as:

$$V_w = \frac{\pi \alpha_{\text{th}}}{D_a} \left(\frac{\rho C (T_w - T)}{\tau_c} \right)^2 \tag{5.14}$$

Like Rice (2006), we consider the same values for the thermal diffusivity, α_{th} and for specific heat capacity, ρC, used before (0.7×10^{-6} m^{2}s^{-1} and 2.7×10^6 PaK^{-1}, respectively) and an ambient temperature of 200 °C. The shear strength of an asperity in brittle materials is taken to be 10% of the shear modulus, which gives a value of ~3 GPa (see Rice, 2006). The weakening temperature is not very well defined, but must be lower than the melting temperature; following Rice (2006), we will assume here 900 °C, with a reference temperature of 20 °C. The displacement over which an asperity exists is most likely related to the size (distribution) of the asperity, similar to the characteristic or critical slip distance, d_c, in rate- and state friction (Section 5.1). A typical value for bare rock friction experiments is 5 μm. For these values, we obtain a weakening velocity of 0.28 m/s, which is remarkably similar to the velocity at which many different rock types show dramatic weakening (see Di Toro et al., 2011). Note that there are multiple parameters poorly constrained, notably the weakening temperature and the displacement over which an asperity exists. Since the mechanisms by which the shear stress thermally degrades cannot be determined, it is difficult to establish the weakening temperature, and it should be different for different materials. For instance, a phyllosilicate such as biotite melts at a much lower temperature (~650 °C) and a weakening temperature of 600 °C is perhaps more appropriate. This change drops the weakening velocity to 0.12 m/s, but at the same time the asperity size in biotite is likely to be very different from that in granite, as also indicated by larger d_c-values in phyllosilicate-bearing samples in quasi-static RSF experiments (see Section 5.1). If we, for example, take a value of 100 μm for D_a, the weakening velocity drops further down to 0.006 m/s.

Experimentally, it is difficult to prove the occurrence of flash heating and weakening because the lifetime of the asperities is so short (less than a microsecond in the first calculation above) and the asperity sizes are small. Inferences for the operation of flash weakening are therefore usually based on the elimination of the other dynamic weakening mechanisms or a comparison of the observed velocity at which weakening occurs with the theoretical predictions.

5.2.4 Thermal pressurization

Most natural faults contain a fluid phase and it is thus appropriate to perform experiments in the presence of a pore fluid. This is not a trivial matter in rotary shear experiments, because there is a competition between the quality of the seal that keeps the fluid inside and the friction between the seal and the rotating piston and sample. The simplest configuration is to use a pre-wetted gouge material, sealed by tightly-fitting PTFE (Teflon) sleeves surrounding a pair of solid cylindrical forcing blocks, typically from rocks such as granite or sandstone (e.g., Mizoguchi et al., 2007). More advanced configurations use metal alloys and dynamic seals (e.g., Smith et al., 2013). The operation of thermal pressurization during fast seismic slip has been suggested and modeled earlier, based on theoretical considerations as well as measurements of the fluid conductivity (permeability) of fault gouges obtained from outcrops (e.g., Sibson, 1973; Andrews, 2002; Wibberley, 2002; Wibberley and Shimamoto, 2005). These calculations demonstrated the possibility of producing high pore fluid pressure, and thus low effective normal stress, from thermal pressurization within a thin gouge layer with low permeability (Platt et al., 2014; Rice et al., 2014).

Experimentally, it is difficult to isolate the effect of thermal pressurization from other thermally-activated dynamic-weakening mechanisms such as flash heating. One option is to contrast wet versus dry behavior (Tanikawa and Shimamoto, 2009; Ujiie and Tsutsumi, 2010; Ferri et al., 2010, 2011; Faulkner et al., 2011), supplemented with microstructural observations of fluidized gouges (e.g., a granular, flow-like microstructure, see e.g., French and Chester, 2018). Another convincing method to highlight the operation of thermal pressurization is to use forcing blocks with different permeabilities, e.g., a permeable sandstone versus an impermeable granite. The operation of thermal pressurization can then be inferred from a more pronounced weakening in the case of an impermeable forcing block.

One caveat of high-velocity friction experiments to date is that the background temperature is low (room temperature) and fluid pressure is low and its spatial and temporal distribution is unknown. Furthermore, thermal pressurization is almost exclusively invoked for clay-rich gouges, because these gouges typically have low permeability ($<10^{-19}$ m^2). Clay minerals lose their surface-bonded water and become mineralogically unstable with increasing temperature, which contributes to potential fluid pressure rise.

5.2.5 Thermal decomposition and pressurization

Temperature rise can induce mineral transformations, which in turn can release fluids in the deforming gouge layer. The local release of fluids can raise the fluid pressure, causing significant weakening. This has been demonstrated convincingly in experiments on solid gypsum cylinders ($CaSO_4 \cdot (H_2O)_2$) which decomposes to basanite ($CaSO_4 \cdot (H_2O)0.5$) at ~100 °C (Brantut et al., 2011a,b) Because the reaction itself is endothermic, i.e., it consumes heat, the progression of the reaction effectively buffers the temperature while water is released. Similar endothermic fluid-releasing reactions occur in many different fault rock mineralogies, such as smectite/illite, kaolinite and calcite. The local permeability is crucial in controlling whether a high fluid pressure and thus significant weakening can be obtained. Thermodynamic models considering the interaction between heat generation from frictional sliding and heat consumption in reactions have been developed and show that important parameters in the amount of weakening that can be achieved, are the thickness of the sliding zone, its permeability and the permeability of the fault surroundings (e.g., Sulem and Famin, 2009; Brantut et al., 2011a,b).

5.2.6 Fluid phase changes

In regular high velocity friction experiments, the fluid is in a liquid form and is expected to undergo a phase transition to vapor (gas) when heated. The vaporization of fluids is another example of an endothermic reaction and can thus buffer the temperature rise induced by frictional heating. This was demonstrated by Chen and coworkers (Chen et al., 2017a,b, see Fig. 5.6) in experiments on calcite (limestone) gouges, with fluid pressure and temperature sensors installed in one of the pistons at different locations, distributed from the center to the edge of the cylinder. These showed that fluid pressure continuously increased, with the highest pressures towards the outer perimeter while at the same time the temperature was buffered to fall along the phase boundary between liquid and vapor for water. Similar fluid phase changes are only expected to occur in nature at relatively shallow conditions, because fluid pressure rises faster than temperature in most settings. A drop of fluid pressure due to dilatation is required to trigger the phase change to vapor.

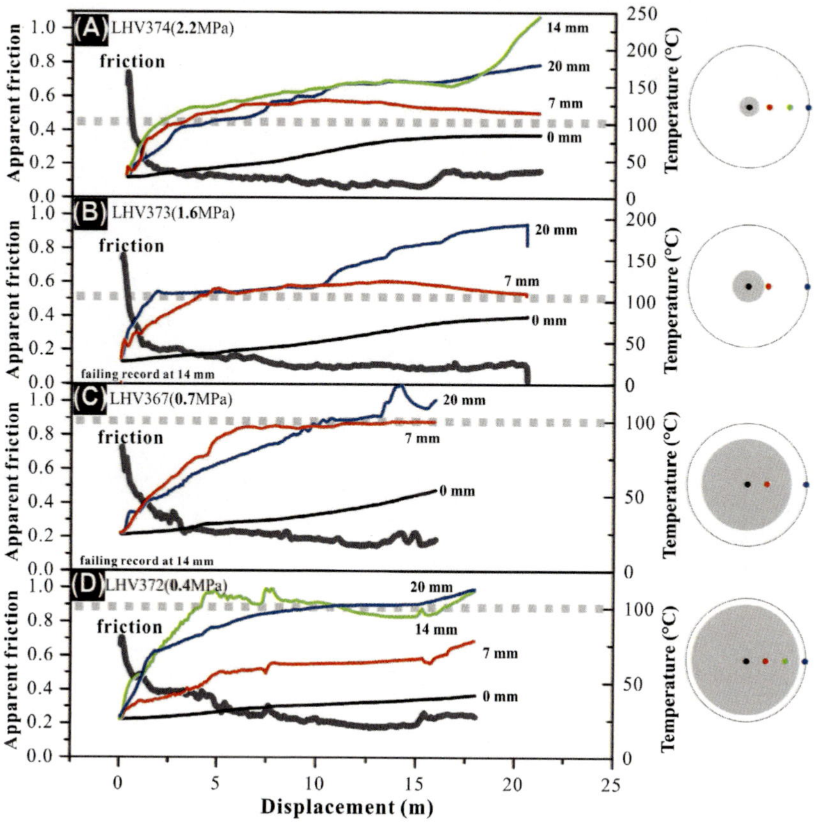

FIG. 5.6 Plots of the evolution of friction (left y-axis) and temperature at different locations (right y-axis) with displacement for four experiments at different normal stresses and 1.3 m/s sliding velocity using a carbonate-rich fault gouge. It is inferred that the temperature is buffered by the endothermic phase transition of water to vapor. The location of the thermocouples is indicated for each curve as the distance to the center of the fault and is sketched on the right hand side. Shading in these sketches indicate the parts of the faults that appeared wet after the experiments. *From Chen et al. (2017a).*

5.2.7 Powder lubrication

It is commonly observed that fast slip generates significant gouge material with a very fine grain size, usually smaller than 1 μm, i.e., nano-sized. Several authors report that the presence of a layer of nanoparticles allows for shear to occur at very low stress level (Han et al., 2007; Reches and Lockner, 2010; Han et al., 2010). The exact mechanism of weakening is unclear, but it is suggested that the shape of nanoparticles is spherical which would allow for grain rolling to be quite efficient, instead of grains sliding over each other and jamming/unjamming the microstructure. A recent study employing

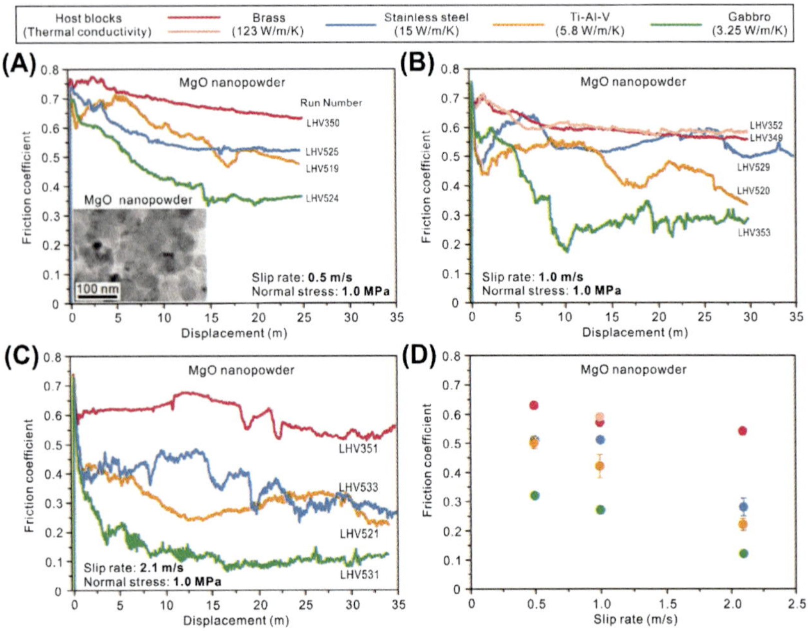

FIG. 5.7 Plots of the evolution of friction with displacement for three different sliding velocities and using four types of host blocks with varying thermal conductivities (A–C) and steady state (final) friction as a function of sliding velocity (D). Experiments were done using a dry gouge consisting of MgO nano-powder and a normal stress of 1.0 MPa. The different host blocks results in different temperature evolutions within the gouge layer and demonstrate that weakening is caused by elevated temperature (compare brass vs. gabbro host blocks). *From Yao et al. (2016a).*

nano-powdered gouges as the starting material suggest that just the presence of nano-particles is not sufficient to cause weakening (Yao et al., 2016a, see Fig. 5.7). In these experiments, a first attempt was made to regulate the fault temperature by using different materials for the host rocks, which, due to their differing thermal conductivities, resulted in different temperatures being achieved during sliding (Yao et al., 2016a,b). These experiments clearly demonstrated the crucial role of temperature on the activation of dynamic weakening mechanisms.

5.2.8 Activation of crystal-plastic (viscous) mechanisms

From the previous sections, it is clear that seismic slip generates heat and reduces the grain size. Crystal-plastic or viscous deformation mechanisms, such as dislocation creep or diffusion creep are thermally activated, and their rates follow an Arrhenius-type relationship with temperature. In addition, diffusion creep rate depends on grain size, i.e., diffusion creep increases with decreasing grain size, with cubic or quadratic dependence (depending on the diffusion pathway). Flow laws describing the dependence of creep rate on temperature, grain size and stress have been established based on mostly uniaxial compression experiments at strain rates of 10^{-4} to 10^{-6} s^{-1}. In contrast, in a localized shear band within a gouge layer, shear strain rates are orders of magnitude larger, e.g., a slip velocity of 1 m/s in a 100 μm thick band translates to a shear strain rate of 10000 s^{-1}. Despite this large difference in strain rate, extrapolation of flow laws to high velocity rotary shear experimental conditions suggests that diffusion creep could have controlled the steady-state strength (De Paola et al., 2015). It is somewhat surprising that the extrapolation yields reasonable values, considering the ten orders of difference in strain rate, but also two to three order change in grain size. Nano-size particles have a large surface area compared to their volume and are thus not expected to be thermodynamically very stable. The properties of nanoparticles of minerals are unknown and a lot more research is needed to better establish differences and similarities in properties.

5.2.9 Dynamic rupture in laboratory experiments

There are essentially three types of dynamic rupture experiments that are performed in laboratory settings, each with slightly different aims:

5.2.9.1 High confinement, small rock sample experiments

The laboratory investigation of rock friction described in the above sections is conducted by imposing the sliding velocity while measuring the friction. The rotary shear configuration used has the advantage of allowing high slip velocity and (almost) unlimited slip. However, it has technical limits in acceleration and normal stress, and does not allow proper confinement of the sample.

Another avenue to investigate friction at high velocity is to conduct experiments where spontaneous rupture propagation is produced. To do this, differential stress is imposed on pre-cut samples in a triaxial machine. The stress level attained is comparable with mid-crustal seismogenic conditions, and, under favourable circumstances, the gradual increase of load triggers a dynamic stick-slip instability that propagates along the slip surface, generating laboratory pico-earthquakes with equivalent magnitudes of minus four.

In this case, dynamic slip velocity is not imposed, but evolves spontaneously from the far-field loading and the interaction of rupture propagation with the surrounding rock. Because of the small sample size, the amount of slip in laboratory earthquakes is limited to a fraction of a millimetre. However, due to the high (up to 200 MPa) normal stress on the sliding surface, the weakening is extremely fast and seismic slip velocities (of the order of 1 m/s) are reached within the very short (~0.1 ms) duration of the slip pulse.

At high sliding velocity, weakening is related to the frictional heating of the sliding surface. The rate of heat production is controlled by the product of shear stress and velocity on the fault surface. Therefore, the increased friction due to higher normal stress, combined with the increased acceleration of slip due to a larger dynamic stress drop, have a tremendous multiplying effect on the heating rate and on the weakening rate. This allows the generation of fully-developed dynamic ruptures, albeit with low slip amounts and reduced duration.

If the shear zone thickness is small, $h \ll 4\sqrt{\alpha\delta/V}$ (where α is thermal diffusivity, δ is slip, see Section 5.2.1) and we consider shear stress τ and slip velocity V as constant for simplicity, the temperature rise T_c can be given as a function of slip time, t_c:

$$T_c = \phi\tau V\sqrt{t_c} \tag{5.15}$$

where $\phi = \frac{1}{\sqrt{\alpha\rho c}}$ and ρ, c are mass density and heat capacity, respectively. Letting T_c represent an indicative critical temperature for weakening of the fault rock, then the time to reach weakening will be

$$t_c = \frac{T_c^2}{\phi^2}\frac{1}{(\tau V)^2} \tag{5.16}$$

Given that the shear stress is proportional to the normal stress during the initial part of the sliding, in triaxial dynamic rupture experiments τ will be initially about ten-fold the value that is typically obtained in rotary shear experiments. Therefore, the weakening time will be reduced by a factor of one hundred with respect to the rotary shear experiments due to the increase in normal stress alone. Furthermore, if the load is mainly supported by contact asperities, which represent only a fraction of the total surface, the asperities will concentrate stress and the weakening process can be described in terms of *flash heating* of asperities, as in the model proposed by Archard (1959) and revised by Rice (2006), in the context of earthquake faulting (see Section 5.2.3). In such a case, weakening will follow the increase in sliding rate almost instantaneously. Therefore, it is expected that the weakening distances and weakening times observed in spontaneous stick-slip experiments, where both slip acceleration and normal stress are substantially higher, are orders of magnitude smaller than those observed in rotary shear experiments.

As a result, dynamic rupture experiments under high confinement can reproduce seismic conditions for fault offsets and slip durations that are orders of magnitude smaller than those produced in rotary shear experiments. The advantages are that (1) stress conditions are closer to mid-crustal faulting and (2) the evolution is due to a spontaneously propagating rupture, which produces a slip pulse probably more similar to that of a real earthquake. The challenge, however, is to obtain a significant measure of stress and slip during the extremely short-lived event.

By instrumenting the sample with small strain gauges and a number of piezoelectric sensors that are monitored at very high frequency (1 MHz or more), it is possible to measure the transient shear stress change during the rupture and the progress of dynamic slip rupture on the simulated fault (Passelègue et al., 2016). This provides a measure of dynamic friction drop as a function of slip and slip velocity, during the laboratory pico-earthquake. As a consequence, it is possible to investigate the extremely fast weakening which can sometimes occur under crustal fault conditions, as inferred by some natural microearthquakes, which show extremely large stress drops, in spite of a very short rupture duration (Nadeau and Johnson, 1998). During large earthquakes, the apparent weakening distance is much longer (of the order of 1 m). In addition, it is observed that the rupture velocity exceeds the shear-wave velocity (i.e., supershear) fairly easily during such experiments, although the propagation distance is of the order of a few centimetres only. Supershear rupture is rarely documented in natural earthquakes, but it does occur after a rupture has propagated on relatively planar sections of a fault for several kilometres (Dunham and Archuleta, 2004). This calls for a number of hypotheses on how to reconcile the difference between small-scale experiments (where supershear is triggered very easily and the stress drop is large) and large-scale earthquake behaviour (where supershear is rare and stress drop appears to be much smaller on average).

One argument is that natural faults have highly heterogeneous structure and frictional properties, and the stress distribution on the fault area is highly inhomogeneous. Extremely fast weakening may occur on only a fraction of the total rupture area, at the site of highly-stressed asperities, where rupture velocity and slip velocity locally accelerate (Passelègue et al., 2013). The bulk of the fault area, however, would be less stressed and less prone to severe velocity-weakening, buffering the rupture process and decelerating the weakening process.

Another argument is that dissipative processes during rupture propagation are not purely frictional. Dissipation is also due to the damage and the anelastic strain that occur within the wider rock volume surrounding the fault surface (Andrews, 2005; Xu et al., 2015). In larger faults, the dissipation affects an increasingly larger volume, therefore the apparent fracture energy also increases (Abercrombie and Rice, 2005; Nielsen et al., 2016). An increase in fracture energy is equivalent to an increase in weakening distance (Spudich and Guatteri, 2000; Peyrat et al., 2004). This possibly explains why large earthquake faults appear to weaken more gradually during rupture than their microscopic laboratory analogues.

The above arguments illustrate that the transposition of results from laboratory dimensions (a few centimetres) to the Earth's crust dimensions ($\sim$10 km) is not trivial. At the laboratory scale, rupture dissipative processes arise from friction (ductile, plastic or viscous shear deformation of grains or contact asperities) combined with surrounding microcracking, within a thickness of a few tens of microns (micro-fault zone). On natural faults, it has been indicated that the thickness of the fault zone increases with the length of the fault (Shipton et al., 2006) and can reach tens of metres. Taking the ratio of an indicative length of 10^4 m for crustal faults (seismogenic crust thickness) and of 10^{-2} m for the laboratory (experimental sample dimensions), we obtain an upscaling factor of one million. The million factor is also an indicative ratio of the thickness of a crustal fault zone (10 m) to the thickness of a laboratory micro-fault zone (10^{-5} m). Because dissipation increases in proportion to the thickness of the fault zone, a similar factor should apply to the dissipative power of crustal vs laboratory faults.

5.2.9.2 Low confinement, large rock sample experiments

Dynamic rupture experiments have also been performed on larger, pre-cut blocks of granite or gabbro, creating simulated faults of a couple of metres long and a few decimetres wide. One advantage is the longer duration of rupture propagation, which makes it easier to monitor its growth. These experiments have been performed using a biaxial geometry, unconfined in the third direction; the large exposed sample face makes it convenient to instrument the fault with all sorts of sensors, such as strain gauges, piezo-electric sensors, micro-accelerometers or optical sensors. One disadvantage is that the normal stress on the fault is only a fraction of that expected on a fault at seismogenic depths.

Okubo and Dieterich (1984) studied the effect of roughness of the fault surface on granite pre-cut faults where roughness was either 0.2 or 80 μm, and normal stress was in the range 0.6–4 MPa. Within this range, they found that slip-weakening distance is insensitive to normal stress, but that the apparent fracture energy and stress drop both increase with normal stress and that rougher faults have larger slip-weakening distances and fracture energies. This experimental setup was later revived by McLaskey and Kilgore (2013), who documented experimental evidence of foreshocks and the nucleation process that takes place on the fault, preceding the onset of dynamic rupture.

A series of experiments were performed by Ohnaka et al. (1987) and subsequent studies (Ohnaka and Shen, 1999 and references therein) who investigated the dynamic friction drop on 40 cm-long pre-cut faults in granite blocks, and normal stress in the range 2–19 MPa. They found that corner frequency (related to maximum slip velocity and acceleration) increased linearly with normal stress to about 17 MPa normal stress. Interestingly, above this normal stress, the linear increase of corner frequency broke down and much higher frequencies were observed, in combination with localised, jerky acceleration of slip on some patches of the simulated fault, but not on others. This can be related to the aforementioned work of Passelègue et al., who observed supershear rupture and extreme weakening in their small-scale experiments and concluded that it may represent high-stress patches on heterogeneous faults during a natural earthquake.

More recently, Fukuyama et al. (2016) investigated super-shear rupture on a pre-cut fault in a gabbro block of 1.5 m × 0.1 m, under normal stress of

2.6 MPa. They found a cohesive zone of about 50 mm at the propagating rupture tip, which decreased with increasing rupture velocity, in particular in the super-shear velocity range.

5.2.9.3 *Low confinement, analogue material experiments*

Analogue material experiments allow the photographic tracking of rupture, stress and slip during dynamic rupture propagation. They are most often performed on transparent materials such as plastic polymers (polycarbonate, allyl diglycol carbonate, or polymethyl methacrylate) which are easy to use for optical techniques and give optimal lighting conditions for high-speed photography. In addition, such materials have elastic stiffnesses about 20 times lower than natural rocks, and consequently the deformation is larger, and easier to measure, even under modest stress values. Another advantage is that the frictional slip weakening distance for these materials is much lower than that measured on natural fault rocks: this allows to produce fully-developed dynamic slip and rupture propagation even on small samples and under modest stress levels.

The frictional behavior of these polymers is not easily compared to that of natural faults; therefore direct characterization and measurement of friction on such materials is of limited interest. However, these experiments provide interesting insights regarding the onset of stick-slip instability at a fundamental level, and regarding the development of rupture nucleation and dynamic rupture propagation processes.

The transition to super-shear rupture velocity, which was documented in several experiments on rocks, has also been the target of detailed photographic and acoustic experiments on analogue materials, for example Xia et al. (2004) used photoelastic interferometry to track the rupture tip, while Rubino et al. (2017) used image-correlation techniques. Use of a high-frequency laser positioning system allowed Lykotrafitis et al. (2006) to measure in the laboratory self-healing pulse-like shear ruptures, which have been predicted by theoretical models (Nielsen and Madariaga, 2003), and inferred in observations of natural earthquakes (Heaton, 1990).

Details of a localised nucleation phase that precedes the dynamic rupture were documented by Nielsen et al. (2010a,b). They created the conditions for spontaneous rupture on a pre-cut fault of 15 cm length, and monitored the growth of an unstable slip patch, and the subsequent acceleration to sub-shear and super-shear rupture velocity using photoelastic interferometry and a high-speed camera. With the same experimental procedure, Guérin-Marthe et al. (2019) showed that the loading rate controlled the length of the nucleation patch on the simulated fault, and discussed implications for precursory slip in natural earthquakes. Rubinstein et al. (2007) and Ben-David et al. (2010) used a grazing laser light sheet to study the evolution of contacts on the sliding interface preceding and during the dynamic propagation of stick-slip events, and documented evidence of a detachment front that precedes the main slip on the fault.

In addition to the mechanics of rupture initiation, the mechanics of rupture arrest have also been investigated using analogue experimental faults (Bayart et al., 2018).

5.2.10 Frontiers

From the previous sections, it should be clear that temperature increase due to frictional heating as a result of fast slip can activate numerous weakening mechanisms. With the continuing development of experimental techniques, the conditions of the experiments will likely approach natural conditions more and more in terms of (effective) normal stress, presence of fluid (pressure) and background temperature. Isolating a unique dynamic weakening mechanism remains challenging, but careful experimentation combined with microstructural observations and numerical modelling should allow the identification of the dominant weakening mechanism under variable conditions. In turn, this will allow us to determine to what extent these mechanisms are active in small-slip events. This will make the link with dynamic rupture experiments (e.g., Okubo and Dieterich, 1984; Nielsen et al, 2010c; Kaneko et al., 2016; McLaskey and Yamashita, 2017), allowing a description of rupture initiation, propagation and arrest and an evaluation of the existing scaling relations with seismology to be made.

5.3 Faults in scaled physical analogue models

5.3.1 Introduction

Following on the pioneering work of Sir James Hall (1815), physical analogue modelling has developed into a powerful tool for studying natural geological phenomena and their underlying processes. In particular, it provides valuable insight in the evolution of deformation structures on various length and time-scales including faults or fault zones. In most cases the incremental stages of their evolution are not preserved and their development is reconstructed based on incomplete data, and only finite geometries are reasonably well constrained by the geological or geophysical record. Analogue modelling overcomes these limitations through designing scale models of the region or object of interest, which are built with materials that exhibit similar behaviour at room temperature and most of the time atmospheric pressures than the natural prototype and thus allow for the study of the natural process at convenient time (hours to days) and length scales (cm−m).

Although subject to limitations when it comes to defining the geometric, kinematic and rheological set-up of an experiment, a detailed qualitative and quantitative analysis of modelling results allows for direct comparison with the natural proto-type, provided that scaling principles have been respected (e.g., Hubbert, 1937; Weijermars and Schmeling, 1986).

5.3.2 Scaling tectonic faulting to the laboratory

Scaling of analogue experiments is based on the principles of geometric, kinematic and dynamic similarity between the model and the natural proto-type. The model to nature proportionality (subscripts m and n in the following, respectively) is parameterized into dimensionless ratios of length ($l^* = l_m/l_n$), density ($\rho^* = \rho_m/\rho_n$) and gravity ($g^* = g_m/g_n$). These ratios allow the derivation of other scaling factors such as stress ($\sigma^* = \sigma_m/\sigma_n$) according to:

$$\sigma^* = \rho^* g^* l^* \tag{5.17}$$

For a dynamically scaled model, the materials used in the experiments must exhibit similar deformation behaviour to the material of the natural proto-type they are to simulate. As such, analogue models are scaled utilizing constitutive equations describing the deformation behaviour of analogue materials and their natural counterparts. These entail the Mohr-Coulomb criterion for deformation under brittle conditions

$$\tau = C + \mu\sigma_n \tag{5.18}$$

and linear

$$\sigma = \eta\dot{\gamma} \tag{5.19}$$

or non-linear flow laws

$$\sigma^n = \eta\dot{\gamma} \tag{5.20}$$

The latter describe the viscous behaviour under constant room temperature, where $\tau, C, \mu, \sigma_n, \eta, \dot{\gamma}$ and n denote the shear stress, cohesion, coefficient of friction, normal stress, viscosity, shear strain rate and the power law exponent, respectively. Scaling of model velocity and forces can be achieved by considering the following equations:

$$\dot{\gamma}^* = \eta^*/\tau^* \tag{5.21}$$

$$\dot{\gamma}^* = V^*/l^* \tag{5.22}$$

where $\dot{\gamma}^*$ $(=\dot{\gamma}^m/\dot{\gamma}^n)$, η^* $(=\eta^m/\eta^n)$, τ^* $(=\tau^m/\tau^n)$ and V^* $(=V^m/V^n)$ are the scaling ratios of strain rate, viscosity, stress and velocity, respectively.

For normal gravity experiments, $g^* = 1$, and ρ^* is dictated by the materials used. As is clear from Eqs (5.17) and (5.18), appropriate scaling within the brittle regime can be obtained by selecting a material with a coefficient of friction similar to those observed in nature and by choosing a correct length scale l_m to obtain a correct model stress. Cohesion has the same dimension as stress and thus must follow a similar scaling, i.e., $C^* = C_m/C_n = \sigma^*$.

From equations (5.17) and (5.18), it is evident that strength in the brittle regime scales with the thickness of the brittle layer and the intrinsic properties of the material, expressed by its friction coefficient, cohesion, and density.

5.3.3 Rock analogue materials and their bulk properties

In most cases, deformation under brittle conditions is simulated using dry granular materials with grain morphologies ranging from almost perfectly spherical (e.g., glass beads) to sub-rounded and angular (e.g., sands of feldspar, quartz, garnet or corundum) and mean grain-sizes typically ranging between 100 and 400 μm (Willingshofer et al., 2018; Klinkmüller et al., 2016 and references therein). Although we are mainly concerned here with the development of faults in pure brittle, granular materials, it should be noted that fault development has also been studied using wet clay. For a comparison of sand and clay models, the reader is referred to Eisenstadt and Sims (2005).

The bulk rheological properties of granular materials have been measured using Hubert-type shear apparatuses as well as ring shear testers. The measurements have yielded friction coefficients at peak strength in the order of 0.4−0.8 and a cohesion between 10 and 100 Pa at normal stresses of a few 100 kPa at most. For the various measurement procedures and an in-depth comparison of the different materials, the reader is referred to Lohrmann et al. (2003), Schellart (2000) and Klinkmüller et al. (2016). Furthermore, these experiments have shown that friction during reactivation and subsequent stable sliding (dynamic friction) is typically 10% and 20%, respectively, below peak friction (Klinkmüller et al., 2016; Ritter et al., 2016). Cohesion is typically not measured directly, but inferred from extrapolation of the Mohr-Coulomb failure envelope to zero normal stress. Therefore, cohesion values are associated with a large error bar, but the values appear to increase with the maturation of the shear zone, i.e., with accumulated displacement. Accordingly, the majority of the measurements suggest that cohesion is higher in reactivated faults than at peak friction for a pristine fault (Klinkmüller et al., 2016).

Overall, friction and cohesion are lowest for perfectly-rounded granular materials, such as glass beads and highest for sands with dominantly-angular grains, i.e., quartz or feldspar sand.

It has been shown that the strength of granular analogue materials scales similar to rocks under natural conditions, with a linear increase of brittle strength, as described by Byerlee's rule (see Section 5.1 of this chapter and Ritter et al., 2016). Some materials may show deviation from a linear Mohr Coulomb failure criterion only at very low normal stresses (<500 Pa) and develop a convex envelope (Schellart, 2000). However, this is difficult to verify because of technical limitations.

5.3.4 Quantifying stress and strain in analogue models

Tracking of deformation in analogue models including the nucleation and evolution of faults is carried out by imaging the deforming model through digital photography, X-ray tomography and subsequent Digital Image Correlation (DIC) on the acquired images (Adam et al., 2005, 2013; Dotare et al., 2016). These methods allow non-intrusive "in-situ" monitoring of the faulting process at high spatial and temporal resolution. Additionally, instrumenting experiments with force sensors allows for linking deformation to the dynamics of the system (e.g., Nieuwland et al., 2000; Herbert et al., 2015; Ritter et al., 2018a,b; McBeck et al., 2018). Digital Image Correlation relies on the tracking of an identifiable area with unique characteristics of the recorded image and performing high resolution cross correlation between time steps. This powerful technique yields the distribution of displacement or strain over time.

5.3.5 Fault formation in analogue models

Faulting in analogue models does not involve the fracturing of material, but is the consequence of rolling and sliding of the sand grains past each other, leading to dilatancy and localization within the zone of deformation (e.g., Mandl et al., 1977). Shear in granular materials typically localizes into 10−15 grain diameter (several mm) wide shear zones (e.g., Panien et al., 2006), rather than thin faults or fractures. Grain-scale analysis of velocity fields within shear bands suggested that deformation entails the build-up and collapse of columns of particles that carry the bulk of the force (Rechenmacher, 2006).

Ring shear tests have shown that the granular material displays a characteristic hardening-softening behaviour that can be correlated with compaction-decompaction cycles of the sand pack (Lohrmann et al., 2003). Typically, shear stress increases quickly until failure occurs close to peak strength (strain hardening part) and decreases upon the formation of faults (strain softening), after which a stable sliding regime under approximately constant shear stress is established (Fig. 5.8, e.g., Klinkmüller et al., 2016). These tests also revealed that peak strength is typically correlated with the maximum rate of volume change (Lohrmann et al., 2003).

A number of recent analogue model experiments provide insight into the fault formation process in various tectonic settings. Ritter et al. (2018a) have shown in their Riedel-type experiments that up to the formation of a continuous strike-slip fault, deformation is diffuse and occurs within a wide shear zone which is correlated with bulk hardening. Activation of individual and partly overlapping fault segments (Riedel shears) subsequently reduced the width of the shear zone, which is accompanied by softening. The shear zone becomes narrowest and weakest once the Riedel shears connect to form a through-going fault zone, with usually an anastomosing geometry. While the

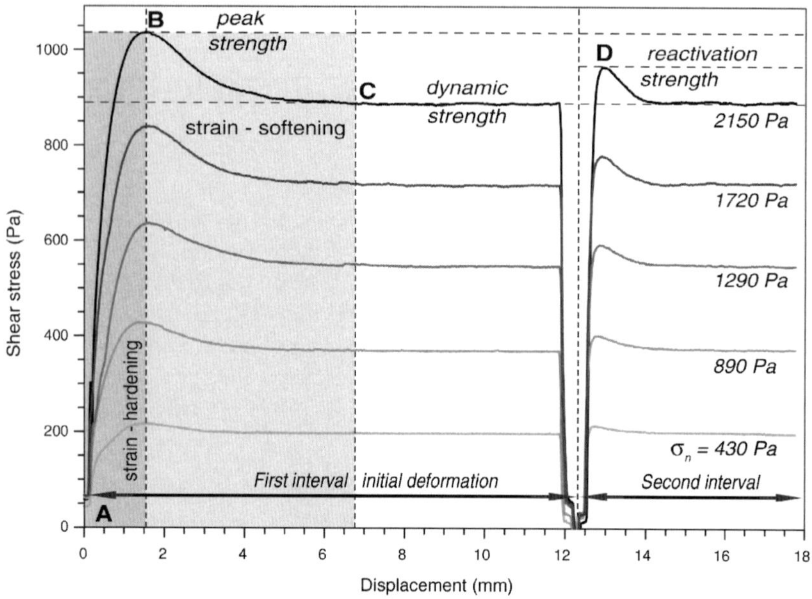

FIG. 5.8 Shear stress evolution with displacement curves for variable normal stresses. B, C and D denote peak-, dynamic-, and reactivation strength. Curve AB is related to strain hardening, whereas BC is related to strain softening. *From Klinkmüller et al. (2016).*

overall width of deformation zones in scale models is defined by spatial relationship of individual faults, the width of single structures is largely controlled by the grain size of the granular material. Similarly, for strike-faults developing in-between two converging indenters (Ritter et al., 2018a), peak strength is reached early during diffuse deformation and dramatic weakening is associated with final fault linkage.

A complementary structural evolution has been documented for accretionary wedge-type experiments, where the localization of deformation through the formation of faults is preceded by an alternating activation of weakly-defined shear bands and the coeval decrease of displacement on already existing faults (Dotare et al., 2016, Fig. 5.9). Similar to strike-slip experiments, this diffuse stage is associated with hardening (Ritter et al., 2018b; McBeck et al., 2018). Once strain localizes on one of the shear bands, the others become inactive, as is also the case for the older thrust faults. Final localization of convergence onto the new frontal thrust is associated with dramatic weakening (Ritter et al., 2018b; McBeck et al., 2018). Data obtained by monitoring of the stress and force build-up with time in thrust wedge experiments (Nieuwland et al., 2000, Ritter et al., 2018b) yield a clear correlation between the fluctuation of stress or force and the activation of new thrusts (Fig. 5.10) and document the development of significant stress variations during shortening (Moulas et al., 2019). An increase of force is measured

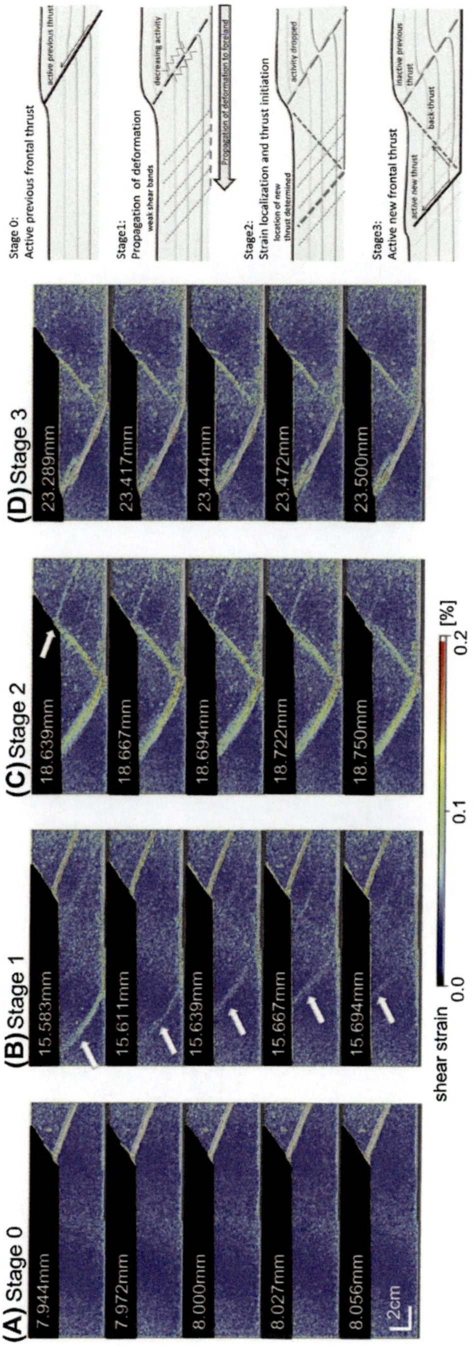

FIG. 5.9 Cross-sections of a sand-box experiment showing the evolution of shear bands through time as determined from the DIC data with a conceptual model of the three stages shown on the right hand side. *From Dotare et al. (2016).*

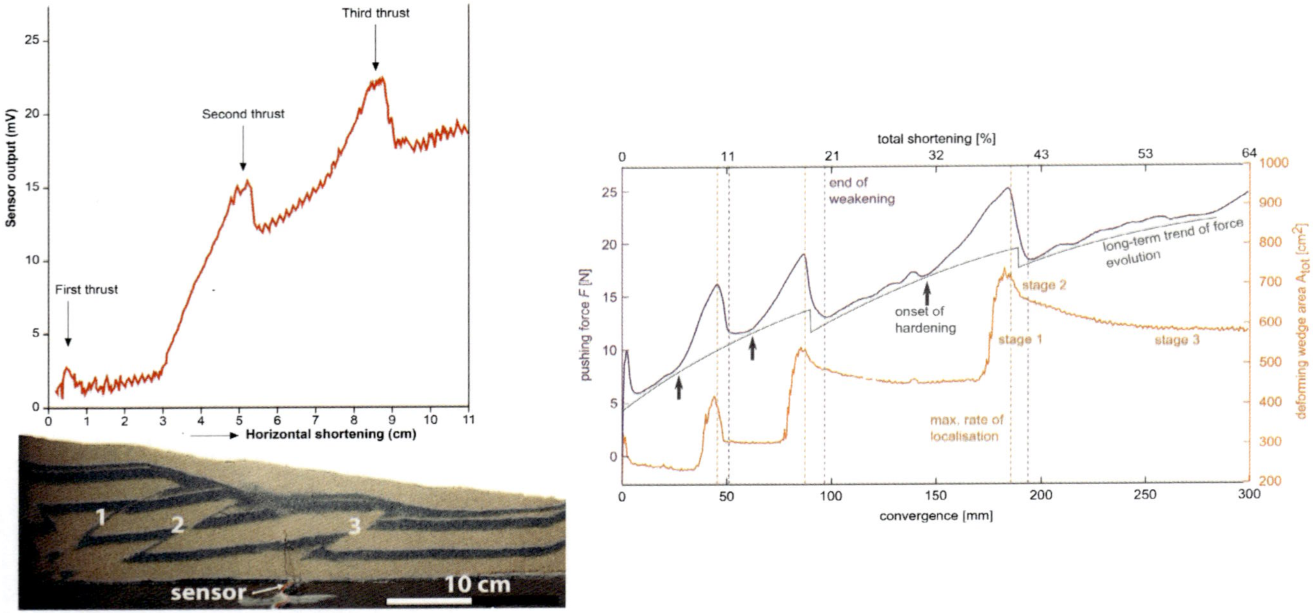

FIG. 5.10 Left: Correlation between thrust activation and in situ-stress evolution. Numbers reflect the sequence of thrust activation. Right: Plot of pushing force against convergence in mm and total shortening in % for an orogenic wedge experiment. Displayed in orange is the change of area affected by deformation as a function of thrust activation. *From Ritter et al. (2018b) and Nieuwland et al. (2000).*

prior to the activation of a new thrust, which is subsequently coeval with a force drop. Additionally, phases of force build-up coincide with an increase in areal extent of deformation, whereas fault activity leads to a decrease in area affected by deformation as thrust sheets move on top of each other.

Integration of data from direct shear tests and analogue experiments, simulating convergent and strike-slip settings such as accretionary prisms, allowed for calculations of the work required for the initiation and propagation of faults (Herbert et al., 2015; Ritter et al., 2018b; McBeck et al., 2018). The results highlight that the work needed for fault propagation scales with the square of the fault length, which is in general agreement with theory (Ritter et al., 2018a). The work, therefore, increases with the thickness of the accreted sand layer in wedges (Herbert et al., 2015). Importantly, modern analogue models, combining kinematic and force measurements, have verified that the fault formation process is dynamically similar in analogue models and nature.

5.3.6 Faulting in single and multi-layer systems

Since the publication of the critical taper theory (Davis et al., 1983), a fair amount of analogue modelling studies have been published addressing the plethora of parameters that influence the overall geometry, internal structure and sequence of deformation of evolving orogenic wedges (e.g., Graveleau et al., 2012 and references therein). Among the studies that use one type of material, faults form on average at angles of ca. 30 degrees with respect to the orientation of maximum compression (σ_1), consistent with their friction angle. Their spacing and geometry (planar vs. listric) is a function of layer thickness and the friction at the base of the sand pack (e.g., Malavieille, 2010). High friction at the base typically yields short, but high wedges that are dominated by forethrusts, whereas wide and low wedges with both fore- and back-thrusts are the result of low friction at the base of the wedges. Additionally, it has been shown that transient changes in fault strength and fault orientation and associated hardening of the rear part of wedges is critical for the wedge taper and wedge segmentation (Lohrmann et al., 2003). Wedge geometry and wedge evolution become increasingly complicated when single or multiple frictional or viscous décollement layers are included. For example, the presence of frictional décollements at the base and within the sand pack leads to simultaneous growth of the wedge at the base and at the front (e.g., Malavieille, 2010). Conversely, the strength ratio between frictional wedge material and a viscous décollement at the base strongly controls the style and sequence of thrusting, such that high strength ratios favour fore-and back-thrusting and oscillating sequences of thrust activation, whereas models with low strength ratios yield results much alike frictional wedges (Smit et al., 2003).

5.3.7 Frontiers

The challenges ahead for the study of fault structures through physical analogue modelling entail bridging time and length-scales. This will lead to an improved understanding of fault slip dynamics including fault formation, fault propagation and fault re-activation in the framework of the seismic cycle (e.g., Rosenau et al., 2017), as well as a better structural characterization of faults and their associated damage zones which will be instrumental for the upscaling of laboratory results to natural conditions.

Additionally, a basic understanding of fault structures that evolve across layers of different rheology and are subject to multiple phases of deformation (e.g., Gabrielsen et al., 2016) is still largely missing, yet of great relevance.

5.4 Microstructures of laboratory faults

5.4.1 Introduction of localization features

It is well known that shearing of gouge-filled faults with artificial rigid side-walls leads to a rotation of the principal compressive stresses ($\sigma_1 \geq \sigma_2 \geq \sigma_3$) within the gouge layer. Initial loading of the shear stress, while keeping the normal stress constant, leads to progressive rotation of σ_1 towards the direction of shear until it is oriented at 45 degrees to the gouge-host rock boundary. This rotation of the stress field activates different localized slip systems, which can be thought of as failure planes or fractures. Fig. 5.11 shows a number of potential orientations and their nomenclature. Recognizing localized features in deformed fault gouges is not always straightforward, particularly in multi-mineralic and clay-rich gouges. Samples of clay-rich gouges are notably difficult to handle because of shrinkage of the gouges upon drying, poor impregnation of epoxy, ripping out of hard(er) clasts, and smearing of the clays. First-order observations of the orientation of preferential planes at which the sample breaks can typically be made when the sample assembly is taken apart (see Fig. 5.11), but these orientations are rarely reported. Slick-enlines are direct evidence of the orientation (but not slip sense) of differential

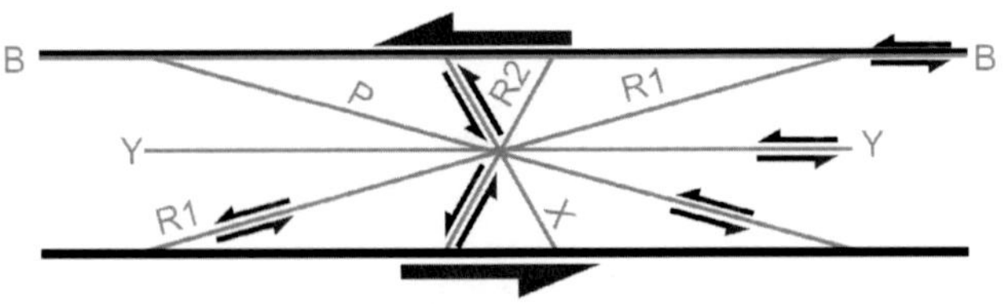

FIG. 5.11 Schematic depiction of a sheared fault gouge showing the nomenclature of different localized shear failure planes according to Logan et al. (1979). Note that a boundary B shear was not defined separately by Logan. Not all localization features have to be present simultaneously. Most commonly only R_1, R_2 and Y or B shears are reported, P shears are sometimes present, X rarely (or not reported). *Figure reproduced after Logan et al. (1992) and Niemeijer et al. (2010).*

displacement, and although these can sometimes be found (Logan et al., 1979), the amount of displacement accommodated cannot be directly estimated from them. Passive strain markers already present in the simulated gouge, such as pyrite clasts, might be used to make quantitative strain estimates (Logan et al., 1979). Alternatively, a passive strain marker might be added to the simulated fault gouge, such as hydrocarbon-coated calcite, making it black (Logan et al., 1992), or dolomite in calcite gouges (Rempe et al., 2017).

5.4.2 Development of gouge microstructure with strain/displacement

In order to study the evolution of gouge microstructure with displacement, it is necessary to perform multiple experiments with the same starting materials under identical conditions and analyze the resulting microstructures. This is very time-consuming, which might be one of the reasons that not many studies have reported on the evolution of fault gouge microstructure as a function of displacement. A general observation, which almost all studies agree on, is that the onset of localization already occurs during the attainment of peak shear stress or slightly thereafter (Fig. 5.12). Gouge layers that exhibit a peak in shear stress show ongoing dilation (increased volume) as the peak in shear stress is approached during initial loading. In these cases, the rate of dilation is greatest as the peak in shear stress is reached. This roll-over in both shear

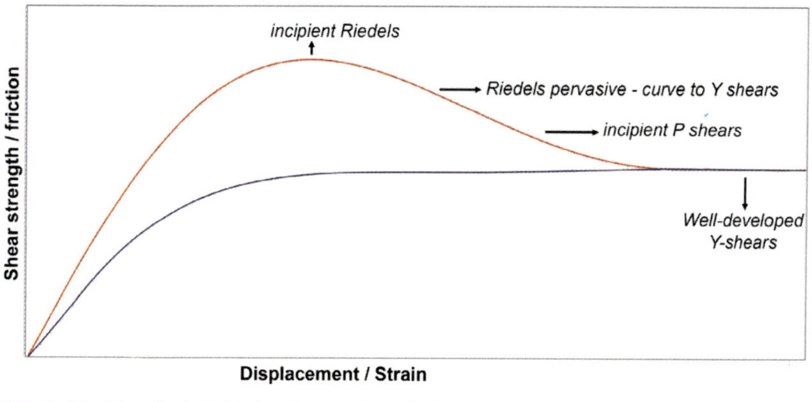

FIG. 5.12 Hypothetical frictional strength evolution curves for gouges deformed at constant sliding velocity. The *red curve* represents an over-compacted gouge sample which dilates as shear stress increases. The dilation rate is maximum at the peak shear stress and this point is associated with the onset of localization (Marone et al., 1990). The scale of the displacement/strain axis has deliberately been omitted, since the strain at which Y or B-shears are well-developed is not constrained and depends on the mineralogy, initial grain size distribution and experimental configuration. The *blue curve* represents a hypothetical curve for a sample that is under-compacted at the onset of shear deformation. In this case, a peak stress is typically absent. The development of localization features in this type of gouge has not been studied sequentially so the labels for the occurrence of these are absent. Note that the experiments of Logan et al. (1992) are of the red type.

stress and volumetric behavior is accompanied by the first occurrence of localized shear bands in the R_1 Riedel orientation (Fig. 5.11). This effect could be even greater in natural faults as an effect of the presence of a pre-existing preferred orientation of phyllosilicates developed during crystallization under a stress field (e.g., Rutter et al., 1986) and/or by dissolution of stronger material and diffusion or advection along potential slip surfaces (e.g., Lacroix et al., 2015).

By far the most comprehensive study of the evolution of gouge microstructure with displacement is that by Logan et al. (1992). These authors earlier reported on the presence of localized slip planes (1979) and identified them as Coulomb failure planes, similar to R_1 Riedel structures reported in the field and in shear box or Riedel experiments (Tchalenko, 1970). The Coulomb failure criterion (Chapter 2) predicts that the angle β between the failure plane and the maximum principal stress, σ_1, should follow the equation:

$$\beta = 45 - \frac{\phi}{2} \tag{5.23}$$

where ϕ is the friction angle, which is related to the static coefficient of friction through $\mu = \tan \phi$. So for a friction of 0.6, the friction angle is about 31 degrees. Assuming that the major principal stress σ_1 is at an angle of 45 degrees to the shear zone boundary, we expect the angle of the R_1 failure plane to the shear zone boundary to be 15.5 degrees. This is surprisingly consistent with observations of R_1 planes over a wide range of scales, from the mm to cm-scale in high normal stress friction experiments, through the cm-m scale in analogue experiments at low normal stress up to the scale of earthquake faults (Tchalenko, 1970). However, departures from this angle do exist, and could be caused by either passive rotation of the shear planes with increased displacement, or progressive stress rotation during shear (Dresen, 1991).

With reference to Fig. 5.12 and the work of Logan et al. (1992, 1979), the evolution of localization can be interpreted as an initial Coulomb failure occurring on R_1 failure planes and their conjugate R_2 planes, both oriented at an angle to the greatest principal stress as determined by the internal friction angle of the material (Eq. 5.23). However, these failure planes cannot accommodate a lot of relative displacement due to the presence of the rigid boundaries. Therefore, with ongoing displacement these initial shear planes deactivate and start to passively rotate, or alternatively, the stress field rotates. If the stresses rotate, new R_1 planes form at a lower angle to the boundaries, or the existing R_1 planes propagate at a lower angle, curving into boundary-parallel Y shears. At the same time, or shortly thereafter, failure planes in the P shear orientation are activated, probably because of a further rotation of the principal stresses in the sense opposite to that of the general movement. Where P-shears crosscut pre-existing Y shears, the latter will grow in length until ultimately a through-going Y shear is formed. Typically, this occurs at one of the two interfaces with the boundary. In rotary shear experiments, this is almost always the rotating side. In some cases, two B-shears develop without any particular identifiable reason. The evolution of this microstructure seems to be fairly universal, and has been reported for most non-phyllosilicate minerals and for some clays.

The experiments by Logan et al. (1992) clearly showed that the evolution of the gouge microstructure is reproducible. For thinner layers, a steady-state microstructure (i.e., the presence of a through-going Y or B shear) is achieved at smaller displacement. This demonstrates that the controlling factor for microstructural development is the distance between the two boundaries and the relative displacement, i.e., shear strain controls structural evolution. Logan also found that a higher confining pressure leads to a faster evolution of the microstructure. This could in part be because gouge layers compact more at high confining pressure, so that the same displacement occurs at a higher shear strain.

5.4.3 Distribution of slip on structural elements

So far, we have only described how gouge microstructure develops in terms of localized shear planes. However, the occurrence of localized shear planes does not necessarily mean that these accommodate all the displacement. As already mentioned, R_1 and R_2 shear planes are not well oriented to accommodate the imposed relative displacement. Logan et al. (1992) created passive markers in their experiments on calcite, using calcite grains coated in hydrocarbons that show up black in plane polarized light. The black marker was deposited at close to 90 degrees to the shear zone boundary, and shear deformation causes the boundary between the black and white to change angle. Fig. 5.13 is an example of a microstructure produced in one of their experiments. There are clear offsets within the boundary between black and white calcite that are related to sliding on R_1 shear planes. Quantitative measures of the angle of the boundary as well as the offsets gives estimates for the percentage of displacement that is accommodated on the different features. Before peak strength is reached, over 90% of the displacement occurs distributed within the gouge volume. After the peak strength, but before the final strength is reached, displacement shifts to the R_1 shears, so that roughly half of the total displacement is accommodated on these. Incipient Y shears accommodate less than 10% of the displacement at this point. Finally, after the maximum displacement of 15 mm or a shear strain of ~25–30 is reached, displacement is partitioned roughly equivalently between distributed shearing, shear on R_1 planes, and shear on Y planes. This indicates that although the bulk of the displacement during steady-state shearing is accommodated by sliding on Y shears, some displacement is still accommodated by distributed shearing and slip on R_1 planes.

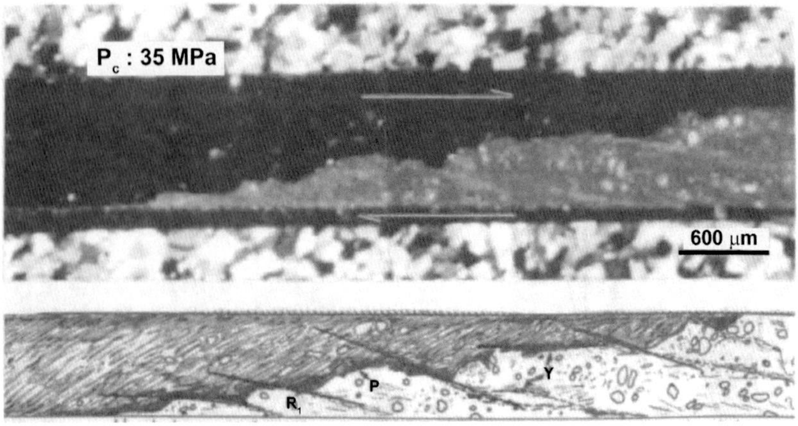

FIG. 5.13 Photomicrograph and interpretive sketch (bottom) of a sheared calcite gouge, deformed at 35 MPa confining pressure and room temperature up to a displacement of 14.5 mm. Loading rate was not directly quoted but was most likely similar to that in earlier experiments at 1 μm/s. The boundary of the black calcite with the normal calcite was perpendicular to the shear zone boundary before shearing. Without shear along localized failure planes, this boundary would have rotated to an angle of ~3 degrees (assuming at thickness of 0.6 mm). The average angle of ~8 degrees here indicates that up to 4.25 mm of displacement was distributed, at least to some extent. *From Logan et al. (1992).*

5.4.4 Role of Y or B shears in generation of unstable slip

Boundary or Y shears have been recognized in samples that experienced unstable sliding or velocity-weakening behavior, leading to suggestions of a link between unstable slip and generation of these structures (e.g., Beeler et al., 1996; Ikari, 2015). Similarly, the presence of a thicker gouge layer appears to correlate with more velocity strengthening behavior (Marone and Scholz, 1988). In fact, Logan et al. (1992) suggested that because most displacement during steady-state sliding is accommodated on boundary-parallel shears, gouge experiments and experiments on bare surfaces would give an identical frictional response. However, the suggestions above seem at odds with a number of other observations. First, it has been demonstrated that frictional wear and gouge accumulation rate increase with increasing normal stress (Scholz, 1987). This suggests that fault gouge thickness should increase with increasing depth, causing faults to become more frictionally stable and indicating increasing $(a - b)$-values with depth. Since the depth at which $(a - b)$ is most negative has the most potential for seismogenesis, this suggests a decrease in seismicity with depth. This contrasts with the common observation and inference that large earthquakes nucleate towards the base of the seismogenic zone, i.e., at the greatest depth (Scholz, 1998). Second, Y and B shears have been reported in fault gouges that show clear velocity-strengthening behavior (Verberne et al., 2010). Third, unstable, velocity-weakening behavior is always reported in experiments employing spherical glass beads as the gouge material, at least as long as the shear zone boundary is rough, even though deformation is distributed in such layers (e.g., Mair et al., 2002; Anthony and Marone, 2005). A common observation of deformed fault gouges in nature and in the lab is that the grain size of the material within the localized B or Y shear is very small, almost always sub-micron (e.g., Friedman et al., 1974; Engelder, 1974; Engelder et al., 1975; Olgaard and Brace, 1983; Power and Tullis, 1989; Verberne et al., 2013). This indicates that even for a very narrow shear band of several microns thickness, there will still be more than 10 grains involved in the shearing process. Therefore, the deformation must involve some part of grain sliding and rotation, the ease of which depends on the properties of the fault material. These properties might be similar for nano- and micron-sized grains, but at present, this is not known, and it remains a challenge for future studies to study the active deformation mechanisms within a gouge volume of nano-crystalline materials. Regardless, we expect the properties of this gouge volume to control the overall frictional characteristics.

5.4.5 Clay-bearing versus non-clay bearing

While the shear planes described in Fig. 5.11 have been observed in both clay- and non-clay-bearing gouges, there are differences between microstructures that develop in the presence and absence of significant clay proportions. At relatively low temperatures, clay-rich gouges develop a pervasive planar fabric with kink bands and/or localized shear zones (Moore and Byerlee, 1989). Moore et al. (Moore and Byerlee, 1989; Moore et al., 1989) further reported that in clay gouges where deformation is localized on boundary-parallel shear bands with R-type shears at high (>14 degrees) angles to the boundaries, stick-slip behavior was observed within otherwise essentially undeformed gouge. On the other hand, pervasively foliated gouge with Riedel shears at less than 10 degrees to the boundaries was found to slide stably. The foliations and kink bands reported in clay and illite gouges were not observed in experiments on quartzofeldspathic gouges, where instead, distributed microcracks were seen without a preferred orientation, but concentrated at grain boundaries. Riedel shears that formed in the quartzofeldspathic gouges were defined by intense grain-size reduction, and similarly to clay samples, stick-slip behavior was correlated with the presence of boundary shears and relatively high-angle Riedel shears accommodating the majority of displacement. Interestingly, high-angle R-type fault traces were reported for locked segments of the San Andreas fault zone whereas fault traces from creeping segments show predominantly P-type fault traces with fewer low angle R-type traces (Moore and Byerlee, 1991, 1992). This suggests that it may be possible to evaluate the seismogenic potential of fault segments by investigating localized structures at all scales, taking into account all features and their relations (see previous discussion on boundary shears).

All observations of foliated clay-rich samples versus non-foliated non-clay gouges were made on samples deformed at laboratory strain rates and temperatures less than 600 °C, where quartzofeldspathic rocks show a macroscopically brittle behavior. At higher temperatures, however, combinations of diffusion and dislocation creep allow foliations to form also in non-clay gouges (Richter et al., 2018). In natural faults, where strain rates are much lower than in the laboratory, extrapolation of laboratory data (Brace and Kohlstedt, 1980), and combinations of observations in natural and experimental faults (Hirth et al., 2001), indicate this brittle-viscous transition occurs at around 350 °C. Richter et al. (2018) report a gradual change from deformation by fracturing and slip localization along Riedel shears at low temperatures, to pervasive foliation development at higher temperatures. Following Kohlstedt et al. (1995), they correlate the onset of viscous creep with the Goetze criterion, where the driving stress for viscous flow is equal to the confining pressure. Cracking may be important for the onset of viscous deformation, because increased surface area and decreased transport distance in finer-grained, intensely-fractured gouges would lead to accelerated solid-state and grain-boundary diffusion creep. Note that in the presence of a fluid, fluid-assisted diffusion creep or pressure solution becomes possible and flow laws of this type of creep suggest weakening is possible at a lower temperatures than for dislocation creep (e.g., Rutter, 1976; Rutter and Mainprice,1979; Angevine et al., 1982).

Where viscous flow in quartz is accommodated by dislocation creep, which is the combination of dislocation glide and climb, three deformation regimes have been identified and correlated to variations in temperature, strain rate, and stress (Hirth and Tullis, 1992; Stipp et al., 2002). These regimes can also be recognized by the recovery and recrystallization mechanisms indicated by their microstructures. Specifically, Regime 1 is distinguished by grain boundary migration, Regime 2 by dominant subgrain rotation, and Regime 3 by a combination of subgrain rotation and grain boundary migration (Hirth and Tullis, 1992). Recognizing the specific dislocation creep regime in which quartz deformed, and by relating quartz grain-size to differential stress through experimentally-determined piezometers (e.g., Stipp et al., 2002; Cross et al., 2017), has therefore become a powerful method to estimate the stress under which rocks deformed, and for extrapolation between natural and experimental condition.

5.4.6 Frontiers

While the paragraphs above highlight that clay-rich gouges develop a penetrative foliation at low temperatures whereas distributed deformation in quartzofeldspathic samples occurs by microcracking, a topic of current development is the microstructural evolution of gouges with mixtures of clays and other soluble minerals (e.g., Bos and Spiers, 2001), including under which conditions and volumetric proportions a mixture behaves more like a clay- or a non-clay-bearing gouge. The location and interconnectivity of the clay or phyllosilicate particles is more important in controlling the frictional behavior than the bulk clay content (Niemeijer, 2018, see also Section 5.1.1). By coupling mechanical and microstructural observations, future experiments are likely to further enhance our ability to determine deformation conditions from natural microstructures, and better extrapolate experiments to nature, as experimentally-derived microphysical models are developed (e.g., Bos and Spiers, 2002; Chen and Spiers, 2016).

References

Abercrombie, R.E., Rice, J.R., 2005. Can observations of earthquake scaling constrain slip weakening? Geophys. J. Int. 162, 406–424.

Adam, J., Urai, J.L., Wieneke, B., Oncken, O., Pfeiffer, K., Kukowski, N., Lohrmann, J., Hoth, S., van der Zee, W., Schmatz, J., 2005. Shear localisation and strain distribution during tectonic faulting – new insights from granular-flow experiments and high-resolution optical image correlation techniques. J. Struct. Geol. 27 (2), 183–301. https://doi.org/10.1016/j.jsg.2004.08.008.

Adam, J., Klinkmüller, M., Schreurs, G., Wieneke, B., 2013. Quantitative 3D strain analysis in analogue experiments simulating tectonic deformation: integration of X-ray computed tomography and digital volume correlation techniques. J. Struct. Geol. 55, 127–149. https://doi.org/10.1016/j.jsg.2013.07.011.

Andrews, D.J., 2002. A fault constitutive relation accounting for thermal pressurization of pore fluid. J. Geophys. Res. 107 (B12), 8. https://doi.org/10.1029/2002JB001942.

Andrews, D.J., 2005. Rupture dynamics with energy loss outside the slip zone. J. Geophys. Res. Solid Earth 110, JB003191. https://doi.org/10.1029/2004JB003191.

Angevine, C.L., Turcotte, D.L., Furnish, M.D., 1982. Pressure solution lithification as a mechanism for the stick-slip behavior of faults. Tectonics 1, 151–160.

Anthony, J.L., Marone, C., 2005. Influence of particle characteristics on granular friction. J. Geophys. Res. B08409. https://doi.org/10.1029/2004JB003399.

Archard, J.F., 1959. The temperature of rubbing surfaces. Wear 2, 438–455. https://doi.org/10.1016/0043-1648(59)90159-0.

Bayart, E., Svetlizky, I., Fineberg, J., 2018. Rupture dynamics of heterogeneous frictional interfaces. J. Geophys. Res. 123 (5), 3828–3848. https://doi.org/10.1002/2018JB015509.

Beeler, N.M., Tullis, T.E., Blanpied, M.L., Weeks, J.D., 1996. Frictional behavior of large displacement experimental faults. J. Geophys. Res. 8697–8715. https://doi.org/10.1029/96JB00411.

Beeler, N.M., Tullis, T.E., Kronenberg, A.K., Reinen, L.A., 2007. The instantaneous rate dependence in low temperature laboratory rock friction and rock deformation experiments. J. Geophys. Res. Solid Earth 112 (B7), B07310.

Ben-David, O., Cohen, G., Fineberg, J., 2010. The dynamics of the onset of frictional slip. Science 330, 211–214.

Blanpied, M.L., Lockner, D.A., Byerlee, J.D., 1995. Frictional slip of granite at hydrothermal conditions. J. Geophys. Res. Solid Earth 100 (B7), 13045–13064.

Blanpied, M.L., Marone, C.J., Lockner, D.A., Byerlee, J.D., King, D.P., 1998a. Quantitative measure of the variation in fault rheology due to fluid-rock interactions. J. Geophys. Res. Solid Earth 103 (B5), 9691–9712.

Blanpied, M.L., Tullis, T.E., Weeks, J.D., 1998b. Effects of slip, slip rate, and shear heating on the friction of granite. J. Geophys. Res. Solid Earth 103 (B1), 489–511.

Bos, B., Spiers, C.J., 2001. Experimental investigation into the microstructural and mechanical evolution of phyllosilicate-bearing fault rock under conditions favouring pressure solution. J. Struct. Geol. 23 (8), 1187–1202.

Bos, B., Spiers, C.J., 2002. Frictional-viscous flow of phyllosilicate-bearing fault rock: microphysical model and implications for crustal strength profiles. J. Geophys. Res. Solid Earth 107 (B2), 2028. https://doi.org/10.1029/2001JB000301.

Bos, B., Peach, C.J., Spiers, C.J., 2000. Frictional-viscous flow of simulated fault gouge caused by the combined effects of phyllosilicates and pressure solution. Tectonophysics 327 (3), 173–194.

Boulton, C., Moore, D.E., Lockner, D.A., Toy, V.G., Townend, J., Sutherland, R., 2014. Frictional properties of exhumed fault gouges in DFDP-1 cores, alpine fault, New Zealand. Geophys. Res. Lett. 41 (2), 356–362.

Brace, W.F., Byerlee, J.D., 1966. Stick-slip as a mechanism for earthquakes. Science 153 (3739), 990–992.

Brace, W.F., Kohlstedt, D.L., 1980. Limits on lithospheric stress imposed by laboratory experiments. J. Geophys. Res. 85, 6248–6252.

Brace, W.F., 1972. Laboratory studies of stick-slip and their application to earthquakes. Tectonophysics 14 (3), 189–200.

Brantut, N., Schubnel, A., Corvisier, J., Sarout, J., 2010. Thermochemical pressurization of faults during coseismic slip. J. Geophys. Res. 115 (B5), 17. https://doi.org/10.1029/2009JB006533.

Brantut, N., Han, R., Shimamoto, T., Findling, N., Schubnel, A., 2011a. Fast slip with inhibited temperature rise due to mineral dehydration: evidence from experiments on gypsum. Geology 39 (1), 59–62. https://doi.org/10.1130/G31424.1.

Brantut, N., Sulem, J., Schubnel, A., 2011b. Effect of dehydration reactions on earthquake nucleation: stable sliding, slow transients, and unstable slip. J. Geophys. Res. 116 (B5), B05304. https://doi.org/10.1029/2010JB007876.

Byerlee, J., 1978. Friction of rocks. Pure Appl. Geophys. 116 (4–5), 615–626.

Carpenter, B.M., Ikari, M.J., Marone, C., 2016. Laboratory observations of time-dependent frictional strengthening and stress relaxation in natural and synthetic fault gouges. J. Geophys. Res. 121 (2), 1183–1201.

Chen, J., Spiers, C.J., 2016. Rate and state frictional and healing behavior of carbonate fault gouge explained using microphysical model. J. Geophys. Res. 121 (12), 8642–8665.

Chen, J., Niemeijer, A., Yao, L., Ma, S., 2017a. Water vaporization promotes coseismic fluid pressurization and buffers temperature rise. Geophys. Res. Lett. 44 (5), 2177–2185. https://doi.org/10.1002/2016GL071932.

Chen, J., Niemeijer, A.R., Fokker, P.A., 2017b. Vaporization of fault water during seismic slip. J. Geophys. Res. 4237–4276. https://doi.org/10.1002/2016JB013824.

Chester, F.M., Higgs, N.G., 1992. Multimechanism friction constitutive model for ultrafine quartz gouge at hypocentral conditions. J. Geophys. Res. Solid Earth 97 (B2), 1859–1870.

Chester, F.M., 1994. Effects of temperature on friction: constitutive equations and experiments with quartz gouge. J. Geophys. Res. Solid Earth 99 (B4), 7247.

Chester, F.M., 1995. A rheologic model for wet crust applied to strike-slip faults. J. Geophys. Res. Solid Earth 100 (B7), 13033–13044.

Cross, A.J., Prior, D.J., Stipp, M., Kidder, S., 2017. The recrystallized grain size piezometer for quartz: an EBSD-based calibration. Geophys. Res. Lett. 44, 6667–6674.

Davis, D., Suppe, J., Dahlen, F.A., 1983. Mechanics of fold-and-thrust belts and accretionary wedges. J. Geophys. Res. 88 (B2), 1153–1172.

De Paola, N., Holdsworth, R.E., Viti, C., Collettini, C., Bullock, R., 2015. Can grain size sensitive flow lubricate faults during the initial stages of earthquake propagation? Earth Planet. Sci. Lett. 431, 48–58. https://doi.org/10.1016/j.epsl.2015.09.002.

del Gaudio, P., Di Toro, G., Han, R., Hirose, T., Nielsen, S., Shimamoto, T., Cavallo, A., 2009. Frictional melting of peridotite and seismic slip. J. Geophys. Res. Solid Earth 114 (B6), B06306. https://doi.org/10.1029/2008JB005990.

den Hartog, S.A.M., Niemeijer, A.R., Spiers, C.J., 2012a. New constraints on megathrust slip stability under subduction zone P–T conditions. Earth Planet. Sci. Lett. 353–354, 240–252.

den Hartog, S.A.M., Peach, C.J., de Winter, D.A.M., Spiers, C.J., Shimamoto, T., 2012b. Frictional properties of megathrust fault gouges at low sliding velocities: new data on effects of normal stress and temperature. J. Struct. Geol. 38, 156–171.

den Hartog, S.A.M., Niemeijer, A.R., Spiers, C.J., 2013. Friction on subduction megathrust faults: beyond the illite–muscovite transition. Earth Planet. Sci. Lett. 373, 8–19.

Di Toro, G., Goldsby, D.L., Tullis, T.E., 2004. Friction falls towards zero in quartz rock as slip velocity approaches seismic rates. Nature 427 (6973), 436–439. https://doi.org/10.1038/nature02249.

Di Toro, G., Han, R., Hirose, T., De Paola, N., Nielsen, S., Mizoguchi, K., Ferri, F., Cocco, M., Shimamoto, T., 2011. Fault lubrication during earthquakes. Nature 471 (7339), 494–498. https://doi.org/10.1038/nature09838.

Dieterich, J.H., 1972. Time-dependent friction in rocks. J. Geophys. Res. Solid Earth 20, 3690–3697.

Dieterich, J.H., Kilgore, B.D., 1994. Direct observation of frictional contacts: new insights for state-dependent properties. Pure Appl. Geophys. 143 (1–3), 283–302.

Dieterich, J.H., Kilgore, B.D., 1996. Imaging surface contacts: power law contact distributions and contact stresses in quartz, calcite, glass and acrylic plastic. Tectonophysics 256 (1–4), 219–239.

Dieterich, J.H., 1978. Time-dependent friction and the mechanics of stick-slip. Pure Appl. Geophys. 116 (4–5), 790–806.

Dieterich, J.H., 1979. Modeling of rock friction 1. Experimental results and constitutive equations. J. Geophys. Res. 2161–2168.

Dieterich, J.H., 1981. Constitutive Properties of Faults with Simulated Gouge, pp. 103–120.

Dotare, T., Yamada, Y., Adam, J., Hori, T., Sakaguchi, H., 2016. Initiation of a thrust fault revealed by analog experiments. Tectonophysics 684, 148–156. https://doi.org/10.1016/j.tecto.2015.12.023.

Dresen, G., 1991. Stress distribution and the orientation of Riedel shears. Tectonophysics 188 (3–4), 239–247.

Dunham, E.M., Archuleta, R.J., 2004. Evidence for a supershear transient during the 2002 Denali Fault earthquake. Bull. Seismol. Soc. Am. 94, 256–268.

Eisenstadt, G., Sims, D., 2005. Evaluating sand and clay models: do rheological differences matter? J. Struct. Geol. 27, 1399–1412.

Engelder, J.T., Logan, J.M., Handin, J., 1975. The sliding characteristics of sandstone on quartz fault-gouge. Pure Appl. Geophys. 113 (1), 69–86. https://doi.org/10.1007/BF01592900.

Engelder, J.T., 1974. Cataclasis and the generation of fault gouge. GSA Bull. 85 (10), 1515.

Faulkner, D.R., Mitchell, T.M., Behnsen, J., Hirose, T., Shimamoto, T., 2011. Stuck in the mud? earthquake nucleation and propagation through accretionary forearcs. Geophys. Res. Lett. 38 (18) https://doi.org/10.1029/2011GL048552 n/a.

Faulkner, D.R., Sanchez-Roa, C., Boulton, C., Hartog, S.A.M., 2018. Pore fluid pressure development in compacting fault gouge in theory, experiments, and nature. J. Geophys. Res. Solid Earth 123 (1), 226–241.

Ferri, F., Di Toro, G., Hirose, T., Shimamoto, T., 2010. Evidence of thermal pressurization in high-velocity friction experiments on smectite-rich gouges. Terra Nova 22 (5), 347–353. https://doi.org/10.1111/j.1365-3121.2010.00955.x.

Ferri, F., Di Toro, G., Hirose, T., Han, R., Noda, H., Shimamoto, T., Quaresimin, M., de Rossi, N., 2011. Low- to high-velocity frictional properties of the clay-rich gouges from the slipping zone of the 1963 vaiont slide, northern Italy. J. Geophys. Res. 116 (B9) https://doi.org/10.1029/2011JB008338.

French, M.E., Chester, J.S., 2018. Localized slip and associated fluidized structures record seismic slip in Clay-Rich fault gouge. J. Geophys. Res. 123 (10), 8568–8588. https://doi.org/10.1029/2018JB016053.

Friedman, M., Logan, J.M., Rigert, J.A., 1974. Glass-indurated quartz gouge in sliding-friction experiments on sandstone. GSA Bull. 85 (6), 937.

Fukuyama, E., Xu, S., Yamashita, F., Mizoguchi, K., 2016. Cohesive zone length of metagabbro at supershear rupture velocity. J. Seismol. 20 (4), 1207–1215. https://doi.org/10.1007/s10950-016-9588-2.

Gabrielsen, R.H., Sokoutis, D., Willingshofer, E., Faleide, J.I., 2016. Fault linkage across weak layers during extension: an experimental approach with reference to the Hoop Fault Complex of the southwestern Barents Sea. Petrol. Geosci. https://doi.org/10.1144/petgeo2015-029.

Giorgetti, C., Carpenter, B.M., Collettini, C., 2015. Frictional behavior of talc-calcite mixtures. J. Geophys. Res. Solid Earth 120 (9), 6614–6633.

Graveleau, F., Malavieille, J., Dominguez, S., 2012. Experimental modelling of orogenic wedges: a review. Tectonophysics 538–540, 1–66. https://doi.org/10.1016/j.tecto.2012.01.027.

Guérin-Marthe, S., Nielsen, S., Bird, B., Giani, S., Di Toro, G., 2019. Earthquake nucleation size: evidence of loading rate dependence in laboratory faults. J. Geophys. Res. Solid Earth 124. https://doi.org/10.1029/2018JB016803.

Hall, J., 1815. On the vertical position and convolutions of certain strata and their relationship with granite. Trans. R. Soc. Edinb. 7, 79–108.

Han, R., Shimamoto, T., Hirose, T., Ree, J., Ando, J., 2007. Ultralow friction of carbonate faults caused by thermal decomposition. Science 316 (5826), 878–881. https://doi.org/10.1126/science.1139763.

Han, R., Hirose, T., Shimamoto, T., 2010. Strong velocity weakening and powder lubrication of simulated carbonate faults at seismic slip rates. J. Geophys. Res. 115 (B3) https://doi.org/10.1029/2008JB006136.

Harbord, C.W.A., Nielsen, S.B., De Paola, N., Holdsworth, R.E., 2017. Earthquake nucleation on rough faults. Geology 45 (10), 931–934.

He, C., Yao, W., Wang, Z., Zhou, Y., 2006. Strength and stability of frictional sliding of gabbro gouge at elevated temperatures. Tectonophysics 427 (1), 217–229.

He, C., Luo, L., Hao, Q., Zhou, Y., 2013. Velocity-weakening behavior of plagioclase and pyroxene gouges and stabilizing effect of small amounts of quartz under hydrothermal conditions. J. Geophys. Res. Solid Earth 118 (7), 3408–3430.

Heaton, T.H., 1990. Evidence for and implications of self-healing pulses of slip in earthquake rupture. Phys. Earth Planet. In. 64, 1–20.

Herbert, J.W., Cooke, M.L., Souloumiac, P., Madden, E.H., Mary, B.C., Maillot, B., 2015. The work of fault growth in laboratory sandbox experiments. Earth Planet. Sci. Lett. 432, 95–102. https://doi.org/10.1016/j.epsl.2015.09.046.

Hirose, T., Shimamoto, T., 2005. Growth of molten zone as a mechanism of slip weakening of simulated faults in gabbro during frictional melting. J. Geophys. Res. 110 (B5), B05202. https://doi.org/10.1029/2004JB003207.

Hirth, G., Tullis, J., 1992. Dislocation creep regimes in quartz aggregates. J. Struct. Geol. 14, 145–159.

Hirth, G., Teyssier, C., Dunlap, W., 2001. An evaluation of quartzite flow laws based on comparisons between experimentally and naturally deformed rocks. Int. J. Earth Sci. 90, 77–87.

Hubbert, M.K., 1937. Theory of scale models as applied to the study of geologic structures. Geol. Soc. Am. Bull. 48, 1459–1520.

Ikari, M.J., Kopf, A.J., 2017. Seismic potential of weak, near-surface faults revealed at plate tectonic slip rates. Sci. Adv. 3 (11), e1701269.

Ikari, M.J., Saffer, D.M., Marone, C., 2009. Frictional and hydrologic properties of clay-rich fault gouge. J. Geophys. Res. Solid Earth 114 (B5), B05409.

Ikari, M.J., Marone, C., Saffer, D.M., 2011. On the relation between fault strength and frictional stability. Geology 39 (1), 83–86.

Ikari, M., Niemeijer, A.R., Spiers, C.J., Kopf, A.J., Saffer, D.M., 2013. Experimental evidence linking slip instability with seafloor lithology and topography at the Costa Rica convergent margin. Geology 41 (8), 891–894.

Ikari, M., Trütner, S., Carpenter, B., Kopf, A., 2015. Shear behavior of DFDP-1 borehole samples from the alpine fault, New Zealand, under a wide range of experimental conditions. Int. J. Earth Sci. 104 (6), 1523–1535.

Ikari, M.J., 2015. Principal slip zones: precursors but not recorders of earthquake slip. Geology 43 (11), 955–958. https://doi.org/10.1130/G37028.1.

Kaneko, Y., Nielsen, S.B., Carpenter, B.M., 2016. The onset of laboratory earthquakes explained by nucleating rupture on a rate-and-state fault. J. Geophys. Res. 121 (8), 6071–6091. https://doi.org/10.1002/2016JB013143.

Klinkmüller, M., Schreurs, G., Rosenau, M., Kemnitz, H., 2016. Properties of granular analogue model materials: A community wide survey. Tectonophysics 684, 23–38.

Kohlstedt, D.L., Evans, B., Mackwell, S.J., 1995. Strength of the lithosphere: constraints imposed by laboratory experiments. J. Geophys. Res. 100, 17587–17602.

Kurzawski, R.M., Stipp, M., Niemeijer, A.R., Spiers, C.J., Behrmann, J.H., 2016. Earthquake nucleation in weak subducted carbonates. Nat. Geosci. 9 (9), 717–722.

Lacroix, B., Tesei, T., Oliot, E., Lahfid, A., Collettini, C., 2015. Early weakening processes inside thrust fault. Tectonics 34, 1396–1411.

Lee, S.K., Han, R., Kim, E.J., Jeong, G.Y., Khim, H., Hirose, T., 2017. Quasi-equilibrium melting of quartzite upon extreme friction. Nat. Geosci. 10 (6), 436–441. https://doi.org/10.1038/ngeo2951.

Leeman, J.R., Saffer, D.M., Scuderi, M.M., Marone, C., 2016. Laboratory observations of slow earthquakes and the spectrum of tectonic fault slip modes. Nat. Commun. 7 (1), 11104.

Logan, J.M., Rauenzahn, K.A., 1987. Frictional dependence of gouge mixtures of quartz and montmorillonite on velocity, composition and fabric. Tectonophysics 144 (1), 87–108.

Logan, J.M., Friedman, M., Higgs, N., Dengo, C., Shimamoto, C., 1979. Experimental studies of simulated gouge and their application to studies of natural fault zones. In: Proceedings of Conference VIII-Analysis of Actual Fault Zones in Bedrock, pp. 305–343.

Logan, J.M., Dengo, C.A., Higgs, N.G., Wang, Z.Z., 1992. Fabrics of experimental fault zones: their development and relationship to mechanical behavior. In: Wong, T.F., Evans, B. (Eds.), Fault Mechanics and Transport Properties of Rocks, A Festschrift in Honor of W.F. Brace, International Geophysics, vol. 51, pp. 33–67.

Lohrmann, J., Kukowski, N., Adam, J., Oncken, O., 2003. The impact of analogue material properties on the geometry, kinematics, and dynamics of convergent sand wedges. J. Struct. Geol. 25 (10), 1691–1711. https://doi.org/10.1016/S0191-8141(03)00005-1.

Lykotrafitis, G., Rosakis, A.J., Ravichandran, G., 2006. Self-healing pulse-like shear ruptures in the laboratory. Science 313, 1765–1768. https://doi.org/10.1126/science.1128359.

Mair, K., Frye, K.M., Marone, C., 2002. Influence of grain characteristics on the friction of granular shear zones. J. Geophys. Res. Solid Earth 107 (B10), 9. https://doi.org/10.1029/2001JB000516.

Malavieille, J., 2010. Impact of erosion, sedimentation, and structural heritage on the structure and kinematics of orogenic wedges: analog models and case studies. GSA Today 20 (1), 4–10.

Mandl, G., de Jong, L.N.J., Maltha, A., 1977. Shear zones in granular material. Rock Mech. 9, 95–144.

Marone, C., Saffer, D.M., 2015. The Mechanics of Frictional Healing and Slip Instability during the Seismic Cycle. Treatise on Geophysics, second ed., pp. 111–138

Marone, C., Scholz, C.H., 1988. The depth of seismic faulting and the upper transition from stable to unstable slip regimes. Geophys. Res. Lett. 15 (6), 621–624. https://doi.org/10.1029/GL015i006p00621.

Marone, C., Raleigh, C.B., Scholz, C.H., 1990. Frictional behavior and constitutive modeling of simulated fault gouge. J. Geophys. Res. 95 (B5), 7007. https://doi.org/10.1029/JB095iB05p07007.

Marone, C., Vidale, J.E., Ellsworth, W.L., 1995. Fault healing inferred from time dependent variations in source properties of repeating earthquakes. Geophys. Res. Lett. 22 (22), 3095–3098.

McBeck, J., Cooke, M., Souloumiac, P., Maillot, B., Mary, B., 2018. The influence of detachment strength on the evolving deformational energy budget of physical accretionary prisms. Solid Earth 9, 1421–1436. https://doi.org/10.5194/se-9-1421-2018.

McLaskey, G.C., Kilgore, B.D., 2013. Foreshocks during the nucleation of stick-slip instability. J. Geophys. Res. 118, 2982–2997. https://doi.org/10.1002/jgrb.50232.

McLaskey, G.C., Yamashita, F., 2017. Slow and fast ruptures on a laboratory fault controlled by loading characteristics. J. Geophys. Res. 122, 3719–3738. https://doi.org/10.1002/2016JB013681.

Mitchell, E.K., Fialko, Y., Brown, K.M., 2013. Temperature dependence of frictional healing of Westerly granite: experimental observations and numerical simulations. Geochem. Geophys. Geosyst. 14 (3), 567–582.

Mitchell, E.K., Fialko, Y., Brown, K.M., 2015. Frictional properties of gabbro at conditions corresponding to slow slip events in subduction zones. Geochem. Geophys. Geosyst. 16 (11), 4006–4020.

Mitchell, E.K., Fialko, Y., Brown, K.M., 2016. Velocity-weakening behavior of westerly granite at temperature up to 600°C. J. Geophys. Res. Solid Earth 121 (9), 6932–6946.

Mizoguchi, K., Hirose, T., Shimamoto, T., Fukuyama, E., 2007. Reconstruction of seismic faulting by high-velocity friction experiments: an example of the 1995 kobe earthquake. Geophys. Res. Lett. 34 (1), L01308. https://doi.org/10.1029/2006GL027931.

Moore, D.E., Byerlee, J.D., 1989. Textural development of clayey and quartzofeldsphatic fault gouges relative to their sliding behavior. Phys. Chem. Earth 17, 1–9.

Moore, D.E., Byerlee, J.D., 1991. Comparative geometry of the San Andreas fault, California, and laboratory fault zones. Geol. Soc. Am. Bull. 103, 762–774.

Moore, D.E., Byerlee, J.D., 1992. Relationship between sliding behavior and internal geometry of laboratory fault zones and some creeping and locked strike-slip faults of California. Tectonophysics 211, 305–316.

Moore, D.E., Lockner, D.A., 2004. Crystallographic controls on the frictional behavior of dry and water-saturated sheet structure minerals. J. Geophys. Res. Solid Earth 109 (B3), B03401.

Moore, D.E., Summers, R., Byerlee, J.D., 1989. Sliding behavior and deformation textures of heated illite gouges. J. Struct. Geol. 11, 329–342.

Morrow, D., Radney, B., Byerlee, J., 1992. Frictional Strength and the Effective Pressure Law of Montmorillonite and Illite Clays, p. 1.

Morrow, C.A., Moore, D.E., Lockner, D.A., 2000. The effect of mineral bond strength and adsorbed water on fault gouge frictional strength. Geophys. Res. Lett. 27 (6), 815–818.

Moulas, E., Sokoutis, D., Willingshofer, E., 2019. Pressure build-up and stress variations within the earth's crust in the light of analogue models. Nat. Sci. Rep. 9 (1) https://doi.org/10.1038/s41598-018-38256-1.

Nadeau, R., Johnson, L., 1998. Seismological studies at Parkfield VI: moment release rates and estimates of source parameters for small repeating earthquakes. Bull. Seismol. Soc. Am. 88 (3), 790–814.

Nielsen, S., Madariaga, R., 2003. On the self-healing fracture mode. Bull. Seismol. Soc. Am. 93, 2375–2388. https://doi.org/10.1785/0120020090.

Nielsen, S., Di Toro, G., Hirose, T., Shimamoto, T., 2008. Frictional melt and seismic slip. J. Geophys. Res. 113 (B1), B01308. https://doi.org/10.1029/2007JB005122.

Nielsen, S., Mosca, P., Giberti, G., Di Toro, G., Hirose, T., Shimamoto, T., 2010a. On the transient behavior of frictional melt during seismic slip. J. Geophys. Res. 115 (B10) https://doi.org/10.1029/2009JB007020.

Nielsen, S., Toro, G.D., Griffith, W.A., 2010b. Friction and roughness of a melting rock surface. Geophys. J. Int. https://doi.org/10.1111/j.1365-246X.2010.04607.x.

Nielsen, S., Taddeucci, J., Vinciguerra, S., 2010c. Experimental observation of stick-slip instability fronts. Geophys. J. Int. 180, 697–702. https://doi.org/10.1111/j.1365-246X.2009.0444.x.

Nielsen, S., Spagnuolo, E., Violay, M., Smith, S., Di Toro, G., Bistacchi, A., 2016. G: fracture energy, friction and dissipation in earthquakes. J. Seismol. 1–19. https://doi.org/10.1007/s10950-016-9560-1.

Niemeijer, A., Collettini, C., 2014. Frictional properties of a low-angle normal fault under in situ conditions: thermally-activated velocity weakening. Pure Appl. Geophys. 171 (10), 2641–2664.

Niemeijer, A.R., Spiers, C.J., 2005. Influence of Phyllosilicates on Fault Strength in the Brittle-Ductile Transition: Insights from Rock Analogue Experiments, vol. 245. Geological Society, London, Special Publications, pp. 303–327, 1.

Niemeijer, A.R., Spiers, C.J., 2006. Velocity dependence of strength and healing behaviour in simulated phyllosilicate-bearing fault gouge. Tectonophysics 427 (1), 231–253.

Niemeijer, A.R., Spiers, C.J., Peach, C.J., 2008. Frictional behaviour of simulated quartz fault gouges under hydrothermal conditions: results from ultra-high strain rotary shear experiments. Tectonophysics 460 (1), 288–303.

Niemeijer, A., Marone, C., Elsworth, D., 2010. Frictional strength and strain weakening in simulated fault gouge: competition between geometrical weakening and chemical strengthening. J. Geophys. Res. Solid Earth 115 (B10). https://doi.org/10.1029/2009JB000838.

Niemeijer, A., Di Toro, G., Nielsen, S., Di Felice, F., 2011. Frictional melting of gabbro under extreme experimental conditions of normal stress, acceleration, and sliding velocity. J. Geophys. Res. 116 (7) https://doi.org/10.1029/2010JB008181, 074044/1.

Niemeijer, A., Di Toro, G., Griffith, W.A., Bistacchi, A., Smith, S.A.F., Nielsen, S., 2012. Inferring earthquake physics and chemistry using an integrated field and laboratory approach. J. Struct. Geol. 39, 2–36. https://doi.org/10.1016/j .jsg .2012 .02.018.

Niemeijer, A.R., Boulton, C., Toy, V.G., Townend, J., Sutherland, R., 2016. Large-displacement, hydrothermal frictional properties of DFDP-1 fault rocks, alpine fault, New Zealand: implications for deep rupture propagation. J. Geophys. Res. 121 (2), 624–647.

Niemeijer, A.R., 2018. Velocity-dependent slip weakening by the combined operation of pressure solution and foliation development. Sci. Rep. 8 (1), 1–10.

Nieuwland, D.A., Urai, J.L., Knoop, M., 2000. In-situ stress measurements in model experiments of tectonic faulting. In: Lehner, F.K., Urai, J.L. (Eds.), Aspects of Tectonic Faulting. Springer Berlin Heidelberg, Berlin, Heidelberg, pp. 155–166. https://doi.org/10.1007/978-3-642-59617-9_8.

Ohnaka, M., Shen, L., 1999. Scaling of the shear rupture process from nucleation to dynamic propagation: implications of geometric irregularity of the rupturing surfaces. J. Geophys. Res. 104, 817. https://doi.org/10.1029/1998JB900007.

Ohnaka, M., Kuwahara, Y., Yamamoto, K., 1987. Constitutive relations between dynamic physical parameters near a tip of the propagating slip zone during stick-slip shear failure. In: Wesson, R.L. (Ed.), Mechanics of Earthquake Faulting, Tectonophysics, vol. 144, pp. 109–125.

Okubo, P.G., Dieterich, J.H., 1984. Effects of physical fault properties on frictional instabilities produced on simulated faults. J. Geophys. Res. 89 (B7), 5817–5827. https://doi.org/10.1029/JB089iB07p05817.

Olgaard, D.L., Brace, W.F., 1983. The microstructure of gouge from a mining-induced seismic shear zone. Int. J. Rock Mech. Min. Sci. Geomech. Abstr. 20 (1), 11–19. https://doi.org/10.1016/0148-9062(83)91610-8.

Panien, M., Schreurs, G., Pfiffner, A.O., 2006. Mechanical behaviour of granular materials, used in analogue modelling: insights from grain characterisation, ring-shear tests and analogue experiments. J. Struct. Geol. 28 (9), 1710–1724.

Passelègue, F.X., Schubnel, A., Nielsen, S., Bhat, H.S., Madariaga, R., 2013. From Sub-Rayleigh to Supershear ruptures during stick-slip experiments on crustal rocks. Science 340, 1208. https://doi.org/10.1126/science.1235637.

Passelègue, F., Schubnel, A., Nielsen, S., Bhat, H., Deldicque, D., Madariaga, R., 2016. Dynamic rupture processes inferred from laboratory micro-earthquakes. J. Geophys. Res. Solid Earth. https://doi.org/10.1002/2015jb012694.

Peyrat, S., Olsen, K.B., Madariaga, R., 2004. Which dynamic rupture parameters can be estimated from strong ground motion and geodetic data ? Pure Appl. Geophys. 161, 2155–2169.

Platt, J.D., Rudnicki, J.W., Rice, J.R., 2014. Stability and localization of rapid shear in fluid-saturated fault gouge: 2. localized zone width and strength evolution. J. Geophys. Res. 119 (5), 4334–4359. https://doi.org/10.1002/2013JB010711.

Power, W.L., Tullis, T.E., 1989. The relationship between slickenside surfaces in fine-grained quartz and the seismic cycle. J. Struct. Geol. 11 (7), 879–893. https://doi.org/10.1016/0191-8141(89)90105-3.

Rechenmacher, A.L., 2006. Grain-scale processes governing shear band initiation and evolution in sands. J. Mech. Phys. Solids 54 (1), 22–45. https://doi.org/10.1016/j.jmps.2005.08.009.

Reches, Z., Lockner, D.A., 2010. Fault weakening and earthquake instability by powder lubrication. Nature 467 (7314), 452–455. https://doi.org/10.1038/nature09348.

Reinen, L.A., Weeks, J.D., 1993. Determination of rock friction constitutive parameters using an iterative least squares inversion method. J. Geophys. Res. Solid Earth 98, 15937–15950.

Rempe, M., Smith, S., Mitchell, T., Hirose, T., Di Toro, G., 2017. The effect of water on strain localization in calcite fault gouge sheared at seismic slip rates. J. Struct. Geol. 97, 104–117. https://doi.org/10.1016/j.jsg.2017.02.007.

Rice, J.R., Rudnicki, J.W., Platt, J.D., 2014. Stability and localization of rapid shear in fluid-saturated fault gouge: 1. linearized stability analysis. J. Geophys. Res. Solid Earth 119 (5), 4311–4333. https://doi.org/10.1002/2013JB010710.

Rice, J.R., 1999. Flash heating at asperity contacts and rate-dependent friction. EOS Trans. AGU 80 (46), F471.

Rice, J.R., 2006. Heating and weakening of faults during earthquake slip. J. Geophys. Res. 111, B05311. https://doi.org/10.1029/2005JB004006.

Richter, B., Stünitz, H., Heilbronner, R., 2018. The brittle-to-viscous transition in polycrystalline quartz: an experimental study. J. Struct. Geol. 114, 1–21.

Ritter, M., Leever, K., Rosenau, M., Oncken, O., 2016. Scaling the sand box – mechanical (dis-) similarities of granular materials and brittle rock. J. Geophys. Res. https://doi.org/10.1002/2016JB012915.

Ritter, M.C., Rosenau, M., Onno, O., 2018a. Growing faults in the lab: insights into the scale dependence of the fault zone evolution process. Tectonics 37. https://doi.org/10.1002/2017TC004787.

Ritter, M.C., Santimano, T., Rosenau, M., Leever, K., Oncken, O., 2018b. Sandbox rheometry: Co-evolution of stress and strain in Riedel– and Critical Wedge–experiments. Tectonophysics 722, 400–409. https://doi.org/10.1016/j.tecto.2017.11.018.

Rosenau, M., Corbi, F., Dominguez, S., 2017. Analogue earthquakes and seismic cycles: experimental modelling across timescales. Solid Earth 8 (3), 597–635. https://doi.org/10.5194/se-8-597-2017.

Rowe, C.D., Griffith, W.A., 2015. Do faults preserve a record of seismic slip: a second opinion. J. Struct. Geol. 78, 1–26. https://doi.org/10.1016/j.jsg.2015.06.006.

Rubin, A.M., Ampuero, J.-P., 2005. Earthquake nucleation on (aging) rate and state faults. J. Geophys. Res. Solid Earth 110 (B11).

Rubino, V., Rosakis, A.J., Lapusta, N., 2017. Understanding Dynamic Friction Through Spontaneously Evolving Laboratory Earthquakes. https://doi.org/10.1038/ncomms15991.

Rubinstein, S.M., Cohen, G., Fineberg, J., 2007. Dynamics of precursors to frictional sliding. Phys. Rev. Lett. 98, 2117–2120.

Ruina, A., 1983. Slip instability and state variable friction laws. J. Geophys. Res. Solid Earth 88 (B12), 10359–10370.

Rutter, E.H., Mainprice, D.H., 1979. On the possibility of slow fault slip controlled by a diffusive mass transfer process. Gerlands Betr. Geophysik, Leipzig 88, 154–162.

Rutter, E.H., Maddock, R.H., Hall, S.H., White, S.H., 1986. Comparative microstructures of natural and experimentally produced clay-bearing fault gouges. Pure Appl. Geophys. 124 (1–2), 3–30.

Rutter, E.H., 1976. The kinetics of rock deformation by pressure solution. Phil. Trans. Roy. Soc. Lond. 283, 203–219.

Saber, O., Chester, F., Alvarado, J., 2016. Development of a material-testing machine for study of friction: experimental analysis of machine dynamics and friction of rock. Exp. Mech. 56 (5), 813–831.

Saffer, D.M., Marone, C., 2003. Comparison of smectite- and illite-rich gouge frictional properties: application to the updip limit of the seismogenic zone along subduction megathrusts. Earth Planet. Sci. Lett. 215 (1–2), 219–235.

Sawai, M., Niemeijer, A.R., Plümper, O., Hirose, T., Spiers, C.J., 2016. Nucleation of frictional instability caused by fluid pressurization in subducted blueschist. Geophys. Res. Lett. 43 (6), 2543–2551.

Schellart, W.P., 2000. Shear test results for cohesion and friction coefficients for different granular materials: scaling implications for their usage in analogue modelling. Tectonophysics 324 (1–2), 1–16.

Scholz, C.H., 1987. Wear and gouge formation in brittle faulting. Geology 15 (6), 493.

Scholz, C.H., 1998. Earthquakes and friction laws. Nature 391, 37–42. https://doi.org/10.1038/34097.

Scruggs, V.J., Tullis, T.E., 1998. Correlation between velocity dependence of friction and strain localization in large displacement experiments on feldspar, muscovite and biotite gouge. Tectonophysics 295 (1), 15–40.

Scuderi, M.M., Collettini, C., Viti, C., Tinti, E., Marone, C., 2017. Evolution of shear fabric in granular fault gouge from stable sliding to stick slip and implications for fault slip mode. Geology 45 (8), 731.

Shimamoto, T., Logan, J.M., 1981. Effects of simulated clay gouges on the sliding behavior of Tennessee sandstone. Tectonophysics 75 (3–4), 243–255.

Shimamoto, T., Logan, J.M., 1986. Velocity-Dependent Behavior of Simulated Halite Shear Zones: an Analog for Silicates, pp. 49–63.

Shimamoto, T., Tsutsumi, A., 1994. A new rotary-shear high-velocity frictional testing machine: its basic design and scope of research. J. Tectonic Res. Group of Jpn. 39, 65–78 (in Japanese with English Abstract).

Shimamoto, T., 1986. Transition between frictional slip and ductile flow for halite shear zones at room temperature. Science 231 (4739), 711–714.

Shipton, Z.K., Soden, A., Kirkpatrick, J.D., Lunn, R., 2006. How thick is a fault? Fault displacement-thickness scaling revisited. In: Earthquakes: Radiated Energy and the Physics of Faulting, pp. 193–198. https://doi.org/10.1029/170GM19. AGU Geophysical Monograph Series 170.

Sibson, R.H., 1973. Interactions between temperature and pore-fluid pressure during earthquake faulting and a mechanism for partial or total stress relief. Nat. Phys. Sci. 243 (126), 66. https://doi.org/10.1038/physci243066a0.

Smit, J.H.W., Brun, J.-P., Soukoutis, D., 2003. Deformation of brittle–ductile thrust wedges in experiments and nature. J. Geophys. Res. 108 (B10) https://doi.org/10.1029/2002JB002190.

Smith, S.A.F., Di Toro, G., Kim, S., Ree, J., Nielsen, S., Billi, A., Spiess, R., 2013. Coseismic recrystallization during shallow earthquake slip. Geology 41 (1), 63–66. https://doi.org/10.1130/G33588.1.

Spray, J.G., 1987. Artificial generation of pseudotachylyte using friction welding apparatus: simulation of melting on a fault plane. J. Struct. Geol. 9 (1), 49–60. https://doi.org/10.1016/0191-8141(87)90043-5.

Spudich, P., Guatteri, M., 2000. What can strong-motion data tell us about slip-weakening fault-friction laws? Bull. Seismol. Soc. Am. 90, 98–116.

Stipp, M., Stünitz, H., Heilbronner, R., Schmid, S.M., 2002. Dynamic recrystallization of quartz: correlations between natural and experimental conditions. In: Meer, S., Drury, M.R., de Bresser, J.H.P., Pennock, G.M. (Eds.), Deformation Mechanisms, Rheology and Tectonics: Current Status and Future Perspectives, vol. 200. Geological Society of London Special Publications, pp. 171–190.

Sulem, J., Famin, V., 2009. Thermal decomposition of carbonates in fault zones: slip-weakening and temperature-limiting effects. J. Geophys. Res. 114 (B3), B03309. https://doi.org/10.1029/2008JB006004.

Takahashi, M., Mizoguchi, K., Kitamura, K., Masuda, K., 2007. Effects of clay content on the frictional strength and fluid transport property of faults. J. Geophys. Res. 112 (B8), B08206.

Tanikawa, W., Shimamoto, T., 2009. Frictional and transport properties of the Chelungpu Fault from shallow borehole data and their correlation with seismic behavior during the 1999 Chi-Chi earthquake. J. Geophys. Res. 114 (B1) https://doi.org/10.1029/2008JB005750.

Tchalenko, J.S., 1970. Similarities between shear zones of different magnitudes. GSA Bull. 81 (6), 1625.

Tembe, S., Lockner, D.A., Wong, T., 2010. Effect of clay content and mineralogy on frictional sliding behavior of simulated gouges: binary and ternary mixtures of quartz, illite, and montmorillonite. J. Geophys. Res. 115 (B3).

Tinti, E., Cocco, M., Fukuyama, E., Piatanesi, A., 2009. Dependence of slip weakening distance (dc) on final slip during dynamic rupture of earthquakes. Geophys. J. Int. 177 (3), 1205–1220. https://doi.org/10.1111/j.1365-246X.2009.04143.x.

Tsutsumi, A., Shimamoto, T., 1997. Temperature measurements along simulated faults during seismic fault motion. In: Hong, Y. (Ed.), Contemporary Lithospheric Motion Seismic Geology: Proceedings of the 30th International Geological Congress. VSP, Beijing, p. 223.

Ujiie, K., Tsutsumi, A., 2010. High-velocity frictional properties of clay-rich fault gouge in a megasplay fault zone, nankai subduction zone. Geophys. Res. Lett. 37, L24310. https://doi.org/10.1029/2010GL046002.

Van Diggelen, E.W.E., De Bresser, J.H.P., Peach, C.J., Spiers, C.J., 2010. High shear strain behaviour of synthetic muscovite fault gouges under hydrothermal conditions. J. Struct. Geol. 32 (11), 1685–1700.

Verberne, B.A., He, C., Spiers, C.J., 2010. Frictional properties of sedimentary rocks and natural fault gouge from the Longmen Shan Fault Zone, Sichuan, China. Bull. Seismol. Soc. Am. 100 (5B), 2767–2790. https://doi.org/10.1785/0120090287.

Verberne, B.A., de Bresser, J.H.P., Niemeijer, A.R., Spiers, C.J., de Winter, D.A.M., Plümper, O., 2013. Nanocrystalline slip zones in calcite fault gouge show intense crystallographic preferred orientation: crystal plasticity at sub-seismic slip rates at 18–150°C. Geology 41 (8), 863–866. https://doi.org/10.1130/G34279.1.

Verberne, B.A., Niemeijer, A.R., De Bresser, J.H.P., Spiers, C.J., 2015. Mechanical behavior and microstructure of simulated calcite fault gouge sheared at 20-600°C: implications for natural faults in limestones. J. Geophys. Res. 12, 8169–8196.

Weeks, J.D., 1993. Constitutive laws for high-velocity frictional sliding and their influence on stress drop during unstable slip. J. Geophys. Res. B10, 17637–17648.

Weijermars, R., Schmeling, H., 1986. Scaling of Newtonian and non-Newtonian fluid dynamics without inertia for quantitative modelling of rock flow due to gravity (including the concept of rheological similarity). Phys. Earth Planet. In. 43, 316–330.

Wibberley, C.A.J., Shimamoto, T., 2005. Earthquake slip weakening and asperities explained by thermal pressurization. Nature 436 (7051), 689–692. https://doi.org/10.1038/nature03901.

Wibberley, C., 2002. Hydraulic diffusivity of fault gouge zones and implications for thermal pressurization during seismic slip. Earth Planets Space 54 (11), 1153–1171. https://doi.org/10.1186/BF03353317.

Willingshofer, E., Sokoutis, D., Beekman, F., Schönebeck, J.-M., Warsitzka, M., Rosenau, M., 2018. Ring Shear Test Data of Feldspar Sand and Quartz Sand Used in the Tectonic Laboratory (TecLab) at Utrecht University for Experimental Earth Science Applications, vol. 1. GFZ Data Services. https://doi.org/10.5880/fidgeo.2018.072.

Xia, K., Rosakis, A.J., Kanamori, H., 2004. Laboratory earthquakes; the sub-Rayleigh-to-supershear rupture transition. Science 303, 1859–1861. https://doi.org/10.1126/science.1094022.

Xu, S., Ben-Zion, Y., Ampuero, J.-P., Lyakhovsky, V., 2015. Dynamic ruptures on a frictional interface with off-fault brittle damage: feedback mechanisms and effects on slip and near-fault motion. Pure Appl. Geophys. 172, 1243–1267. https://doi.org/10.1007/s00024-014-0923-7.

Yao, L., Ma, S., Niemeijer, A.R., Shimamoto, T., Platt, J.D., 2016a. Is frictional heating needed to cause dramatic weakening of nanoparticle gouge during seismic slip? insights from friction experiments with variable thermal evolutions. Geophys. Res. Lett. 43 (13), 6852–6860. https://doi.org/10.1002/2016GL069053.

Yao, L., Ma, S., Platt, J.D., Niemeijer, A.R., Shimamoto, T., 2016b. The crucial role of temperature in high-velocity weakening of faults: experiments on gouge using host blocks with different thermal conductivities. Geology 44 (1), 63–66. https://doi.org/10.1130/G37310.1.

Chapter 6

The growth of faults

Andrew Nicol[a], John Walsh[b], Conrad Childs[b], Tom Manzocchi[b]
[a]*Department of Geological Sciences, University of Canterbury, Christchurch, New Zealand;*
[b]*Fault Analysis Group, School of Geological Sciences, UCD, Dublin, Ireland*

Chapter outline

6.1 Introduction

Fault growth is achieved by increases of both surface dimensions and displacement (e.g., Elliott, 1976; Watterson, 1986; Muraoka and Kamata, 1983; Walsh and Watterson, 1987, 1988; Cowie and Scholz, 1992a; Dawers et al., 1993; Walsh et al., 2002; Schultz et al., 2008; Nicol et al., 2010, 2016; Rotevatn et al., 2019) (Fig. 6.1). These changes accrue in incremental slip events that can be accompanied by fault linkage and slip localisation for both individual faults and fault systems (Cartwright et al., 1995; Meyer et al., 2002; Walsh et al., 2002, 2003a; Childs et al., 2009; Fossen and Rotevatn, 2016). In the 1980s and 1990s the majority of studies used mainly 2D map-view data in combination with finite fault dimensions and displacements to constrain fault

Understanding Faults. https://doi.org/10.1016/B978-0-12-815985-9.00006-0

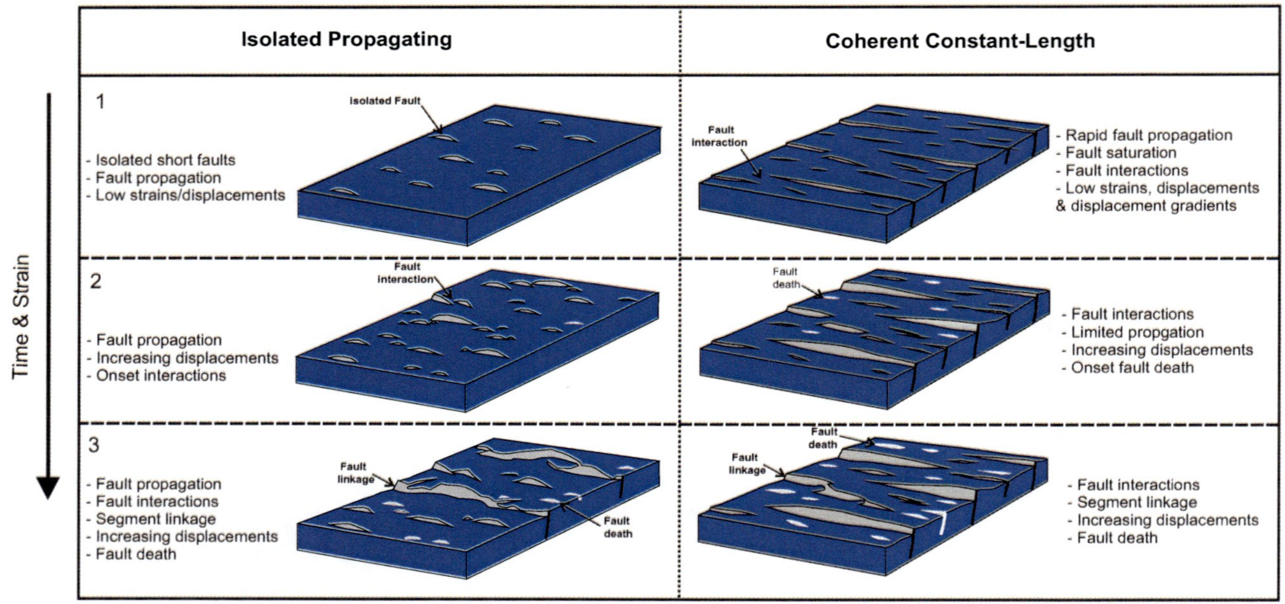

FIG. 6.1 Schematic block diagrams showing the evolution of normal fault systems in the 'isolated propagating' (left) and 'coherent constant-length' (right) growth models. Isolated propagating model modified from Cowie (1998) and the coherent constant-length model adapted from Walsh et al. (2002). Text adjacent to each block diagram describes the salient components of each stage of fault evolution.

growth. A positive correlation between fault length and maximum displacement (see Fig. 6.2 for definitions) was interpreted to indicate that both increase progressively during growth (e.g., Elliott, 1976; Watterson, 1986; Gillespie et al., 1992; Cowie and Scholz, 1992a,b; Cartwright et al., 1995; Peacock and Sanderson, 1996) (Fig. 6.1 left). For such a model increase in fault length is achieved by propagation of individual faults (Walsh and Watterson, 1987, 1988), with or without the linkage of initially isolated faults (Cartwright et al., 1995; Ellis and Dunlap, 1988; Cowie et al., 2000; Bergen and Shaw, 2010). More recently, the increased availability of 3D seismic reflection datasets

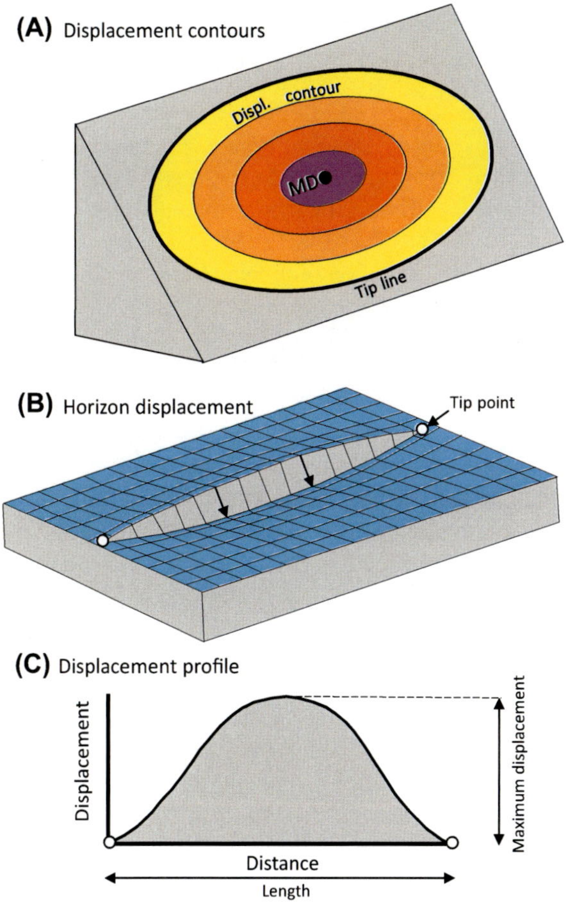

FIG. 6.2 Schematic diagram of the ideal isolated normal fault showing (A) displacement contours on the fault surface, (B) oblique view of horizon displacement and, (C) strike-parallel displacement profile. MD in (A) refers to maximum displacement, *arrows* on the fault plane in (B) indicate fault slip vectors, and open circles in (B) and (C) show the locations of fault tips. Measures of fault maximum displacement and length plotted in Fig. 6.14 are illustrated in (C).

permits the use of growth strata to chart the evolution of fault lengths and displacements throughout the history of faulting (Petersen et al., 1992; Morley, 1999, 2016; Walsh et al., 2002; Childs et al., 2003; Paton, 2006; Tvedt et al., 2013, 2016; Rotevatn et al., 2019). These studies, however, provide few examples of propagating faults and instead support a growth model in which fault displacements accrue progressively for the duration of faulting and dimensions are achieved early in the history of faulting (Morley, 1999; Poulimenos, 2000; Meyer et al., 2002; Walsh et al., 2002; Nicol et al., 2005, 2010, 2016; Manzocchi et al., 2006; Schlagenhauf et al., 2008; Jackson and Rotevatn, 2013; Whipp et al., 2016; Finch and Gawthorpe, 2017; Rotevatn et al., 2019) (Fig. 6.1 right). Fault interactions are generally inferred to prevent, or at least retard, fault propagation, with interaction between faults from an early stage of faulting being a key element of this growth model (Fig. 6.1 right).

Fault growth models proposed in the literature differ primarily in terms of two basic criteria that define the extent to which individual faults, or fault segments, were in kinematic 'isolation' from one another in their early stages of growth and to their 'propagation' histories (see discussion in Childs et al., 2017). The fault isolation criterion defines a continuum from faults that remain isolated throughout their growth to faults that were interacting and part of a coherent array from the onset of faulting. Few believe that faults remain isolated for the duration of their growth, with isolated fault models generally requiring initial isolation followed by later interactions: differences in fault growth are therefore defined along an isolation-interaction spectrum, which we discuss in the next section. The propagation criterion also defines a continuum, the limits of which are characterised by models where the final length and maximum finite displacement are only achieved at the cessation of faulting (Fig. 6.1 left) and, at the other extreme, where the final (maximum) length is reached geologically instantaneously at the start of faulting and displacements accrue throughout deformation (Fig. 6.1 right). As we will see later in this chapter, these phenomena are often related because the degree to which faults interact generally has a first-order impact on their propagation. Determining where faults are on the isolation/interaction and propagation spectrums can be challenging with the currently available datasets. It can, for example, be very difficult, if not impossible, to discriminate between faults that interacted from their initiation and those that did so from an early stage in their faulting history, because the required time-series information from growth strata are not available or poorly resolved.

Here we briefly present geometric and kinematic constraints on fault growth, in an outline that partly reflects the progression of understanding over the past five decades. This outline will provide the essential backdrop to a more detailed consideration of existing growth models for tectonic faults followed by their reconciliation with constraints derived from earthquake studies. As we progress through our description, it will become clear that faults range in dimensions from millimetres to 100s of kilometres, often with fractal scaling properties, and are components of complex systems (Walsh et al., 1991; Marrett and Allmendinger, 1992; Schultz et al., 2010; Torabi and Berg, 2011). In these circumstances the distinction between individual faults, fault zones (discrete zones in which displacement is distributed onto two or more slip-surfaces) and fault systems can be subjective and dependent on the scale of observation, which is ultimately controlled by the methods of data acquisition. Despite the limitations imposed by the scale of observation, the nucleation, propagation, interaction and linkage of individual faults can be interpreted with reference to the isolated/interaction and propagation criteria. Since the most complete datasets constraining fault growth are available for normal faults (Childs et al., 2017), they are the main focus of this chapter, however, data for reverse and strike slip faults suggest that the interpreted patterns of growth may be independent of fault-slip type (Elliott, 1976; Ellis and Dunlap, 1988; Peacock, 1991; Bergen and Shaw, 2010; Nicol et al., 2016).

6.2 Geometric indicators of fault growth

Finite fault displacement patterns at the cessation of faulting have been used to constrain the factors that control fault growth and the models that best account for fault evolution. These displacement patterns are typically represented as contours on the fault surface, horizon separation diagrams and displacement profiles (Fig. 6.2). An important underpinning principle of fault growth studies is that displacements can include discrete (fault displacement) and ductile or continuous (fault-related folding and bed rotation) components (Fig. 6.3). Continuous deformation for normal and reverse faults is interpreted to primarily occur as folds related to fault-tip propagation, taking the form of normal drag and as bed rotations within relay zones that separate adjacent fault

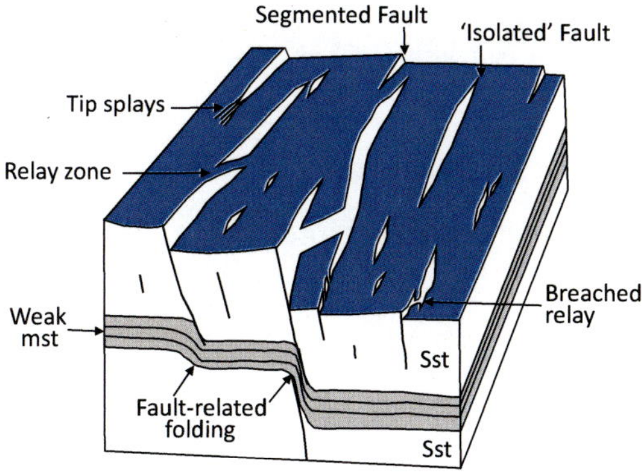

FIG. 6.3 Schematic block diagram showing the main geometric characteristics of normal faults in a system. The descriptions and terminology are also applicable to strike-slip and thrust faults. *Diagram modified from Schultz et al. (2010).*

segments observed in maps or cross sections (Fig. 6.3; Dahlstrom, 1970; Williams and Chapman, 1983; Walsh and Watterson, 1987; Peacock, 1991; McConnell et al., 1997; Nicol et al., 2002; Bergen and Shaw, 2010; Khalil and McClay, 2016; Lăpădat et al., 2016; Childs et al., 2017; Conneally et al., 2017). These studies, and many others, highlight the temporal and spatial complementarity of fault-related folding and fault displacements, indicating that they are related components of the same structure and must be considered together when using displacements to draw conclusions about fault growth.

6.2.1 Conceptual 'ideal isolated fault' model

Variations in fault displacements over normal and reverse fault surfaces have been documented from seismic reflection and coal-mine datasets (Rippon et al., 1985; Barnett et al., 1987; Gillespie, 1991; Nicol et al., 1996; Torabi et al., 2019). These data help define displacements on the "ideal isolated" fault which has elliptical displacement contours parallel to the tip-line loop and concentric about a centrally located maximum (Figs 6.2A and 6.4A and B; Barnett et al., 1987; Walsh and Watterson, 1987). As is the case for the shapes of fault surfaces, the geometries and spacings of displacement contours (i.e., displacement gradients) typically depart from the ideal for a range of reasons which are considered below. The displacement contour patterns for the isolated fault and more complex fault surfaces shown in Fig. 6.4 highlight, for example, that the maximum displacements are often not located centrally on the fault surface and, as a consequence, displacement gradients can be variable. Furthermore, faults that terminate along their branch-line (i.e., intersection line) with other faults (i.e. those that are physically linked, or hard-linked, to others) can have displacement contours that are either locally sub-parallel to the branch-line along which the throw is low or zero (Fig. 6.4C), or trend at a high angle to the branch line along which the maximum throw is located (Fig. 6.4D; Yielding, 2016). Despite these caveats, the concept of an 'ideal isolated' fault and its characteristics are valuable for understanding fault geometry, processes and evolution (Barnett et al., 1987; Walsh and Watterson, 1987).

6.2.2 Mechanical layering and displacement variations

In heterogenous multi-layered rocks the ideal isolated normal fault has an approximately elliptical tip-line with a sub-horizontal, bedding parallel, major axis and an average aspect ratio of about two (e.g., Barnett et al., 1987; Nicol et al., 1996) (Figs 6.2A and 6.4A). Few faults documented in the literature display all of the properties of the ideal fault (e.g., see Fig. 6.4), as the shape of fault surfaces is influenced by a number of factors, including the type of fault (normal, reverse or strike slip), host rock mechanical properties, geometries of pre-existing faults in the faulted sequence or underlying basement, and

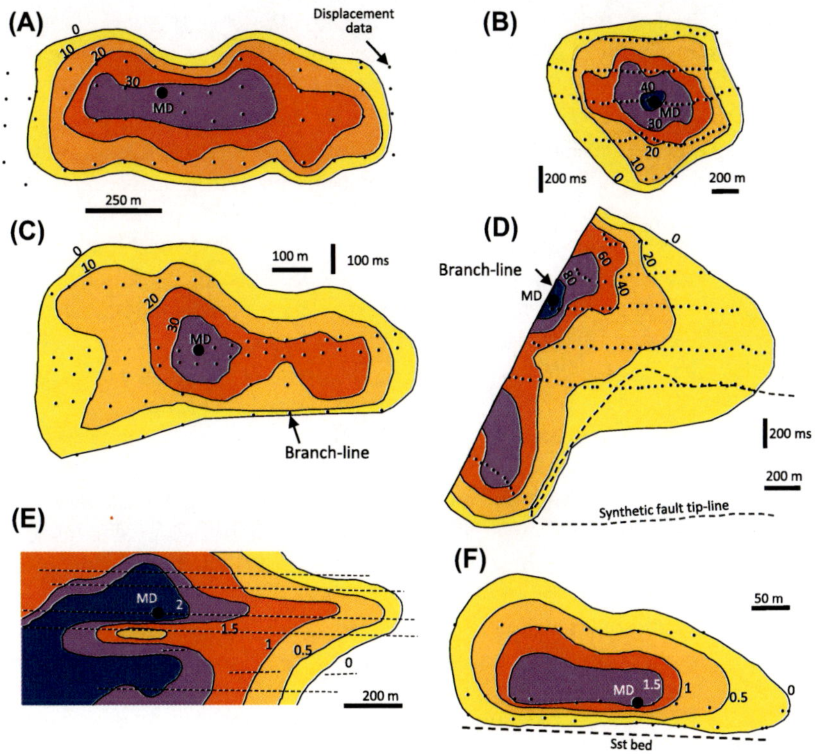

FIG. 6.4 Selection of fault surface displacement diagrams for hard- and soft-linked normal faults: hard-linked faults are physically linked with another fault, whereas soft-linked faults transfer displacement without physical linkage at the scale of observation. (A) Displacement contour diagram for a fault constructed using 3D seismic data from the North Sea (modified from Fig. 3 of Barnett et al., 1987). Contours approximate those of the ideal isolated fault. (B) Displacement contour diagram of an equant or penny shaped fault surface from the Gulf of Mexico (Fig. 1 of Nicol et al., 1996). (C) Displacement contour diagram for a fault, imaged in a 3D seismic reflection survey in the Timor Sea. The contoured fault abuts a contemporaneous antithetic fault along a sub-horizontal branch-line (Fig. 10 in Nicol et al., 1996). (D) Displacement contour diagram for a fault abutting a contemporaneous antithetic fault along a sub-vertical branch-line, with a second fault in the hangingwall of the contoured fault surface (tip line indicated by dashed line). Fault from a 3D seismic reflection dataset in the Timor Sea (Fig. 8 in Nicol et al., 1996). (E) Partial displacement contour diagram from the Derbyshire coalfields constructed from coal-seam throws. Displacement low coincides with weak mudstones and seatearth layers (Fig. 6 in Rippon, 1985). (F) Displacement contour diagram from the Derbyshire coalfields constructed from coal-seam throws (Fig. 3 in Rippon, 1985). Propagation of the lower tip-line of this fault is interpreted by Rippon to have been retarded by a thick sandstone bed. *Small black filled circles* on the contour diagrams show the locations of displacement measurements. *Large black filled circles labelled MD* show the location of the maximum displacement.

interactions with nearby faults (e.g., Walsh and Watterson, 1987; Nicol et al., 1996, 2016; Peacock and Sanderson, 1996; Peacock, 2002; Schultz et al., 2008; Giba et al., 2012; Whipp et al., 2014; Ghalayini et al., 2016) (Fig. 6.4). Data constraining the shapes of faults in homogeneous isotropic rocks are rare. Crystal dislocation theory suggests the tip-line loops of glide planes should be elongate parallel to the Burgers vector or slip direction (Hull and Bacon, 1984). If the same is also true of normal and reverse faults, we would expect these surfaces to be elliptical with a horizontal to dip dimension aspect ratio of 0.5. However, few rocks are isotropic and the disparity between theoretical prediction and normal fault surfaces, elongate perpendicular to the slip vector, is attributed to the influence of bedding anistropy on fault propagation (Nicol et al., 1996; Ghalayini et al., 2016). The influence of anistropy on fault shape is most readily demonstrated in sedimentary sequences where the layers are thick compared to the fault dimensions or the contrast in layer mechanical properties is high (e.g., Muraoka and Kamata, 1983; Peacock and Sanderson, 1992; Nicol et al., 1996; Manighetti et al., 2001; Wilkins and Gross, 2002; Schöpfer et al., 2006; Schultz et al., 2008; Ghalayini et al., 2016; Roche et al., 2016). The influence of mechanical layering on displacement patterns is usually reflected in a decrease in fault displacement either within or towards weaker layers. For example, displacement contour patterns on faults from the British Coalfields highlight the presence of rapid displacement changes into coal seams immediately underlain by relatively thick and weak clay-rich seatearths (Fig. 6.4E). These seams are characterised by an increased amount of folding and normal drag, with an associated decrease in the component of discontinuous fault displacement (which is what was measured by mine surveyors). In some cases upper and lower parts of tip-lines can be located at bed boundaries or within adjacent weak beds (e.g., Muraoka and Kamata, 1983; Rippon et al., 1985; Bürgmann et al., 1994; Nicol et al., 1996, 2016; Wilkins and Gross, 2002; Schöpfer et al., 2006), with the fault surface often having an approximately rectangular shape, a large aspect ratio (e.g., >10) and a major axis parallel to bedding (Nicol et al., 1996; Schultz et al., 2008). In the case of Fig. 6.4F, for example, the relatively straight lower tip line of the fault is interpreted to terminate within weak shales immediately above a thick strong sandstone bed. In other cases, displacement variations suggest that individual faults were segmented along bedding planes, at least in the earliest stages of localisation, with fault displacements in strong beds giving way to fault-related folding (i.e., normal drag) in weak layers, but with the segments collectively representing a single kinematic structure. Such coherence is illustrated in Fig. 6.5, which shows that the aggregate displacement profile for all segments is less variable than those for the individual segments. The available observations indicate that the rheology and geometry of layering in the faulted rock volume can influence the geometry and growth of faults over a wide range of scales (Muraoka and Kamata, 1983; Chapman and Williams, 1984; Eisenstadt and DePaor, 1987; Ellis and Dunlap, 1988; Peacock, 2002, 1991; Peacock and Sanderson, 1992; Nicol et al., 1996; Schultz and Fossen, 2002; Wilkins and Gross, 2002). Mechanical layering is a key determinant of propagation-related segmentation and associated displacement partitioning along profiles parallel to the slip direction of normal and reverse faults. Therefore, investigations of fault propagation and

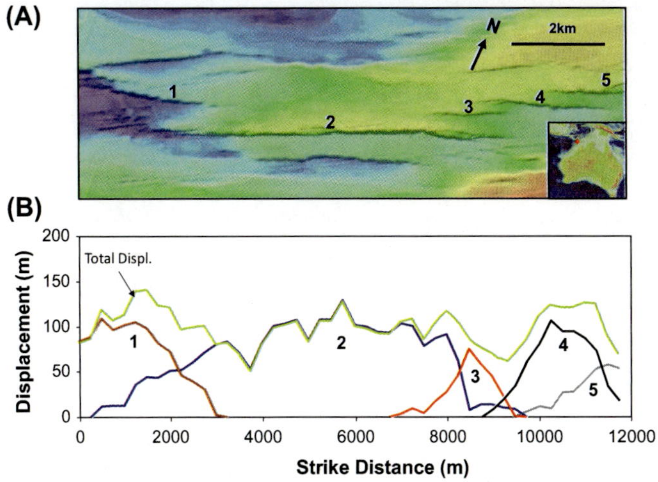

FIG. 6.5 (A) Map of segmented array of normal faults from the Timor Sea (see Meyer et al., 2002). Colours represent two-way travel time (TWTT) to the base of growth strata; cold colours indicate the largest TWTT. (B) Displacement profiles for each fault in the array. Numbers on the map and profiles indicate faults for which displacement profiles are presented in (B). *Green line* in (B) is the total displacement profile aggregated for all segments. Inset map shows location of the map.

interaction using sub-horizontal lateral displacement profiles along individual layers are less susceptible to complications arising from changes in mechanical properties.

6.2.3 'Isolated' fault lateral displacement profiles

Given the paucity of 3D displacement data for many fault surfaces, interpretation of fault displacements and growth models is generally conducted using the shapes of finite displacement profiles (e.g., Walsh and Watterson, 1987; Peacock, 1991; McConnell et al., 1997; Manighetti et al., 2001; Davis et al., 2005) (see Figs. 6.2C and 6.6). The detailed shapes of these profiles have the potential to constrain whether faults lengthen as displacement increases and the rates of fault propagation (Williams and Chapman, 1983; Watterson, 1986; Peacock and Sanderson, 1996; Manighetti et al., 2001; Childs et al., 2003; Manzocchi et al., 2006). In the absence of strong fault interactions, finite displacement profiles are often approximately triangular (also referred to as C-type profiles, Fig. 6.6 top) with varying degrees of asymmetry and near-constant displacement gradients from the maximum displacement to fault tip (e.g., Walsh and Watterson, 1987; Nicol et al., 1996, 2016; Cowie and Shipton, 1998; Manighetti et al., 2001, 2004; Manzocchi et al., 2006; Nixon et al., 2014) (Fig. 6.6 top). The triangular-shaped finite profiles are most often observed for strike-slip and normal faults and indicate that many faults have approximately constant displacement gradients (Fig. 6.6 top). Near-triangular displacement profiles do not display the low displacement tails

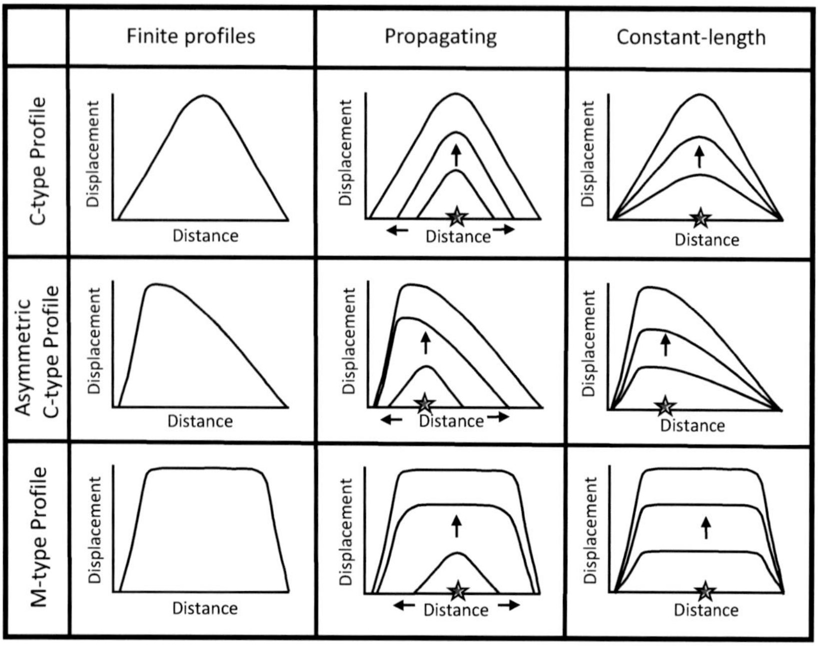

FIG. 6.6 Normalized displacement profiles for normal faults with length/maximum length (L/LMax) along the x axis versus displacement/maximum displacement (D/Dmax) on the y axis. Finite profile shapes are indicated in the left column and include; C-type (*triangular*), asymmetric C-type and M-type finite profile shapes are shown on the left side of the diagram. Profiles on the right side of the diagram show profile growth for propagating (centre right) and constant-length (far right) fault models. *Red star* shows the nucleation point of the fault which is assumed to be in the plane of observation. *Concept of the diagram from Nixon et al. (2014).*

predicted for faults by post-yield fracture mechanics (Cowie and Scholz, 1992b) or for rapidly-propagating faults (Peacock and Sanderson, 1996). Similarly, the observed triangular profiles do not replicate the semi-elliptical incremental profiles modelled for linear elastic materials (e.g., Pollard and Segall, 1987). Departure from these finite semi-elliptical displacement profiles and the production of near-linear profiles has been interpreted to arise due to off-fault deformation (see Manighetti et al., 2004). Stochastic models suggest that near-triangular finite displacement profiles can also be produced using triangular, circular or rectangular incremental slip profiles when these slip events follow a Gutenburg-Richter size distribution and the fault tips do not propagate as displacement accrues (Manzocchi et al., 2006).

6.2.4 Interaction and lateral displacement profiles

Fault interactions can provide a primary control on the shapes of displacement profiles and elevated displacement gradients are inferred to indicate interactions between faults (e.g., Peacock, 1991; Walsh and Watterson, 1991; Nicol et al., 1996; Manighetti et al., 2001; Nixon et al., 2014) (Fig. 6.6), and associated soft-linkage accommodated by deformation of the intervening host rock (e.g. see relay zones below). Elevated displacement gradients may arise because fault interaction retards the rate of fault propagation without impacting the rate of maximum displacement increase (Peacock and Sanderson, 1996). This retardation is generally attributed to the interference of elastic strain fields (e.g., Willemse, 1997), which can take place well before the surfaces intersect within the plane of inspection. Extreme forms of fault interactions are inferred from flat topped or highly asymmetric profiles displaying high tip displacement gradients which, in many cases, can be demonstrated to reflect the transfer of displacement to near-by faults (e.g., Fig. 6.6 middle and lower profiles). By contrast, faults with symmetrical profiles (Fig. 6.6 upper profiles) tend to have low displacement gradients and be removed from other faults of similar size in the system. In such cases, fault interactions are difficult to unequivocally demonstrate, in part because the faults are small with subtle changes in displacement and low displacement gradients (typically <0.01 for faults that are soft-linked). Despite these sampling issues, we argue that the observations are consistent with the notion that all faults interact and the degree of interaction is positively related to the lateral displacement gradients (e.g., Walsh and Watterson, 1991; Nicol et al., 1996, 2010; Nixon et al., 2014).

6.2.5 Relay zones and lateral interactions

The best documented examples of fault interaction are those at lateral overlaps, or relay zones, between en échelon pairs of synthetic normal faults (Figs 6.5, 6.7–6.9; Morley et al., 1990; Trudgill and Cartwright, 1994; Childs et al., 1995; Fossen and Rotevatn, 2016). In these circumstances, interaction and displacement transfer between fault segments is marked by high displacement gradients (up to 0.25) associated with bed rotation in relay ramps (Fig. 6.7). Faults can generally be demonstrated to interact when the ratio of the distance between faults to fault displacement is less than c. 10 (Rippon et al., 1985; Barnett et al., 1987; Nicol et al., 1996; Childs et al., 2003). At higher ratios, the associated displacement gradients become difficult to distinguish from those of isolated faults. Relay zones arise from the interaction of either fault segments that were related from their inception or from the interaction of originally isolated faults (Fig. 6.8). The precise shapes of displacement profiles can be used to distinguish between these end-member models (Fig. 6.9). Fault segments that were kinematically related from their initiation have complementary finite displacement distributions, which produce a smooth and regular aggregate displacement distribution that resembles that of a single isolated fault (Fig. 6.7B, C). Arrays of faults with complementary displacements that sum to produce a profile comparable to that

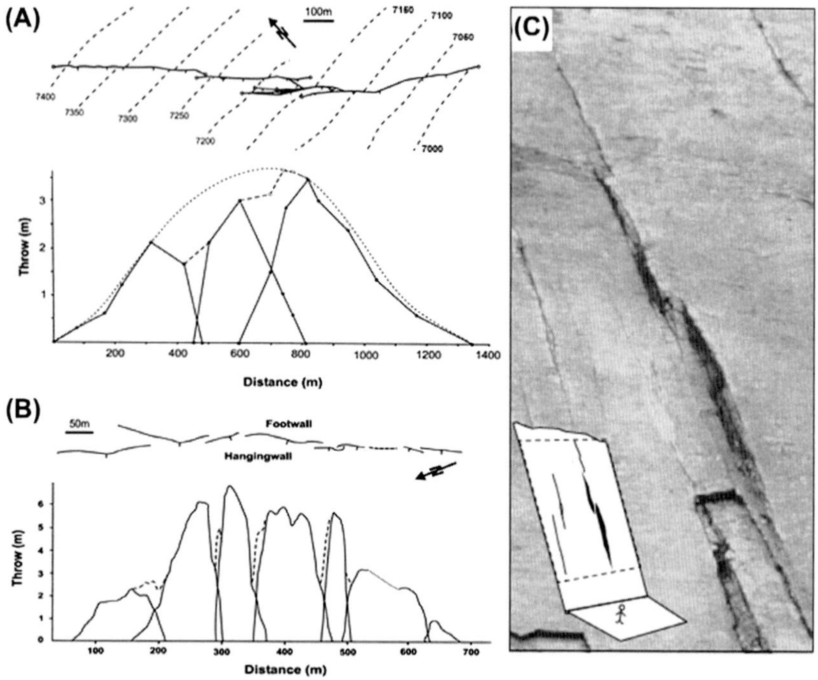

FIG. 6.7 (A) Map (top) and throw profiles (bottom) of a segmented normal fault recorded on coal seam mine plans from Lancashire, UK (reproduced from Walsh and Watterson, 1990). The map shows coal seam height contours (*dashed*) and fault trace tip locations (*open circles*). Displacement profiles show throws for individual segments and summed displacements (*dashed line*) for all three segments. Summed profile is comparable to the profile for an idealized profile (*dotted line*) and is interpreted to have formed as one kinematically coherent structure. (B) Segmented fault array in the Bishop Tuff, California, showing a systematic distribution of throw along seven fault segments (*solid lines*) (reproduced from Willemse, 1997). The aggregate throw profile, shown as dashed lines, does not include the strain accommodated by rotation of bedding within relay zones. (C) Photograph of fault segments separated by relay ramps on a bedding surface dipping at ~70 degrees at Jaca, Pyrenees, Spain (modified from Fig. 1 of Walsh et al., 2003a). The relationships between the bed and fault are comparable to those of a normal fault displacing horizontal beds. Inset sketch shows person for scale.

of an ideal single fault are referred to as geometrically coherent (Walsh and Watterson, 1991). For previously isolated faults the aggregate displacement distribution will display local maxima on each fault segment separated by displacement lows at the intervening relay zones (Fig. 6.9B), which are not compensated by increases in continuous displacements (e.g., folding or bed rotations). In most cases, however, displacement lows are apparent, and simply reflect the local accommodation of displacements by bed rotations and small-scale faulting within relay zones. The preservation of displacement along arrays of fault segments indicates that they were coherent and formed interacting structures from the outset of faulting; a scenario that may apply in

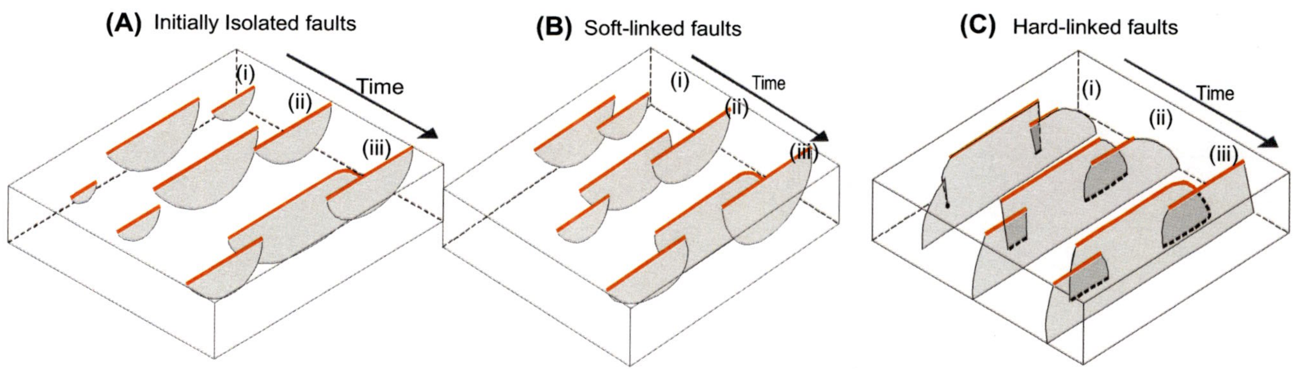

FIG. 6.8 Schematic illustrations of three end-member models for the formation of segmented fault arrays: (A) isolated faults, (B) soft-linked fault segments arising from 3D segmentation or (C) hard-linked segments due to fault surface bifurcation. Each block diagram shows three stages in the growth of a segmented fault array (i–iii). Initially isolated faults (A) require early fault propagation and independence followed by coherent growth, while the soft-linked (B) and hard-linked (C) diagrams both require coherent fault growth from the onset of faulting. *Redrawn from Fig. 2a–d of Walsh et al. (2003a).*

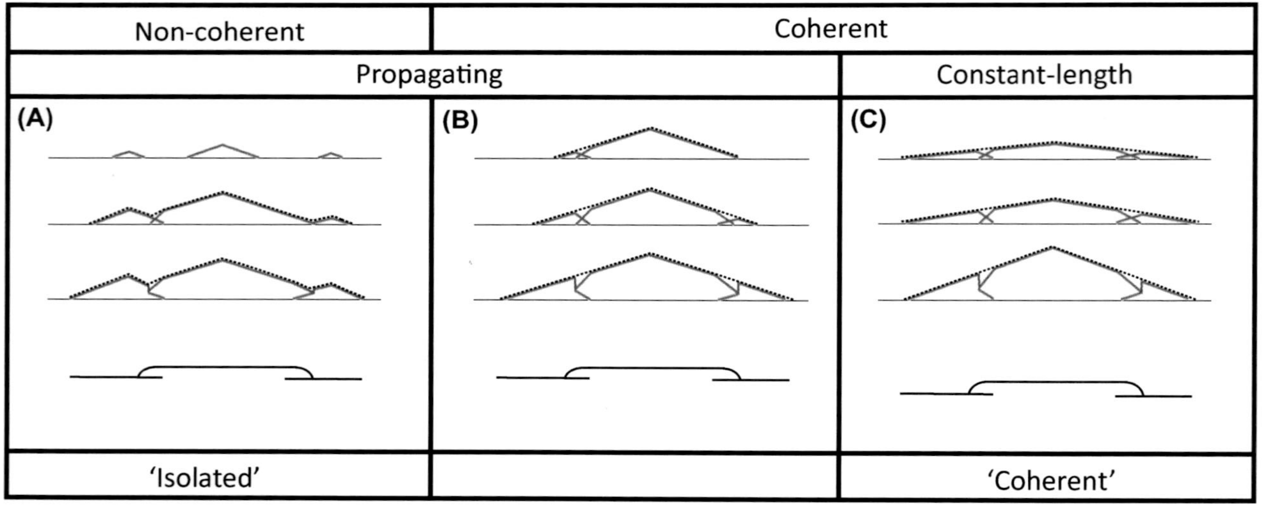

FIG. 6.9 Displacement-length profiles illustrating three potential growth histories leading to the same segmented fault trace pattern (shown in the lower part of the figure). The *solid blue lines* are the displacement profiles for individual faults and the *dashed lines* are profiles of aggregate displacement for all segments. The three cases are distinguished on the basis of whether the faults are coherent and whether the faults propagate continuously as displacement increases. The non-coherent propagating case is here referred to as the isolated propagating model and the constant-length coherent case is here referred to as the coherent constant-length model. The propagating coherent case is also possible, but is less commonly invoked. The profiles shown are for normal faults in map view, but would be equally applicable for strike slip and reverse faults in map view or cross-section. *Diagram from Childs et al. (2017).*

the vast majority of cases (Walsh and Watterson, 1991; Morley, 2002; Walsh et al., 2003a; Giba et al., 2012). Our perception is that the appeal of the isolated model for relay formation often reflects the intrinsically 2D considerations of most studies, which usually assume that propagation is within the plane of inspection, whether it is perpendicular or parallel to the slip direction (i.e., map or cross-section) (Walsh et al., 2003a).

6.2.6 Damage zones and lateral growth

Fault branch-lines and tip lines are often characterised by elevated densities of small-scale faulting and fracturing within fault zones or fault damage zones, and can provide indirect information about fault propagation (McGrath and Davison, 1995; Kim et al., 2000; Perrin et al., 2015; Nicol et al., 2016). Structures, such as fault-tip splays (Fig. 6.3), horsetail faults, wing cracks or tail cracks (Chapter 2), occur where slip on a primary fault surface locally produces stress concentrations at fault tips which are relieved by distributed strains in the adjacent rock volume (Chinnery, 1996). In cases where faults propagate progressively during growth, we would expect to see secondary faults formed at the locations of previous tips along the entire length of faults (Fig. 6.10 left). For many faults however, such secondary faulting is restricted to the region surrounding the present fault tip (McGrath and Davison, 1995; Kim et al., 2000; Perrin et al., 2015; Nicol et al., 2016) (Fig. 6.10 right). The observed fault-tip geometries can be explained if faults initially propagate rapidly with secondary tip structures only forming when the fault reaches its final length (Fig. 6.10 right). Secondary faulting restricted to the fault-tip zone have been replicated by numerical models in which tip stresses and strains are localised by a pre-existing plane of weakness that does not propagate (e.g., Pollard and Segall, 1987; Bürgmann et al., 1994). Therefore, both outcrop and models do not indicate progressive tip propagation and are suggestive of the coherent constant-length model, although in the absence of growth strata neither the constant length nor propagating faults can be unequivocally demonstrated from inspection of damage zones.

6.3 Direct kinematic indicators of fault growth

The most direct geological evidence for fault growth is provided by displacement variations derived from seismic reflection surveys that image growth strata deposited as faults accrued displacement (Fig. 6.11A). In circumstances where sedimentation rates exceed fault displacement rates, faults are buried and aspects of their growth can be recovered from analysis of growth strata using either, expansion index, growth index or displacement backstripping techniques (Petersen et al., 1992; Childs et al., 1993, 2003). These methods provide a means of defining the timing of faulting,

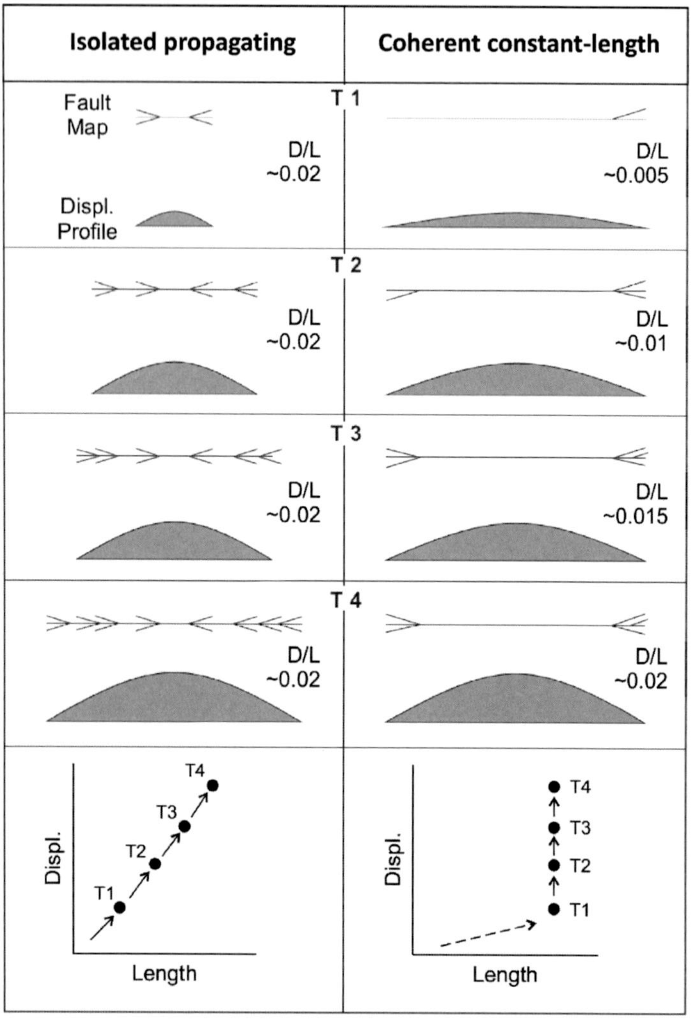

FIG. 6.10 Schematic diagram illustrating two potential fault-growth models. Diagram shows fault length, tip-zone secondary faulting (fault-trace maps) and along-strike displacement profiles at four times (T1–T4) for isolated propagating (left) and coherent constant-length (right) growth models. Displacement–length (D–L) ratios are indicated; they are approximately constant for the isolated propagating model and increase with displacement for the constant-length model. Evolution of displacement and length for each model is shown in the lower schematic graphs. The horizontal and vertical scales are different for both displacement profiles and displacement–length graphs. Diagram from *Nicol et al. (2016)*.

displacement rates, and fault lateral propagation. Growth strata analysis can also be used to examine the upward propagation of faults within a system dominated by the reactivation of underlying pre-existing faults, circumstances that are relatively common given the importance of structural inheritance.

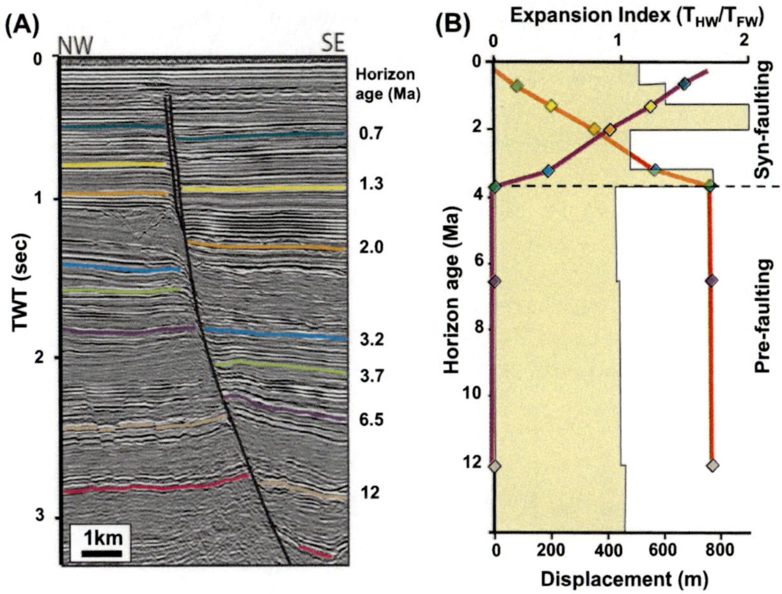

FIG. 6.11 (A) 3D seismic reflection profile showing the upper part of the Parihaka Fault from the Taranaki Basin, New Zealand. Interpreted horizons and fault trace from Giba et al. (2012). (B) Graph showing horizon age vs expansion index (top of the *yellow polygon*), present displacement (*red line*) and backstripped displacement (*purple line*) profiles. Colours of datapoints on the displacement *curves* indicate the horizon from which the displacement was measured. Near-linear displacement increases from the onset of faulting indicate that changes of expansion index within the growth sequence are most likely due to changes in sediment supply arising from eustatic sea-level fluctuations (see Thorsen, 1963).

6.3.1 Displacement through time

In circumstances where sedimentation rates outstrip fault displacement rates, fault activity has been defined from expansion index (EI = Hangingwall thickness/Footwall thickness) curves, which record across-fault thickness changes though time (Fig. 6.11). Expansion index values greater than one indicate fault displacement with the development of accommodation space and strata thickening on the downthrown side (i.e. hangingwall) of the fault. A common misconception is that expansion indices relate directly to displacement rates. This is not the case because an expansion index incorporates both fault displacement rates and sedimentation rates, so that for the same displacement rate, higher expansion index values merely reflect lower sedimentation rates. The relationships between sedimentation and displacement rates are highlighted for the Parihaka Fault in the Taranaki Basin, which shows highly variable expansion indices and approximately constant displacement rates derived from displacement backstripping (Fig. 6.11). The principal requirement of backstripping is that sedimentation rates are equal to, or greater than, fault displacement rates and that the amount of incremental slip (e.g., for individual slip events) is uniform within syn-faulting strata for a given cross-section. Under these conditions displacement that accrues during the period between the deposition of two horizons is equal to the difference in displacement between those horizons. Previous work has shown that this relatively simple approach permits definition of fault displacement increase through time, with no requirement for sediment decompaction which has a subordinate effect on the resulting displacement curves (Taylor et al., 2008). Displacement backstripping of the Parihaka Fault in Fig. 6.11, indicates approximately constant fault displacement rates, albeit with the potential for slightly elevated displacement rates in the first resolvable time interval of fault growth (Giba et al., 2012). Similar studies in other basins suggest that approximately linear displacement vs time curves are common (Fig. 6.12A),

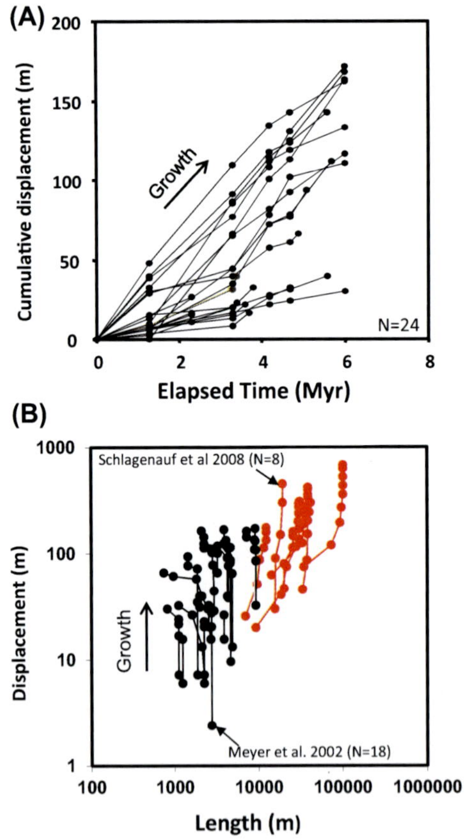

FIG. 6.12 (A) Cumulative displacement versus time plot of faults in the Timor Sea, offshore Australia (for location see Fig. 6.5). *Black lines* show the accumulation of displacement from the inception of faulting at time 0 to the cessation of faulting after ~6 Myr. (B) Displacement-length plot showing the evolution of normal faults from the Timor Sea (*black lines,* Walsh et al., 2002) and from an analogue model (*red lines*, Schlagenhauf et al., 2008). Model results have been rescaled assuming that 10 mm in the model is equal to 1.5 km in nature (see Schlagenhauf et al., 2008). The trends of the growth lines, which are independent of the precise scaling relationships for the analogue model, are steeper than the slope of the D-L finite data.

and, perhaps more importantly, that fault displacement rates are generally greater for faults with the highest finite displacements and lengths (Nicol et al., 1997; Walsh et al., 2002; Rotevatn et al., 2019; Fig. 6.12).

6.3.2 Fault lateral propagation

Displacement backstripping analysis performed along the length of a fault can be used to produce incremental displacement profiles which chart changes in fault displacements and lengths through time (Fig. 6.13). The different shapes of incremental displacement profiles in Fig. 6.13 arise because older horizons accumulated larger maximum displacements and because the fault tip locations were either stationary or migrated with time. In each of the diagrams, the coloured polygons between displacement profiles for successively older horizons represent the incremental displacement that accrued along the length of the fault; the dark green polygon representing the first or oldest increment and the dark brown the last increment (Fig. 6.13). The displacement profiles represent conceptual models (Fig. 6.13 left) and real faults (Fig. 6.13 right) with stationary tips (non-propagating fault, upper profiles), retreating tips (dying fault, middle profiles) and propagating tips (propagating fault, lower profiles). Many of the incremental profiles derived from seismic reflection data

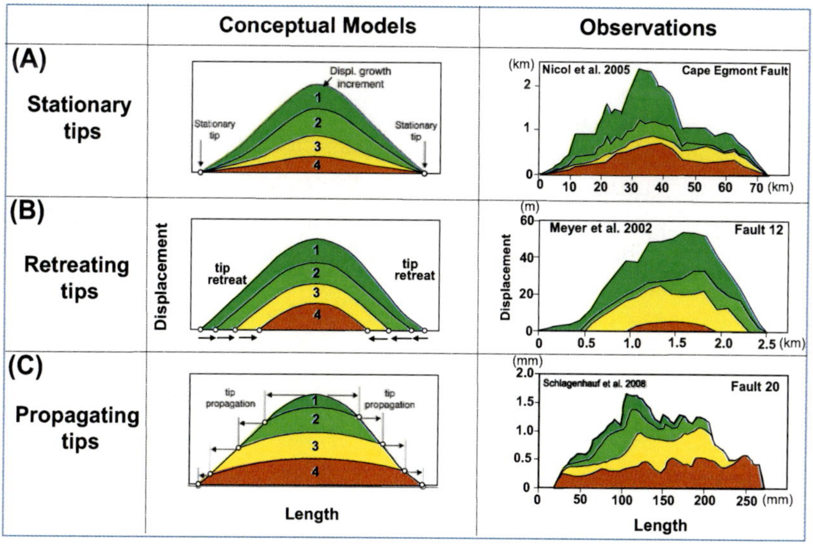

FIG. 6.13 Strike-parallel displacement profiles showing conceptual models and natural examples of fault growth. (A) constant length fault characterised by stationary tips and increasing displacement on older horizons (natural profile from Nicol et al., 2005), (B) retreating fault with decreasing fault length and displacements on younger horizons (natural profile from Meyer et al., 2002), and (C) propagating fault with increasing length and decreasing displacement on younger horizons (profile from analogue modelling of Schlagenhauf et al., 2008).

in the literature resemble the stationary tip profiles with all increments of displacement intersecting the x-axis at the same distances (e.g., Meyer et al., 2002; Childs et al., 2003; Nicol et al., 2005; Rotevatn et al., 2019), a feature we will return to later in the end-member growth models section. Propagating faults, where younger increments of displacement define longer faults than older increments of displacement, are rarely observed from seismic reflection datasets, but are encountered in the early stages of faulting in analogue models (Schlagenhauf et al., 2008). Retreating fault tips are characterized by shorter profile lengths for younger displacement increments, a change in profile geometry that is rarely reported and may be associated with fault death in some fault systems (Meyer et al., 2002).

6.3.3 Fault upward propagation and reactivation

Reactivation of pre-existing faults can influence the location, orientation and size (both length and maximum displacement) of younger faults (Giba et al., 2012; Morley, 2016; Reilly et al., 2016; Whipp et al., 2016; Withjack et al., 2017), together with the locations and geometries of structural domains (Fossen et al., 2016). In many cases reactivated faults appear to have nucleated along the upper tip line of pre-existing faults and propagated upwards to form geometrically coherent arrays of fault segments. The segments become progressively hard linked with depth, often forming a single, through-going fault immediately above the pre-existing structure. In cases where normal faults are reactivated in extension and the principal stress axes are not parallel, the new fault surfaces can rotate in strike with increasing vertical distance above the pre-existing structure, with shallower parts of the surface forming sub-normal to the new σ_1 direction (Frankowicz and McClay, 2010; Giba et al., 2012). These circumstances promote the formation of arrays of fault segments, where the degree of segmentation increases with a rise in obliquity between the two extension directions. Pre-existing crustal structure can have a profound effect on fault growth histories because reactivated faults usually propagate upwards, and may play an important role in controlling the locations and lengths of the largest faults in a fault system (Kelly et al., 1999; Paton, 2006; Jackson and Rotevatn, 2013; Whipp et al., 2014; Morley, 2016; Reilly et al., 2016). Fault reactivation causes faults to reach their final lengths rapidly and promotes the early formation of fault size hierarchy in some fault systems (Walsh et al., 2002). However, fault reactivation is not a pre-requisite for the coherent constant-length fault growth model or establishing fault hierarchy (Walsh et al., 2002; Schlagenhauf et al., 2008). Material science literature suggests that longer defects established at low strains are generally the loci for higher stresses (Gordon, 1978) and such stress localisation may help explain why the longest faults in a system move fastest from an early stage of deformation (Walsh et al., 2001).

6.4 Displacement-length relations and fault growth

A positive relationship between maximum displacement and length underpinned early studies of fault growth, which advocated for the isolated propagating model (Elliott, 1976; Watterson, 1986; Walsh and Watterson, 1988; Cowie and Scholz, 1992b). The maximum displacement (D) and fault length (L) relationship is expressed as $D = cL^n$, where the parameter c defines the vertical position of the data and is referred to as the "scaling factor" (Cowie and Scholz, 1992b) or the "characteristic shear strain" (Watterson, 1986), and the exponent n defines the slope of the distribution and describes the rate of displacement accumulation relative to the fault length. D-L data have been compiled for many fault systems worldwide with the entire dataset extending over eight orders of magnitude and exponents for individual systems ranging from 0.5 to 2.0 (e.g., Schlische et al., 1996; Fossen and Hesthammer, 1997; Bailey et al., 2005; Kim and Sanderson, 2005; Schultz et al., 2008; Nicol et al., 2010; Torabi and Berg, 2011) (Fig. 6.14, light grey symbols). For individual fault systems the data form a cloud with displacement varying by up to an order of magnitude for a given displacement (e.g., Fig. 6.14, dark grey,

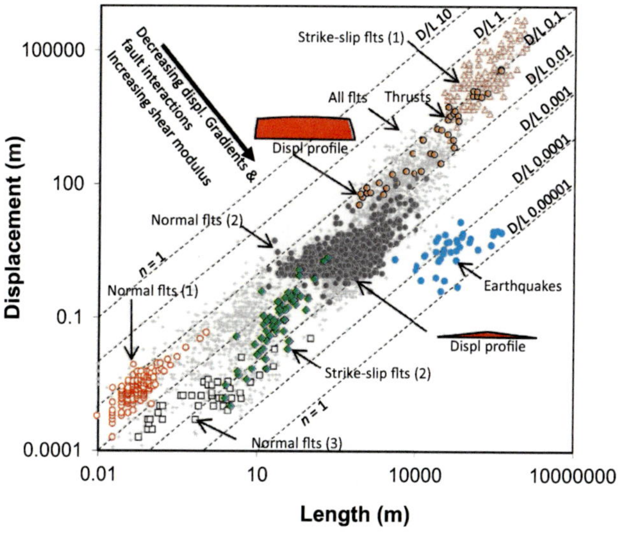

FIG. 6.14 Displacement (D)-length (L) plot showing global data for ancient fault systems (*light grey filled circles*) together with normal faults from (1) Solite Quarry (Schlische et al., 1996), (2) the East Peninnes coalfield (Bailey et al., 2005), (3) Utah (Fossen and Hesthammer, 1997), thrusts from the Canadian Rockies (Elliott, 1976), strike-slip faults from (1) oceanic transforms (MacMillian, 1975), (2) the eastern North Island, New Zealand (Nicol et al., 2016) and historical earthquakes (Wesnousky, 2008). *Dashed lines* have a slope of 1 and show equal displacement gradients. Decreases in D/L ratios are associated with decreases in fault displacement gradients and fault interactions, and with increasing shear moduli. Changes in the shapes of displacement profiles for different displacement gradients are indicated by the *red polygons* which represent schematic displacement profiles.

green and red symbols). The spread of the data cloud is even greater when all fault systems are plotted together (Fig. 6.14, light grey symbols) and changes in the D:L ratio may reflect a combination of natural differences between individual faults and fault systems together with uncertainties in the fault lengths and maximum displacements arising from incomplete sampling (Gillespie et al., 1992; Watterson et al., 1996; Bailey et al., 2005; Kim and Sanderson, 2005; Manzocchi et al., 2009; Nicol et al., 2010). The value of the exponent n can vary depending on the technique used to produce the line of best fit through the data, with the Ordinary Least Squares technique routinely producing a lower n than the techniques using regressions on both x- and y-axes (e.g., Reduced Major Axis technique; Smith, 2009) which may be most appropriate for D-L data (Walsh and Watterson, 1988). Irrespective of the precise value of n, the positive correlation between D and L has been widely interpreted to represent a growth trend with progressive increases in both displacement and length (Watterson, 1986; Walsh and Watterson, 1988; Cowie and Scholz, 1992b).

The positive relationship between D and L is generally interpreted to provide evidence for the isolated propagating growth model, however, alternative interpretations are possible. In recent years it has been proposed that faults mainly grow in the D-L distribution along steeply-inclined paths rather than evolving along the data trend (Walsh et al., 2002; Nicol et al., 2010; Rotevatn et al., 2019). Such steep or vertical growth trends require that, after initial rapid propagation, which typically occurs on timescales shorter than the resolution of natural growth strata, faults do not grow significantly in length (Walsh et al., 2002; Schlagenhauf et al., 2008; Rotevatn et al., 2019). Such growth paths are depicted by the data from a natural normal fault system (black lines) and for an analogue model (red lines) in Fig. 6.12B. These growth trends are consistent with the coherent constant-length model of Walsh et al. (2002) (Fig. 6.1 right).

In circumstances where individual faults grow along trends steeper than an exponent of 1, they could do so with, or without, a change in the value of the exponent of the system as a whole (Schultz, 1999; Poulimenos, 2000; Walsh et al., 2002; Schlagenhauf et al., 2008) (Fig. 6.12B). Fault growth associated with proportional increases in average displacement gradients for all faults in a system will produce upward migration of the D-L distribution as strain accrues without a change in n. In the case where n is equal to 1 from the outset of faulting and all faults experience proportional increases in displacement gradients, a slope of 1 will be maintained, with increasing values of the scaling factor c. However, the D-L exponent for single events or historical earthquakes may be 0.5−1, with the data plotting below ancient systems (e.g., Cowie and Scholz, 1992a; Wells and Coppersmith, 1994; Fossen and Hesthammer, 1997; Wesnousky, 2008; Schultz et al., 2008; Nicol et al., 2010; Rotevatn and Fossen, 2012) (Fig. 6.14, blue symbols), and it is possible that n for immature fault systems are also <1. In such cases n may increase with strain accumulation (Nicol et al., 2010; Rotevatn and Fossen, 2012), and be consistent with fault system models in which displacements are progressively localised onto the largest faults (Meyer et al., 2002; Walsh et al., 2003b). Strain localisation and associated changes in n during fault growth could be achieved by, for example; (1) preferential linkage of smaller faults or fault segments to form longer faults on which displacements continue to accrue (e.g., Cartwright et al., 1995; McLeod et al., 2000; Mansfield and Cartwright, 2001; Fossen and Rotevatn, 2016), and/or (2) preferential death or temporary abandonment of the smallest faults in a system (e.g., Meyer et al., 2002; Walsh et al., 2002, 2003b; Nicol et al., 2010) (Fig. 6.1). More data and analyses are required to test the hypothesis that n increases with fault system maturity and to examine how factors such as fault inheritance and rock properties might impact the evolution of n during fault growth (Walsh and Watterson, 1988; Cowie and Scholz, 1992c; Walsh et al., 2002).

Fault interactions have a profound impact on the dimensions and displacement accumulation of faults over timescales of millions of years (Figs 6.1 and 6.14). Increases in the D:L ratio for larger displacements, with or without changes in n, reflect increases in average displacement gradients which may indicate that fault interactions become increasingly important as fault systems evolve. These changes in displacement gradients can be accompanied by changes in the shape of displacement profiles, with flat-topped profiles being more common on faults with high D:L ratios (Fig. 6.14, see red profiles). Fault interactions are an essential feature of all fault systems (Walsh and Watterson, 1991), as indicated by complementary displacement patterns on the component faults (e.g., Childs et al., 1995; Cartwright et al., 1995; Dawers and Anders, 1995; Nicol et al., 1996; Walsh et al., 2003a). This interaction may occur early in the growth of a fault system and many faults have near-constant lengths from an early stage in their history, due to retardation of lateral propagation by interaction between fault tips.

6.5 End-member fault growth models

The previous sections outline some of the principal lines of evidence for fault growth models. Here we present the two fault growth models that are most widely supported in the literature. The first of these is the 'isolated propagating fault model' (Figs. 6.1 left and 6.10 left), a model which has previously been referred to as the 'conventional fault model' (Walsh et al., 2002), the 'increasing-length fault model' (Nicol et al., 2016), the 'isolated fault model' (e.g., Walsh et al., 2003a; Ghalayini et al., 2016; Morley, 2016), the 'non-coherent propagating model' (Fig. 6.9; Childs et al., 2017), and the 'propagating fault model' (Rotevatn et al., 2019). In this model, faults nucleate randomly in space and grow by progressive propagation or by the interaction and linkage of fortuitously aligned smaller faults to form a single through-going structure (Fig. 6.1 left). The displacement distribution on the resultant through-going fault is irregular, with local displacement lows recording the locations of boundaries between previously separate faults or fault segments (Ellis and Dunlap, 1988; Cartwright et al., 1995; Cowie et al., 2000; Davis et al., 2005; Bergen and Shaw, 2010) (Fig. 6.1 left). The second model is the 'coherent constant-length model', which has previously been referred to as the 'alternative model' (Walsh et al., 2002), the 'constant-length model' (Curry et al., 2016; Nicol et al., 2016; Jackson et al., 2017; Rotevatn et al., 2019) and the 'coherent model' (Walsh et al., 2003a; Morley, 2016) (Figs. 6.1 right and 6.10 right). In the 'coherent constant-length model', individual faults or fault segments initiate as interacting elements within a larger structure or fault system. The lengths of coherent faults are established early in the history of faulting with further propagation restricted by their interactions. These interactions are a key component of coherent fault arrays and, as we have seen in earlier sections, are most clearly illustrated for fault segments in 2D, where displacements on these segments are complementary and produce an aggregated displacement pattern similar to that of a single isolated fault (Walsh and Watterson, 1989; Childs et al., 1995; Walsh et al., 2003a) (Figs 6.5B, 6.7A,B and 6.9B,C). For such coherent arrays the faults are not initially hard-linked (i.e., physically connected), with soft-linkage of displacements achieved by distributed strain in the rock between the faults (Walsh and Watterson, 1991). These off-fault strains are most easily recognised at relay ramps between the ends of fault segments, where bed rotations and splay faults are common (Figs 6.3 and 6.7).

The two primary growth models are end-members in a range of potential behaviours and, although useful conceptually, reality is likely to lie somewhere between these end members. Within the isolated propagating model, initially isolated faults will eventually interact with one another as strain increases. Similarly, for the coherent constant-length model, there must be a period, however short, during which faults propagate through the rock volume. The key distinction between the models is therefore, at what strains, and how rapidly, are fault lengths established? Determining the answer to this question from natural and analogue data is hampered by limitations in the temporal resolution of growth strata, the displacement resolution limits for analogue models or the 2D form of many fault growth datasets. Where the first increment of faulting resolved by growth strata typically accounts for as much as ~20% of the duration of deformation (e.g., Meyer et al., 2002; Nicol et al., 2005), faults exhibit only coherent constant-length behaviour and appear to have established their lengths instantaneously on geological timescales, but may have initiated as isolated propagating structures on temporal scales less than the first growth interval (Walsh et al., 2002; Rotevatn et al., 2019). Analogue models appear to support this mixed-mode model; however, in some cases the measured lengths of traces may be impacted by the resolution limit of the displacement measurements. Faults could also propagate to form new segments that initiate as elements of a coherent array (i.e., the 'coherent propagating' model depicted in Fig. 6.9B), a scenario which is between our end-member models. In circumstances where fault lengths are established very quickly and are restricted by processes other than fault interactions (e.g., intersection with weak beds, salt diapirs and pre-existing faults that do not appear to be reactivated), it is, in principle, also possible for a system to subscribe to a 'non-coherent constant length' model, although few of these faults have been described in the literature.

The majority of fault-growth analyses of geological and analogue data are conducted using fault traces in map view, a perspective that influences the way we both think about, and choose to represent, fault growth models (Walsh et al., 2003a). In particular, fault growth models are often constrained using segmented fault arrays comprising two or more fault segments that are observed in map view (Figs 6.7–6.9) at a range of scales (Walsh and Watterson, 1989, 1990, 1991; Peacock and Sanderson, 1991; Trudgill and Cartwright, 1994; Cartwright et al., 1995; Childs et al., 1995; Dawers and Anders, 1995; Willemse, 1997; Morley, 2002; Walsh et al., 2003a; Fossen and Rotevatn, 2016). As we have already discussed, segmented arrays show aggregate displacement variations that are similar to those of a single isolated fault (e.g., Walsh and Watterson, 1989, 1990; Peacock and Sanderson, 1991; Dawers and Anders, 1995; Willemse, 1997), suggesting that they are (and were always) elements of a single coherent structure (e.g., Fig. 6.8C). However, interpretations of the growth of segmented fault arrays most often involve nucleation and propagation of kinematically independent segments, followed by incidental overlap, interaction and the formation of fault relay zones (Fig. 6.8A and B; e.g., Ellis and Dunlap, 1988; Morley et al., 1990; Trudgill and Cartwright, 1994; Cartwright et al., 1995, 1996; Dawers and Anders, 1995; Morley, 2002; Bergen and Shaw, 2010). One explanation for this anomaly is that when interpreting fault-trace maps or cross sections it is common to invoke a 2D growth model and assume that the propagation direction of each fault or fault segment is in the plane of inspection (Cowie, 1998; Cowie et al., 2000). A 3D perspective of these faults is more likely to lead to identification of fault surface propagation and fault segment linkage out of the plane of inspection, observations that will promote recognition of coherent constant-lengths in map view (Walsh et al., 2003a).

A variety of lines of evidence favour fault growth that is closer to the coherent constant-length model than to the isolated propagating model, the main evidence being the following. Firstly, most aggregate displacement profiles for segmented fault arrays are geometrically coherent. Secondly, kinematic constraints suggest that at the temporal scale of existing datasets, the lengths of individual faults are established instantaneously and thereafter accumulate displacements with relatively limited changes in length. Some of the best studied datasets, such as those from the East African Rift, Inner Moray Firth, Timor Sea and Taranaki Basin, suggest that fault localisation occurs at very low strains (i.e., <5%; Morley, 1999; Walsh et al., 2002, 2003b; Giba et al., 2012). Thirdly, when segmented fault arrays are mapped in 3D, they frequently merge into a single fault and many large segmented faults predominantly form by upward propagation from underlying pre-existing faults. Whilst there are well defined differences between the two leading fault growth models, they refer only to the mechanism by which faults and fault systems are established. They are not, for example, distinguished in terms of other aspects of the evolution of faults, such as the hard-linkage of segments bounding relays with increasing displacements, the instantaneous increase in fault length achieved by the coalescence and linkage of previously isolated faults (e.g., Fig. 6.9B) or the decrease in active fault lengths as deformation progresses and strain becomes localised onto fewer and larger faults. In such cases, fault growth is more complicated than is described by the end-member models, and is accompanied by a gradual localisation of displacements onto fewer and larger faults with the progressive death of smaller faults. Thus, fault evolution is dependent on the scale of individual faults and on the maturity of the fault system in which it occurs.

6.6 Earthquakes and incremental growth

Fault growth is generally considered to be achieved by discrete increments of slip which accrue during earthquakes (Gilbert, 1884; Reid, 1910; Walsh and Watterson, 1987; Stein et al., 1988; Cowie and Scholz, 1992a; Nicol et al., 2005). Historical and paleo-earthquakes provide information about fault growth processes which, in many cases, could not be retrieved by analysis of finite displacements and lengths or from growth strata dated on million-year timescales. In areas of active faulting it is often possible to characterise fault growth arising from small numbers of earthquakes (e.g., <10) which displaced the ground surface. Comparison of short-term displacements that produce active fault scarps crossing the landscape (or seabed) and long-term fault displacements on individual faults offer the opportunity to characterise the temporal and spatial changes in fault growth on timescales ranging from thousands to millions of years (Wallace, 1987; Stein et al., 1988; Cowie and Scholz, 1992a; Nicol et al., 1997, 2005, 2010; Friedrich et al., 2003; Manzocchi et al., 2006; Mouslopoulou et al., 2009; Cowie et al., 2012). A primary conclusion from these earthquake studies is that fault growth during repeated earthquakes (short-term) is more variable than might be predicted from million-year (long-term) displacements. An increasing body of evidence suggests that both the short-term variability and the long-term stability of fault growth are at least partly due to fault interactions.

Aperiodic recurrence intervals and non-characteristic slip earthquake behaviour observed for many active faults are widely thought to reflect fault interactions. In central Italy these interactions may produce a temporal clustering of paleoearthquakes across the entire width of a normal fault system

(Benedetti et al., 2013), alternatively in other fault systems interactions have been interpreted to produce migration of earthquake activity associated with near-constant regional strain rates (Wallacc, 1987; Rockwell et al., 2000; Nicol et al., 2006; Dolan et al., 2007). For both models the accumulation of displacements during earthquakes on separate faults are interdependent, with faults interacting over length scales of up to 100s of kilometres. Whether these fault systems are hard-linked or soft-linked, fault interaction reflects stress concentrations and shadows arising from short-term stress-dependent behaviour of earthquakes (e.g., Harris, 1998; Stein, 1999; King and Cocco, 2000). Dynamic, static and Coulomb stress modelling suggests that large magntiude earthquakes can increase or decrease stress on nearby faults, which advances or delays future events on time scales of seconds to thousands of years. The most extreme forms of fault interaction occur on time scales of seconds, where dynamic stress changes produce co-seismic rupture of multiple faults during the same event, particularly where the faults are closely spaced (e.g., separation <5 km) or intersect one another (Kurushin et al., 1998; Hamling et al., 2017). The recent Kaikōura Earthquake in New Zealand is an extreme form of multi-fault event, rupturing 10s of faults with variable orientation, slip sense and slip rates along a distance of hundreds of kilometres (Hamling et al., 2017). These multi-fault events, and earthquake data in general, indicate that faults interact over length scales far greater than could be demonstrated from fault finite displacement data and support the view that all faults in systems interact, forming coherent arrays for much of the duration of faulting. Fault interactions observed over timescales of millions of years ultimately reflect stress transfers between faults, which can happen on length scales of 100s of kms and instantaneously on geological timescales.

6.7 Concluding remarks

This chapter provides an overview of the datasets and approaches that have been brought to bear in the study of fault growth. The emphasis has been mainly on normal faults, because those are the structures for which relevant data are most readily obtained, although the conclusions are likely to apply to other fault types and over a range of scales. Each of the various data types provide insights into different aspects and time intervals. The key objective of this chapter is to rationalise all of the available data with a comprehensive model of fault growth that incorporates the short term erratic behaviour of faults, due to continually changing stress distributions, with the long term systematic behaviour of fault systems that reflect the generally simple boundary conditions of diverging basin margins.

This chapter was introduced in terms of two end-member models of fault growth, the isolated propagating and the coherent constant-length models, and most sections described data that can be used to discriminate between them. The key question that distinguishes the models is, at what strain does the transition occur from a system in which faults are mechanically isolated from one another and continue to propagate, to a system in which all faults interact so that tip-line propagation ceases and throw accumulates on an existing fault pattern? While we have attempted to maintain a balanced description of the relevant data, our favoured model lies close to the coherent constant-length model and we consider that the available geological data, including displacement profiles, displacement backstripping, displacement vs length relationships and fault damage zone structure support this view. Geological data are however subject to uncertainties and issues of temporal resolution and discussions on these data will continue into the future as models of fault growth develop further. Independent of the precise form of the fault growth model it must be acknowledged that interaction between faults over a range of spatial and temporal scales is critical for their evolution.

References

Bailey, W.R., Walsh, J.J., Manzocchi, T., 2005. Fault populations, strain distribution and basement reactivation in the East Pennines Coalfield, U.K. J. Struct. Geol. 27, 913–928.

Barnett, J.A.M., Mortimer, J., Rippon, J.H., Walsh, J.J., Watterson, J., 1987. Displacement geometry in the volume containing a single normal fault. Am. Assoc. Pet. Geol. Bull. 71, 925–937.

Benedetti, L., Manighetti, I., Gaudemer, Y., Finkel, R., Malavieille, J., Pou, K., Arnold, M., Aumaître, G., Bourlès, D., Keddadouche, K., 2013. More than 30 large earthquakes broke the Fucino faults (Central Italy) in synchrony in the last 12 ka, as revealed from in situ ^{36}Cl exposure dating. J. Geophys. Res. 118, 4948–4974.

Bergen, K.J., Shaw, J.H., 2010. Displacement profiles and displacement-length scaling relationships of thrust faults constrained by seismic-reflection data. Geol. Soc. Am. Bull. 122 (7/8), 1209–1219. https://doi.org/10.1130/B26373.1.

Bürgmann, R., Pollard, D.D., Martel, S.J., 1994. Slip distributions on faults: effects of stress gradients, inelastic deformation, heterogeneous host-rock stiffness, and fault interaction. J. Struct. Geol. 16, 1675–1690.

Cartwright, J.A., Trudgill, B., Mansfield, C.S., 1995. Fault growth by segment linkage: an explanation for scatter in maximum displacement and trace length data from the Canyonlands Grabens of S.E. Utah. J. Struct. Geol. 17, 1319–1326.

Chapman, T.J., Williams, G.D., 1984. Displacement-distance methods in the analysis of fold-thrust structures and linked fault systems. J. Geol. Soc. Lond. 141, 121–128.

Childs, C., Easton, S.J., Vendeville, B.C., Jackson, M.P.A., Lin, S.T., Walsh, J.J., Watterson, J., 1993. Kinematic analysis of faults in a physical model of growth faulting above a viscous salt analogue. Tectonophysics 228, 313–329.

Childs, C., Watterson, J., Walsh, J.J., 1995. Fault overlap zones within developing normal fault systems. J. Geol. Soc. Lond. 152, 535–549.

Childs, C., Nicol, A., Walsh, J.J., Watterson, J., 2003. The growth and propagation of synsedimentary faults. J. Struct. Geol. 25, 633–648.

Childs, C., Manzocchi, T., Walsh, J.J., Bonson, C.G., Nicol, A., Schöpfer, M.P.J., 2009. A geometric model of fault zone and fault rock thickness variations. J. Struct. Geol. 31, 117–127.

Childs, C., Holdsworth, R.E., Jackson, C.A.L., Manzocchi, T., Walsh, J.J., Yielding, G., 2017. Introduction to the geometry and growth of normal faults. Geol. Soc. Lond. Spec. Publ. 439, 1–9.

Chinnery, M.A., 1996. Secondary faulting: II. geological aspects. Can. J. Earth Sci. 3, 175–190. https://doi.org/10.1139/e66-014.

Conneally, J., Childs, C., Nicol, A., 2017. Monocline formation during growth of segmented faults in the Taranaki Basin, offshore New Zealand. Tectonophysics 721, 310–321. https://doi.org/10.1016/j.tecto.2017.06.036.

Cowie, P.A., 1998. A healing-reloading feedback control on the growth rate of seismogenic faults. J. Struct. Geol. 20, 1075–1087.

Cowie, P.A., Scholz, C.H., 1992a. Growth of faults by accumulation of seismic slip. J. Geophys. Res. 97, 11085–11096.

Cowie, P.A., Scholz, C.H., 1992b. Physical explanation for the displacement-length relationship of faults using a post-yield fracture mechanics model. J. Struct. Geol. 14, 1133–1148.

Cowie, P.A., Scholz, C.H., 1992c. Displacement-length scaling relationships for faults: data synthesis and discussion. J. Struct. Geol. 14, 1149–1156.

Cowie, P.A., Shipton, Z.K., 1998. Fault tip displacement gradients and process zone dimensions. J. Struct. Geol. 20 (8), 983–997.

Cowie, P.A., Gupta, S., Dawers, N.H., 2000. Implications of fault array evolution for synrift depocentre development: insights from a numerical fault growth model. Basin Res. 12, 241–261.

Cowie, P.A., Roberts, G.P., Bull, J.M., Visini, F., 2012. Relationships between fault geometry, slip rate variability and earthquake recurrence in extensional settings. Geophys. J. Int. 189, 143–160. https://doi.org/10.1111/j.1365-246X.2012.05378.x.

Curry, M.A., Barnes, J.B., Colgan, J.P., 2016. Testing fault growth models with low-temperature thermochronology in the northwest Basin and Range, USA. Tectonics 35, 2467–2492.

Dahlstrom, C.D.A., 1970. Structural geology in the eastern margin of the Canadian Rocky Mountains. Bull. Can. Pet. Geol. 18, 332–406.

Davis, K., Burbank, D.W., Fisher, D., Wallace, S., Nobes, D., 2005. Thrust-fault growth and segment linkage in the active Ostler fault zone, New Zealand. J. Struct. Geol. 27, 1528–1546.

Dawers, N.H., Anders, M.H., Scholz, C.H., 1993. Growth of normal faults: displacement-length scaling. Geology 21, 1107–1110.

Dawers, N.H., Anders, M.H., 1995. Displacement-length scaling and fault linkage. J. Struct. Geol. 17, 607–614.

Dolan, J.F., Bowman, D.D., Sammis, C.G., 2007. Long-range and long-term fault interactions in Southern California. Geology 35, 855–858. https://doi.org/10.1130/G23789A.1.

Elliott, D., 1976. The energy balance and deformation mechanisms of thrust sheets. Phil. Trans. Roy. Soc. Lond. A 283, 289–312.

Ellis, M.A., Dunlap, W.J., 1988. Displacement variation along thrust faults: implications for the development of large faults. J. Struct. Geol. 10, 183–192.

Eisenstadt, G., DePaor, D.G., 1987. Alternative model of thrust-fault propagation. Geology 15, 630–633.

Finch, E., Gawthorpe, R., 2017. Growth and interaction of normal faults and fault network evolution in rifts: insights from three-dimensional discrete element modelling. In: Childs, C., Holdsworth, R.E., Jackson, C.A.-L., Manzocchi, T., Walsh, J.J., Yielding, G. (Eds.), The Geometry and Growth of Normal Faults, vol. 439. Geological Society, London, Special Publications. https://doi.org/10.1144/SP439.23.

Fossen, H., Hesthammer, J., 1997. Geometric analysis and scaling relations of deformation bands in porous sandstone. J. Struct. Geol. 19, 1479–1493.

Fossen, H., Rotevatn, A., 2016. Fault linkage and relay structures in extensional settings— a review. Earth Sci. Rev. 154, 14–28.

Fossen, H., Fazli Khani, H., Faleide, J.I., Ksienzyk, A.K., Dunlap, W.J., 2016. Post-Caledonian extension in the West Norway–northern North Sea region: the role of structural inheritance. In: Childs, C., Holdsworth, R.E., Jackson, C.A.-L., Manzocchi, T., Walsh, J.J., Yielding, G. (Eds.), The Geometry and Growth of Normal Faults, vol. 439. Geological Society, London, Special Publications. https://doi.org/10.1144/SP439.6.

Frankowicz, E., McClay, K.R., 2010. Extensional fault segmentation and linkages, Bonaparte Basin, outer North West Shelf, Australia. Am. Assoc. Pet. Geol. Bull. 94 (7), 977–1010. https://doi.org/10.1306/01051009120.

Friedrich, A.M., Wernicke, B.P., Niemi, N.A., Bennett, R.A., Davis, J.L., 2003. Comparison of geodetic and geologic data from the Wasatch region, Utah, and implications for the spectral character of Earth deformation at periods of 10 to 10 million years. J. Geophys. Res. 108 https://doi.org/10.1029/2001JB000682.

Ghalayini, R., Homberg, C., Daniel, J.M., Nader, F.H., 2016. Growth of layer-bound normal faults under a regional anisotropic stress field. In: Childs, C., Holdsworth, R.E., Jackson, C.A.-L., Manzocchi, T., Walsh, J.J., Yielding, G. (Eds.), The Geometry and Growth of Normal Faults, vol. 439. Geological Society, London, Special Publications. https://doi.org/10.1144/SP439.13.

Giba, M., Walsh, J.J., Nicol, A., 2012. Segmentation and growth of an obliquely reactivated normal fault. J. Struct. Geol. 39, 253–267.

Gilbert, G.K., 1884. A theory of earthquakes of the Great Basin, with practical application (from the Salt Lake Tribune of September 30, 1883). Am. J. Sci. 27, 49–53, 3rd Series.

Gillespie, P.A., 1991. Structural Analysis of Faults and Folds with Examples from the South Wales Coalfield and the Ruhr Coalfield (Unpublished Ph.D. thesis). University of Wales.

Gillespie, P.A., Walsh, J.J., Watterson, J., 1992. Limitations of dimension and displacement data from single faults and the consequences for data analysis and interpretation. J. Struct. Geol. 14, 1157–1172.

Gordon, J.E., 1978. Structures, or Why Things Don't Fall Down. Plenum, New York, 420 p.

Hamling, I.J., Hreinsdóttir, S., Clark, K., Elliott, J., Liang, C., Fielding, E., Litchfield, N., Villamor, P., Wallace, L., Wright, T.J., D'Anastasio, E., Bannister, S., Burbidge, D., Denys, P., Gentle, P., Howarth, J., Mueller, C., Palmer, N., Pearson, C., Power, W., Barnes, P., Barrell, D.J.A., Van Dissen, R., Langridge, R., Little, T., Nicol, A., Pettinga, J., Rowland, J., Stirling, M., 2017. Complex multi-fault rupture during the 2016 Mw 7.8 Kaikōura earthquake, New Zealand. Science 356 (6334), eaam7194. https://doi.org/10.1126/science.aam7194.

Harris, R., 1998. Introduction to special section: stress triggers, stress shadows, and implications for seismic hazard. J. Geophys. Res. 103 (B10), 24347–24358.

Hull, D., Bacon, D.J., 1984. Introduction to Dislocations. In: International Series on Materials Science and Technology, 3rd edition, vol. 37. https://doi.org/10.1016/C2009-0-64358-0. 272 p.

Jackson, C.A.-L., Rotevatn, A., 2013. 3D seismic analysis of the structure and evolution of a salt-influenced normal fault zone: a test of competing fault growth models. J. Struct. Geol. 54, 215–234. https://doi.org/10.1016/j.jsg.2013.06.012.

Jackson, C.A.-L., Bell, R.E., Rotevatn, A., Tvedt, A.B.M., 2017. Techniques to determine the kinematics of synsedimentary normal faults and implications for fault growth models. In: Childs, C., Holdsworth, R.E., Jackson, C.A.-L., Manzocchi, T., Walsh, J.J., Yielding, G. (Eds.), The Geometry and Growth of Normal Faults. Geological Society, London, Special Publications, p. 439. https://doi.org/10.1144/SP439.22.

Khalil, S.M., McClay, K.R., 2016. 3D geometry and kinematic evolution of extensional fault-related folds, NW Red Sea, Egypt. In: Childs, C., Holdsworth, R.E., Jackson, C.A.-L., Manzocchi, T., Walsh, J.J., Yielding, G. (Eds.), The Geometry and Growth of Normal Faults. Geological Society, London, Special Publications, p. 439. https://doi.org/10.1144/SP439.11.

Kelly, P.G., Peacock, D.C.P., Sanderson, D.J., McGuirk, A.C., 1999. Selective reverse reactivation of normal faults, and deformation around reverse reactivated faults in the Mesozoic of the Somerset coast. J. Struct. Geol. 21, 493–509.

Kim, Y.-S., Andrews, J.R., Sanderson, D.J., 2000. Damage zones around strike–slip fault systems and strike–slip fault evolution, Crackington Haven, southwest England. Geosci. J. 4, 53–72.

Kim, Y.-S., Sanderson, D.J., 2005. The relationship between displacement and length of faults. Earth Sci. Rev. 68, 317–334.

King, G., Cocco, M., 2000. Fault interaction by elastic stress changes: new clues from earthquake sequences. Adv. Geophys. 44, 1–36.

Kurushin, R.A., Bayasgalan, A., Ölziybat, M., Enhtuvshin, B., Molnar, P., Bayarsayhan, C., Hudnut, K.W., Lin, J., 1998. The Surface Rupture of the 1957 Gobi-Altay, Mongolia, Earthquake, vol. 320. Geological Society of America Special Publication, 143 p. https://doi.org/10.1130/SPE320.

Lăpădat, A., Imber, J., Yielding, G., Iacopini, D., McCaffrey, K.J.W., Long, J.J., Jones, R.R., 2016. Occurrence and development of folding related to normal faulting within a mechanically heterogeneous sedimentary sequence: a case study from Inner Moray Firth, UK. In: Childs, C., Holdsworth, R.E., Jackson, C.A.-L., Manzocchi, T., Walsh, J.J., Yielding, G. (Eds.), The Geometry and Growth of Normal Faults, vol. 439. Geological Society, London, Special Publications, pp. 373–394. https://doi.org/10.1144/SP439.18.

McConnell, D.A., Kattenhorn, S.A., Benner, L.M., 1997. Distribution of fault slip in outcrop-scale fault related folds, Appalachian Mountains. J. Struct. Geol. 19, 257–267.

McGrath, A.G., Davison, I., 1995. Damage zone geometry around fault tips. J. Struct. Geol. 17 (7), 1011–1024.

McLeod, A.E., Dawers, N.H., Underhill, J.R., 2000. The propagation and linkage of normal faults: insights from the Strathspey-Brent-Statfjord fault array, northern North Sea. Basin Res. 12, 263–284.

MacMillian, R.A., 1975. The Orientation and Sense of Displacement of Strike-Slip Faults in Continental Crust (Unpublished B.S. thesis). Carleton University, Ottawa, Ontario.

Manighetti, I., King, G.C.P., Gaudemer, Y., Scholz, C.H., Doubre, C., 2001. Slip accumulation and lateral propagation of active normal faults in Afar. J. Geophys. Res. 106 (B7), 13667–13696. https://doi.org/10.1029/2000jb900471.

Manighetti, I., King, G., Sammis, C.G., 2004. The role of off-fault damage in the evolution of normal faults. Earth Planet. Sci. Lett. 217, 399–408.

Mansfield, C., Cartwright, J.A., 2001. Fault growth by linkage: observations and implications from analogue models. J. Struct. Geol. 23, 745–763.

Manzocchi, T., Walsh, J.J., Nicol, A., 2006. Displacement accumulation from earthquakes on isolated normal faults. J. Struct. Geol. 28, 1685–1693.

Manzocchi, T., Walsh, J.J., Bailey, W.R., 2009. Population scaling biases in map samples of power-law fault systems. J. Struct. Geol. 31, 1612–1626.

Marrett, R., Allmendinger, R.W., 1992. Amount of extension on 'small' faults: an example from the Viking graben. Geology 20, 47–50.

Meyer, V., Nicol, A., Childs, C., Walsh, J.J., Watterson, J., 2002. Progressive localisation of strain during the evolution of a normal fault system. J. Struct. Geol. 24, 1215–1231.

Morley, C.K., Nelson, R.A., Patton, T.L., Munn, S.G., 1990. Transfer zones in the East African rift system and their relevance to hydrocarbon exploration in rifts. Bull. Am. Assoc. Pet. Geol. 74, 1234–1253.

Morley, C.K., 1999. Patterns of displacement along large normal faults: implications for basin evolution and fault propagation, based on examples from East Africa. Bull. Am. Assoc. Pet. Geol. 83, 613–634.

Morley, C.K., 2002. Evolution of large normal faults: evidence from seismic reflection data. Bull. Am. Assoc. Pet. Geol. 86, 961–978.

Morley, C.K., 2016. The impact of multiple extension events, stress rotation and inherited fabrics on normal fault geometries and evolution in the Cenozoic rift basins of Thailand. In: Childs, C., Holdsworth, R.E., Jackson, C.A.-L., Manzocchi, T., Walsh, J.J., Yielding, G. (Eds.), The Geometry and Growth of Normal Faults, vol. 439. Geological Society, London, Special Publications. https://doi.org/10.1144/SP439.3.

Mouslopoulou, V., Walsh, J.J., Nicol, A., 2009. Fault displacement rates on a range of timescales. Earth and Planetary Science Letters 278, 186–197.

Muraoka, H., Kamata, H., 1983. Displacement distribution along minor fault traces. J. Struct. Geol. 5, 483–495. https://doi.org/10.1016/0191-8141 (83)90054-8.

Nicol, A., Watterson, J., Walsh, J.J., Childs, C., 1996. The shapes, major axis orientations and displacement patterns of fault surfaces. J. Struct. Geol. 18 (2/3), 235–248.

Nicol, A., Walsh, J.J., Watterson, J., Underhill, J.R., 1997. Displacement rates of normal faults. Nature 390, 157–159.

Nicol, A., Walsh, J.J., Berryman, K., Nodder, S., 2005. Growth of a normal fault by the accumulation of slip over millions of years. J. Struct. Geol. 27, 327–342.

Nicol, A., Gillespie, P.A., Walsh, J., Childs, C., 2002. Thrust relays and associated folding in layered sequences. Journal of Structural Geology 24, 709–727.

Nicol, A., Walsh, J., Berryman, K., Villamor, P., 2006. Interdependence of fault displacement rates and paleoearthquakes in an active rift. Geology 34 (10), 865–868. https://doi.org/10.1130/G22335.1.

Nicol, A., Walsh, J.J., Villamor, P., Seebeck, H., Berryman, K.R., 2010. Normal fault interactions, paleoearthquakes and growth in an active rift. J. Struct. Geol. 32, 1101–1113. https://doi.org/10.1016/j.jsg.2010.06.018.

Nicol, A., Childs, C., Walsh, J.J., Manzocchi, T., Schöpfer, M.P.J., 2016. Interactions and growth of faults in an outcrop-scale system. In: Childs, C., Holdsworth, R.E., Jackson, C.A.-L., Manzocchi, T., Walsh, J.J., Yielding, G. (Eds.), The Geometry and Growth of Normal Faults, vol. 439. Geological Society, London, Special Publications. https://doi.org/10.1144/SP439.9.

Nixon, C.W., Bull, J.M., Sanderson, D.J., 2014. Localized vs distributed deformation associated with the linkage history of an active normal fault, Whakatane Graben, New Zealand. J. Struct. Geol. 69, 266–280.

Paton, D.A., 2006. Influence of crustal heterogeneity on normal fault dimensions and evolution: southern South Africa extensional system. J. Struct. Geol. 28, 868–886.

Peacock, D.C.P., 1991. Displacements and segment linkage in strike-slip fault zones. Journal of Structural Geology 11 (9), 1025–1035.

Peacock, D.C.P., 2002. Propagation, interaction and linkage in normal fault systems. Earth Sci. Rev. 58, 121–142.

Peacock, D.C.P., Sanderson, D.J., 1991. Displacements, segment linkage and relay ramps in normal fault zones. J. Struct. Geol. 13, 721–733.

Peacock, D.C.P., Sanderson, D.J., 1992. Effects of layering and anisotropy on fault geometry. J. Geol. Soc. Lond. 149, 793–802.

Peacock, D.C.P., Sanderson, D.J., 1996. Effects of propagation rate on displacement variations along faults. J. Struct. Geol. 18 (2/3), 311–320.

Perrin, C., Manighetti, I., Gaudemer, Y., 2015. Off-fault tip splay networks: a genetic and generic property of faults indicative of their long-term propagation. Compt. Rendus Geosci. https://doi.org/10.1016/j.crte.2015.05.002.

Petersen, K., Clausen, O.R., Korstgård, J.A., 1992. Evolution of a salt-related listric growth fault near the D-1 well, block 5605, Danish North Sea: displacement history and salt kinematics. J. Struct. Geol. 14, 565–577.

Pollard, D.D., Segall, P., 1987. Theoretical displacements and stresses near fractures in rock: with applications to faults, joints, veins, dikes, and solution surfaces. In: Atkinson, B.K. (Ed.), Fracture Mechanics of Rock. Academic Press, New York, pp. 277–349.

Poulimenos, G., 2000. Scaling properties of normal fault populations in the western Corinth Graben, Greece: implications for fault growth in large strain settings. J. Struct. Geol. 22, 307–322.

Reid, H.F., 1910. Mechanics of the Earthquake, the California Earthquake of April 18, 1906; Report of the State Investigation Commission, vol. 2. Carnegie Institution of Washington, Washington, DC.

Rippon, J.H., 1985. Contoured patterns of the throw and hade of normal faults in the coal Measures (Westphalian) of north-east Derbyshire. Proc. Yorks. Geol. Soc. 45 (3), 147–161.

Reilly, C., Nicol, A., Walsh, J.J., 2016. The importance of fault size for the evolution of an inverted fault system. In: Childs, C., Holdsworth, R.E., Jackson, C.A.-L., Manzocchi, T., Walsh, J.J., Yielding, G. (Eds.), The Geometry and Growth of Normal Faults, vol. 439. Geological Society, London, Special Publications, pp. 447–463. https://doi.org/10.1144/SP439.2.

Rippon, J.H., 1985. Contoured patterns of throw and hade of normal faults in the Coal Measures (Westphalian) of north-east Derbyshire. Proceedings of the Yorkshire Geological Society 45, 147–161.

Roche, V., Homberg, C., van der Baan, M., Rocher, M., 2016. Widening of normal fault zones due to the inhibition of vertical propagation. In: Childs, C., Holdsworth, R.E., Jackson, C.A.-L., Manzocchi, T., Walsh, J.J., Yielding, G. (Eds.), The Geometry and Growth of Normal Faults, vol. 439. Geological Society, London, Special Publications, pp. 271–288. https://doi.org/10.1144/SP439.5.

Rockwell, T.K., Lindvall, S., Herzberg, M., Murbach, D., Dawson, T., Berger, G., 2000. Paleoseismology of the Johnson Valley, Kickapoo, and Homestead Valley faults: clustering of earthquakes in the Eastern California Shear Zone. Bull. Seismol. Soc. Am. 90, 1200–1236.

Rotevatn, A., Fossen, H., 2012. Soft faults and hard tips: magnitude-order displacement gradient variations controlled by strain softening versus hardening; complications for fault scaling. J. Geol. Soc. Lond. 169, 123–126. https://doi.org/10.1144/0016-76492011.108.

Rotevatn, A., Jackson, C.A., Tvedt, A.B.M., Bell, R., Blækkan, I., 2019. How do normal faults grow? J. Struct. Geol. 125, 174–184. https://doi.org/10.1016/j.jsg.2018.08.005.

Schlagenhauf, A., Manighetti, I., Malavieille, J., Dominguez, S., 2008. Incremental growth of normal faults: insights from a laser-equipped analog experiment. Earth Planet. Sci. Lett. 273, 299–311.

Schlische, R.W., Young, S.S., Ackermann, R.V., Gupta, A., 1996. Geometry and scaling relations of a population of very small rift-related normal faults. Geology 24, 683–686.

Schöpfer, M.P.J., Childs, C., Walsh, J.J., 2006. Localisation of normal faults in multilayer sequences. J. Struct. Geol. 28, 816–833.

Schultz, R.A., 1999. Understanding the process of faulting: selected challenges and opportunities at the edge of the 21st century. J. Struct. Geol. 21, 985–993.

Schultz, R.A., Fossen, H., 2002. Displacement-length scaling in three dimensions: the importance of aspect ratio and application to deformation bands. J. Struct. Geol. 24, 1389–1411. https://doi.org/10.1016/S0191-8141(01)00146-8.

Schultz, R.A., Soliva, R., Fossen, H., Okubo, C.H., Reeves, D.M., 2008. Dependence of displacement–length scaling relations for fractures and deformation bands on the volumetric changes across them. J. Struct. Geol. 30, 1405–1411.

Schultz, R.A., Soliva, R., Okubu, C., Mége, D., 2010. Fault populations. In: Watters, T.R., Schultz, R.A. (Eds.), Planetary Tectonics. Cambridge University Press, pp. 457–510.

Smith, R.J., 2009. Use and misuse of the reduced major Axis for line-fitting. Am. J. Phys. Anthropol. 140, 476–486.

Stein, R.S., 1999. The role of stress transfer in earthquake occurrence. Nature 402, 605–609.

Stein, R.S., King, G.C.P., Rundle, J.B., 1988. The growth of geological structures by repeated earthquakes. 2 Field examples of continental dip-slip faults. J. Geophys. Res. 93, 13319–13331. https://doi.org/10.1029/JB093iB11p13319.

Taylor, S.K., Nicol, A., Walsh, J.J., 2008. Sediment compaction of normal displacements on growth faults. J. Struct. Geol. 30, 394–405.

Thorsen, C.E., 1963. Age of growth faulting in southeast Louisiana, vol. 13. Gulf Coast Association of Geological Societies, pp. 103–110.

Torabi, A., Berg, S.S., 2011. Scaling of fault attributes: a review. Mar. Pet. Geol. 28 (8), 1444–1460.

Torabi, A., Alaei, B., Libak, A., 2019. Normal fault 3D geometry and displacement revisited: insights from faults in the Norwegian Barents Sea. Mar. Pet. Geol. 99, 135–155.

Trudgill, B., Cartwright, J., 1994. Relay-ramp forms and normal-fault linkages, Canyonlands National Park, Utah. Geological Society of America Bulletin 106, 1143–1157.

Tvedt, A.B.M., Rotevatn, A., Jackson, C.A.-L., Fossen, H., Gawthorpe, R.L., 2013. Growth of normal faults in multilayer sequences: a 3D seismic case study from the Egersund Basin, Norwegian North Sea. J. Struct. Geol. 55, 1–20. https://doi.org/10.1016/j.jsg.2013.08.002.

Tvedt, A.B.M., Rotevatn, A., Jackson, C.A.-L., 2016. Supra-salt normal fault growth during the rise and fall of a diapir: perspectives from 3D seismic reflection data, Norwegian North Sea. J. Struct. Geol. 91, 1–26.

Wallace, R.E., 1987. Grouping and migration of surface faulting and variations in slip rates on faults in the Great Basin Province. Bull. Seismol. Soc. Am. 77, 868–876.

Walsh, J.J., Watterson, J., 1987. Distributions of cumulative displacement and seismic slip on a single normal fault surface. J. Struct. Geol. 9 (8), 1039–1046.

Walsh, J.J., Watterson, J., 1988. Analysis of the relationship between displacements and dimensions of faults. J. Struct. Geol. 10, 239–247.

Walsh, J.J., Watterson, J., 1989. Displacement gradients on fault surfaces. Journal of Structural Geology 11 (3), 307–316.

Walsh, J.J., Watterson, J., 1990. New methods of fault projection for coalmine planning. Proceedings of the Yorkshire Geological Society 48, 209–219.

Walsh, J.J., Watterson, J., 1991. Geometric and kinematic coherence and scale effects in normal fault systems. In: Roberts, A.M., Yielding, G., Freeman, B. (Eds.), The Geometry of Normal Faults, vol. 56. Geological Society Special Publication, pp. 193–203.

Walsh, J.J., Watterson, J., Yielding, G., 1991. The importance of small-scale faulting in regional extension. Nature 351, 391–393.

Walsh, J.J., Childs, C., Meyer, V., Manzocchi, T., Imber, J., Nicol, A., Tuckwell, G., Bailey, W.R., Bonson, C.G., Watterson, J., Nell, P.A.R., Strand, J.A., 2001. Geometrical controls on the evolution of normal fault systems. In: Holdsworth, R.E., Strachan, R.A., Magloughlin, J.F., Knipe, R.J. (Eds.), The Nature of the Tectonic Significance of Fault Zone Weakening, vol. 186. Geological Society of London, Special Publication, pp. 157–170.

Walsh, J.J., Nicol, A., Childs, C., 2002. An alternative model for the growth of faults. J. Struct. Geol. 24, 1669–1675.

Walsh, J.J., Bailey, W.R., Childs, C., Nicol, A., Bonson, C.G., 2003a. Formation of segmented normal faults: a 3-D perspective. J. Struct. Geol. 25, 1251–1262.

Walsh, J.J., Childs, C., Imber, J., Manzocchi, T., Watterson, J., Nell, P.A.R., 2003b. Strain localisation and population changes during fault system growth within the inner Moray Firth, northern north sea. J. Struct. Geol. 25, 1897–1911.

Watterson, J., 1986. Fault dimensions, displacements and growth. Pure Appl. Geophys. 124, 365–373.

Watterson, J., Walsh, J.J., Gillespie, P.A., Easton, S., 1996. Scaling systematics of fault sizes on a large scale range fault map. J. Struct. Geol. 18, 199–214.

Wells, D.L., Coppersmith, K.J., 1994. New empirical relationships among magnitude, rupture length, rupture width, rupture area, and surface displacement. Bull. Seismol. Soc. Am. 84, 974–1002.

Wesnousky, S.G., 2008. Displacement and geometrical characteristics of earthquake surface ruptures: issues and implications for seismic-hazard analysis and the process of earthquake rupture. Bull. Seismol. Soc. Am. 98, 1609–16332. https://doi.org/10.1785/0120070111.

Whipp, P.S., Jackson, C.A.-L., Gawthorpe, R.L., Dreyer, T., Quinn, D., 2014. Normal fault array evolution above a reactivated rift fabric; a subsurface example from the northern Horda Platform, Norwegian North Sea. Basin Res. 26, 523–549.

Whipp, P.S., Jackson, C.A.-L., Schlische, R.W., Withjack, M.O., Gawthorpe, R.L., 2016. Spatial distribution and evolution of fault-segment boundary types in rift systems: observations from experimental clay models. In: Childs, C., Holdsworth, R.E., Jackson, C.A.-L., Manzocchi, T., Walsh, J.J., Yielding, G. (Eds.), The Geometry and Growth of Normal Faults, vol. 439. Geological Society, London, Special Publications, pp. 79–107. https://doi.org/10.1144/SP439.7.

Willemse, E.J.M., 1997. Segmented normal faults: correspondence between three-dimensional mechanical models and field data. J. Geophys. Res. 102, 675–692.

Wilkins, S.J., Gross, M.R., 2002. Normal fault growth in layered rocks at Split Mountain, Utah: influence of mechanical stratigraphy on dip linkage, fault restriction and fault scaling. J. Struct. Geol. 24, 1413–1429.

Williams, G., Chapman, T., 1983. Strains developed in the hangingwalls of thrusts due to their slip/propagation rate: a dislocation model. J. Struct. Geol. 5, 563–571.

Withjack, M.O., Henza, A.A., Schlishe, R.W., 2017. Three-dimensional fault geometries and interactions within experimental models of multiphase extension. Am. Assoc. Pet. Geol. Bull. 101 (11), 1767–1789. https://doi.org/10.1306/02071716090.

Yielding, G., 2016. The geometry of branch lines. In: Childs, C., Holdsworth, R.E., Jackson, C.A.-L., Manzocchi, T., Walsh, J.J., Yielding, G. (Eds.), The Geometry and Growth of Normal Faults. Geological Society, London, Special Publications, p. 439. https://doi.org/10.1144/SP439.1.

Chapter 7

Direct dating of fault movement

Sumiko Tsukamoto[a], Takahiro Tagami[b], Horst Zwingmann[c]

[a]*Leibniz Institute for Applied Geophysics (LIAG), Hannover, Germany;* [b]*Division of Earth and Planetary Sciences, Graduate School of Science, Kyoto University, Kyoto, Japan;* [c]*Department of Geology and Mineralogy, Faculty of Science, Kyoto University, Kyoto, Japan*

Chapter outline

7.1 Dating of authigenic clay minerals in brittle faults

7.1.1 Outline of the concept and the analytical method

Numerous authors (e.g., Hobbs et al., 1976; Holdsworth et al., 1997; Scholz, 2002, Fig. 7.1) have described the geological nature of faults in relation to their tectonic environments in the upper crust. Some of the geological processes in active fault zones involve fragmentation of host rocks, grain-size reduction and subsequent authigenesis of clay minerals such as illite, illite-smectite mixed-layers and chlorite. Fault zones often contain synkinematic phyllosilicates (such as muscovite), and authigenic illite, a general term for a 10Å, potassium-rich, di-octahedral, mica-like clay mineral (Velde, 1992; Meunier, 2005). Illite has a di-octahedral mineral structure, such as $K(Mg,Fe,Al)_2(Si,Al)_4O_{10}$ (Velde, 2004) (Fig. 7.2). As it contains potassium, it can be dated by the K-Ar (e.g. Lyons and Snellenberg, 1971) and $^{40}Ar/^{39}Ar$

Understanding Faults. https://doi.org/10.1016/B978-0-12-815985-9.00007-2

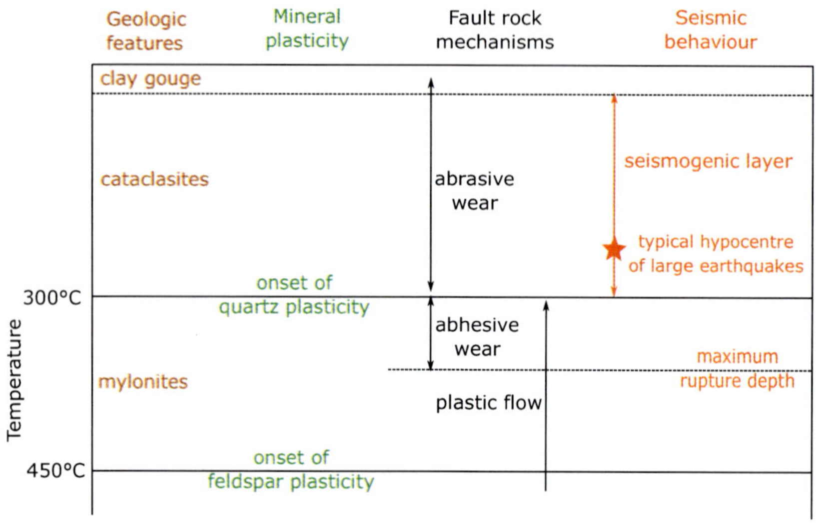

FIG. 7.1 Gouge formation areas in the crust.

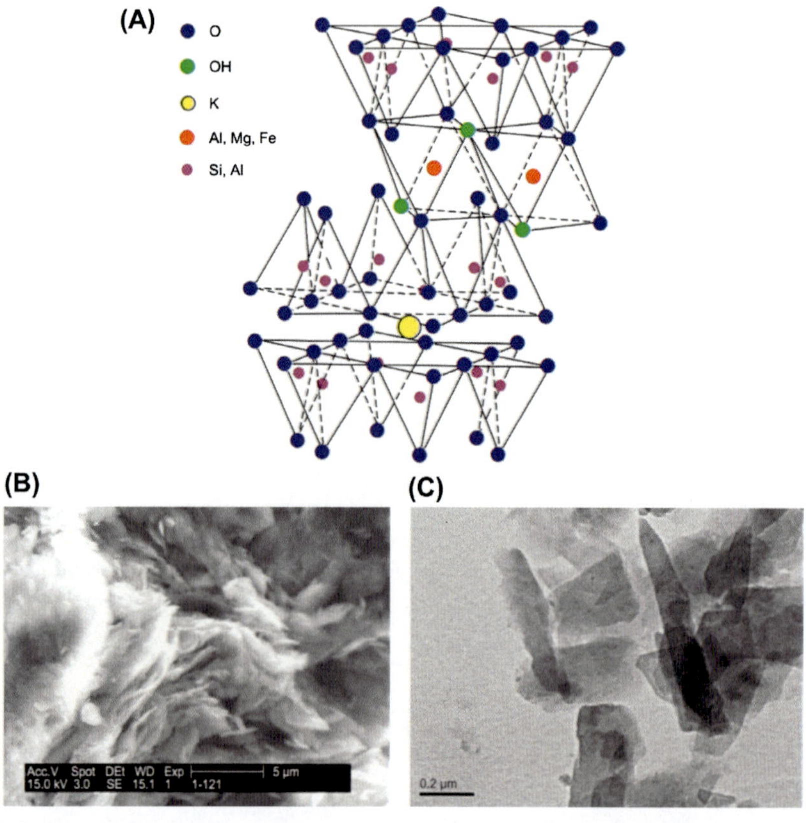

FIG. 7.2 (A) Illite crystallographic structure, (B) SEM and (C) TEM examples, after Zwingmann and Mancktelow (2004).

dating methods (e.g., van der Pluijm et al., 2001). Vrolijk et al. (2018) provide a comprehensive summary and highlight applications within prograde and retrograde geological environments to constrain the absolute timing of brittle fault formation, fault gouge dating and its scientific history.

7.1.2 K-Ar and $^{40}Ar/^{39}Ar$ clay dating principles

The age of illite concentrates can be determined by measuring the amount of the ^{40}Ar isotope relative to the K content, as ^{40}Ar is produced by the radioactive decay of the ^{40}K isotope. General backgrounds to geochronological methods, applications and technical requirements of K-Ar and $^{40}Ar/^{39}Ar$ dating have been detailed by Dalrymple and Lanphere (1969) and McDougall and Harrison (1999). Zwingmann (2014) summarizes the three major clay-dating methods that involve (1) classical spiked K-Ar; (2) unspiked K-Ar, and (3) $^{40}Ar/^{39}Ar$ methods, all of which are applicable to fault zones. Minerals usually contain negligible amounts of atmospheric (= common) argon when they are first formed, although small amounts of atmospheric argon may remain in the samples or can be added during preparation and of the size separates, which needs to be corrected using the amount and the known $^{40}Ar/^{36}Ar$ ratio of 295.5 (Steiger and Jäger, 1977). Measuring the ratio of ^{40}Ar to ^{40}K and knowing the decay rate of ^{40}K in selected samples enables the calculation of the time elapsed since the rock or mineral crystallized.

In the case of $^{40}Ar/^{39}Ar$ dating, the samples are irradiated by fast neutrons in a nuclear reactor forming ^{39}Ar, a daughter isotope of the ^{39}K isotope. The age of the rock is thus directly related to the $^{40}Ar/^{39}Ar$ ratio, which can be measured directly in a mass spectrometer. However, $^{40}Ar/^{39}Ar$ dating of fine-grained clay minerals has a major drawback, namely the distance of the recoil resulting from irradiation is systematically larger than the grain size, which requires sample encapsulation during irradiation as clay minerals are small ($<2\ \mu m$) (Foland et al., 1992; van der Pluijm et al., 2001; Clauer et al., 2012). Precise determination of this recoiled ^{39}Ar trapped in the vial, is analytically challenging in the case of dating illite separates (Clauer et al., 2012). Due to the hygroscopic nature of fine-grained clays, special care must be taken in the preparation of sample splits, clay fraction separation and storage prior to analysis for K-Ar or $^{40}Ar/^{39}Ar$ dating (Clauer and Chaudhuri, 1995; Zwingmann et al., 2017).

7.1.3 Fault gouge dating constraints

In order to obtain reliable authigenic fault clay minerals ages by the K-Ar or $^{40}Ar/^{39}Ar$ method, several prerequisites need to be fulfilled:

1. The clay mineral must behave as a closed system. The system should not have lost radiogenic ^{40}Ar produced by decay of ^{40}K, which requires a closed system with respect to radiogenic ^{40}Ar and K since clay authigenesis.

This aspect questions the $^{40}Ar/^{39}Ar$ method because of the systematic recoil effect and potential loss of ^{39}Ar, which remains an analytical challenge.

2. The radiogenic ^{40}Ar should be produced only by the decay of ^{40} K. No ^{40}Ar should have been incorporated into the rock or mineral before or after its formation or recrystallization.
3. Atmospheric ^{40}Ar should be accounted for by utilizing the present-day $^{40}Ar/^{36}Ar$ value of 295.5.

In nature, clay minerals can roughly be organized into two groups depending on their origin: either.

(1) detrital clays, mainly as weathering products (2 M polytypes, Velde, 1965), or
(2) newly-formed clays (1 M and 1Md polytypes, Velde, 1965; Moore and Reynolds, 1997; Clauer and Lerman, 2012).

The isotopic dating of newly-formed authigenic clay minerals and their crystallization in fault zones requires a special approach (Clauer and Chaudhuri, 1995; Meunier et al., 2004). Consideration of initial sample disintegration methods and clay separation processes are important prior to age dating (Zwingmann et al., 2017). For instance, coarse detrital minerals should not be ground to less than the size of the clay crystals. Observation of additional analytical aspects when dating authigenic clay minerals is essential and might involve: (1) contamination effects and potential mixing with other K-bearing phases (i.e. K-feldspars, micas), (2) Ar closure and loss or capture of radiogenic ^{40}Ar during crystallization, (3) duration of K accumulation, and (4) initial clay mineralogy and, in particular, the presence of detrital illite.

Fault clay mineral authigenesis can occur over short, prolonged, or successive geological time intervals. The process is fundamentally different from classical mineral cooling model, where the Ar closure temperature is considered to be of short duration relative to the mineral lifetime (Faure and Mensing, 2005). The mineral crystallization process can be constrained independently of the concept of a closure temperature as different timescales are involved. If the time interval over which authigenic fault clay minerals crystallize is below the analytical precision of the method (1 or 2 sigma values), an "age" is measured. Conversely, in some fault zones, mineral authigenesis and crystallization processes can extend over extensive geological time periods, larger than the analytical uncertainties, and an "integrated age" must be considered (e.g. Meunier and Velde, 2004). Clauer et al. (2012) emphasized that illite nucleation processes are not temperature dependent during a crystallization event, and closure of the illite K-Ar system does not occur when the temperature reaches a theoretical closure value. Illite-type minerals stop crystallizing at a given temperature ($\sim$60–140 °C), when the "Ar retention temperature" is reached, that is to say when illitization stops, the K-Ar chronometer will start. Instead of a simple cooling model, the authigenic clays in fault zones can, after initial fault formation, be subjected to recrystallization episodes during fault reactivation, if the temperature exceeds the crystallization temperature (e.g., Clauer and Chaudhuri, 1995; Sasseville et al., 2008; Meunier et al., 2004; Vrolijk et al., 2018). Each subsequent event of higher intensity might then partially or fully reset the previous isotopic equilibrium between ^{40}K and radiogenic ^{40}Ar, resulting in a new "integrated age" and thus careful evaluation of the fault gouge material is required to enable geological interpretation of fault formation (Viola et al., 2016).

7.1.4 Authigenic clay gouge age interpretation

Fault gouge illite of authigenic origin may consist of mixtures of illite particles, which formed at different episodes during growth. The growth history can be evaluated by dating a range of different grain-size fractions, the finer separates being younger. Special clay-gouge sample-preparation methods and extensive clay mineral separation (e.g. <0.02, < 0.1, <0.4, <2, 2–6 μm etc.) and characterization techniques (thin-section petrography, scanning electron microscopy (SEM), transmission electron microscopy (TEM), X-ray diffraction (XRD), laser particle size analysis, and microprobe in-situ analysis are important to identify and distinguish suitable clay mineral fractions prior to age dating. Jackson (1979) provided a summary of technical processes that help to improve the clay size separation. Fault gouge illite ages are often interpreted as "maximum" ages because the neo-formed illite, i.e., the finest separated particle size (<0.1 μm or finer), crystallized at the edges of authigenic clay particles and is the most recently crystallized. ^{40}Ar/^{39}Ar dating of coarse-grained mica is a valuable tool in dating higher-grade deformation processes; see, for example, Sherlock et al. (2003). However, as stated before, dating fine-grained μm size fault-gouge illite mineral fractions is complicated due to ^{39}Ar recoil and the necessary consequent corrections (Clauer et al., 2012). For instance, the typical inverse relationship between grain size and amount of contamination can therefore make ^{40}Ar/^{39}Ar dating of illite fault gouges problematic: the finest grain sizes are normally least contaminated, but are susceptible to highest ^{39}Ar recoil and are, therefore, not applicable for ^{40}Ar/^{39}Ar dating without encapsulation (e.g. Foland et al., 1992).

The illite age analysis approach (IAA, Pevear, 1999) can theoretically also provide additional information by distinguishing between authigenic and detrital illite clay mineral phases. Pevear (1999) discussed the influence of detrital mineral phases on authigenic illite formation for sedimentary systems in detail. Detrital contamination and authigenic end member ages can be investigated by estimating the ages of these end-members by:

(1) quantitatively determining (by XRD) the proportions of the end members in each of the size fractions on the basis of their relative 2 M/1 M ratios,
(2) plotting the points (normalized to 100%) as apparent K-Ar age versus percent of detrital illite, and

(3) linearly extrapolating to a 0% and 100% detrital component to obtain the end-member ages. Pevear (1999) proposed that the extrapolated diagenetic age should represent the mean (integrated) age of the time interval over which illite grew, a concept which has been extended to fault-gouge dating studies (e.g. van der Pluijm et al., 2001; Zwingmann et al., 2010a; Mancktelow et al., 2015) (Fig. 7.3).

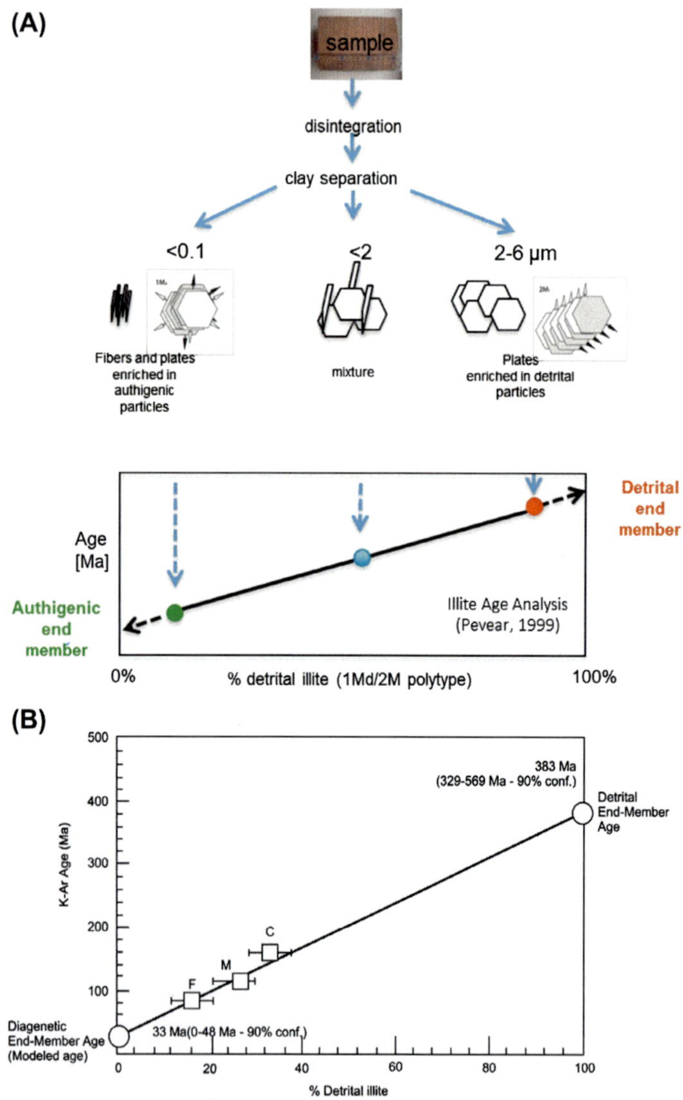

FIG. 7.3 (A) Sample separation and characterization summary with simplified IAA approach. Polytype cartoons redrawn after Moore and Reynolds (1997). (B) Example IAA plot (Illite Age Analysis).

7.1.5 Case studies

Vrolijk et al. (2018) summarize an approach to constrain timing of complex fault isotopic records involving initiation and reactivation over Earth's long history, ranging from the Mesoproterozoic (e.g. Viola et al., 2013) to the Neogene (e.g. Zwingmann et al., 2010a). Vrolijk and van der Pluijm (1999) highlighted, in a pioneering study, the complexities of clay gouge investigations of the Lewis Thrust, Canada, in a prograde sedimentary environment. The study explores processes that occur in clay gouge and discusses detailed mineralogical, isotopic and textural observations in fault zones. The fault gouge samples were dated using the K-Ar method. Younger apparent K-Ar ages within the fault zone imply advanced mineralogical reactions, compared to adjacent undeformed, shale host rocks. In a follow-up study, van der Pluijm et al. (2001) extended the dating of shallow crustal faults of the Lewis Thrust, Canada. They combined $^{40}Ar/^{39}Ar$ clay dating using clay fraction encapsulation with X-ray modeling (WILDFIRE; Grathoff and Moore, 1996) to investigate detrital and authigenic clay polytypes (2M1 vs 1 M). The combined $^{40}Ar/^{39}Ar$ total fusion age data and clay polytype results of fractions indicate, via use of the IAA approach (Pevear, 1999), that the Lewis Thrust was formed ~52 Ma ago, as documented by an extrapolated authigenic clay end member. Another implication of the IAA approach allows to constraint uplift and exhumation ages of the sedimentary source of these rocks to be ~172 Ma, i.e. via its extrapolated detrital-clay end member (Fig. 7.4). Numerous additional studies applied this approach, for example Haines and van der Pluijm (2008, 2010, 2012) investigated the Miocene Sierra Mazatan detachment fault, Sonora in Mexico and Ruby Mountains, Nevada, US and Canada (Sasseville et al., 2008).

Zwingmann and Mancktelow (2004) focused on a different approach by investigating retrograde mineral reactions in fault zones. This work was subsequently extended to the European Alps by Zwingmann et al. (2010a), Surace et al. (2011), Pleuger et al. (2012) and Mancktelow et al. (2015). Similar studies have been reported from Mongolia (Guy et al., 2014), Korea (Zwingmann et al., 2011) and Japan (Zwingmann et al., 2010b; Yamasaki et al., 2013). The main advantage of this approach is that the high-grade metamorphic host rock is primarily void of clay minerals and that all authigenic, fault-gouge clay material was formed by retrograde reactions during brittle deformation. Zwingmann et al. (2010a) report that authigenic illite ages from fresh, non-weathered tunnel samples, collected from the Alptransit construction, document a homogenous age range of 6–7 Ma, which can be related to subsequent closure of the connection with the Atlantic during subduction and slab rollback in the Alps. Additional advantages of these studies are that, given the abundance of geological constraints and additional thermo-chronological information, such as apatite fission track analysis (AFTA) or zircon fission track analysis (ZFTA) data sets, the active retrograde

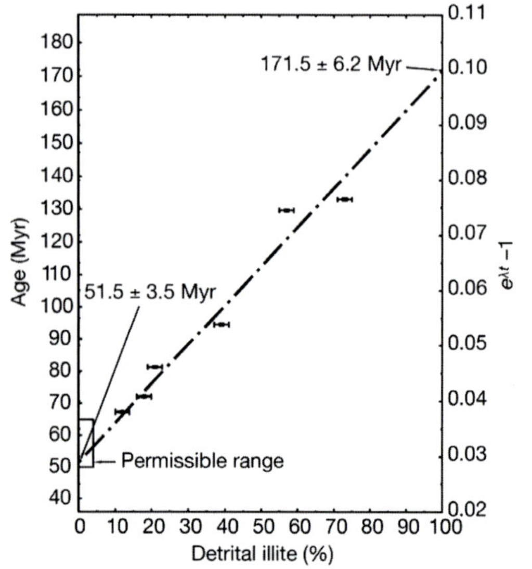

FIG. 7.4 IAA summary after van der Pluijm et al. (2001).

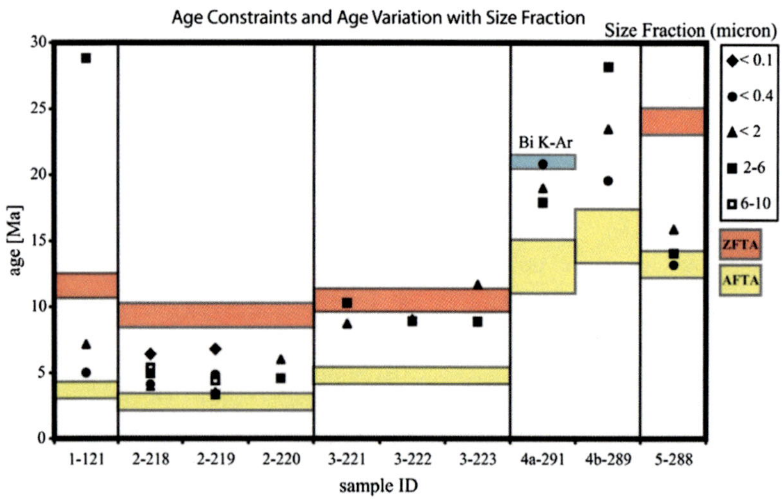

FIG. 7.5 Illite fault gouge ages in relation to AFTA and ZFTA age ranges, after Zwingmann and Mancktelow (2004).

faulting processes are well constrained. Brittle faulting and authigenic illite growth occur under conditions bracketed by the partial annealing zones of apatite and zircon. Ages of retrograde illite fault-gouge plot between the passive uplift-related AFTA and ZFTA ages. Then the illite ages provide absolute time constraints on the youngest retrograde neotectonic part of the Alpine Orogeny (Fig. 7.5).

Viola et al. (2013) extended the complexities of fault zone dating to the Mesoproterozoic in a study of fault core material from two drill-core gouge samples in the basement of Finland that record two distinct faulting events. The size fractions of an older sample yield apparent ages from 1240 to 1006 ± 21 Ma. A second sample is structurally younger and yields statistically distinct ages, ranging from 978 to 886 ± 18 Ma. The ages of the finest <0.1 μm fractions, which are the youngest, are interpreted to document the actual time of faulting. These results provide an example of meaningful isotopic dating of illite-type clay material formed during Precambrian faulting; they also demonstrate and constrain fault reactivation, and provide evidence for brittle Sveconorwegian Mesoproterozoic shortening and orogenic growth, Neoproterozoic extension and orogen collapse in Fennoscandia. Although petrographic and textural evidence might not be as clear as with younger faults, K-Ar illite age data of well-characterised gouge samples may also provide reliable insights into geologically-old, low-temperature faulting histories. This is important because apatite- and zircon-fission-track data, which are common tools to constrain the timing of shallow faulting histories, might be inconclusive or questionable in the case of geological old and complex terranes (e.g., Viola et al., 2013, Fig. 7.6).

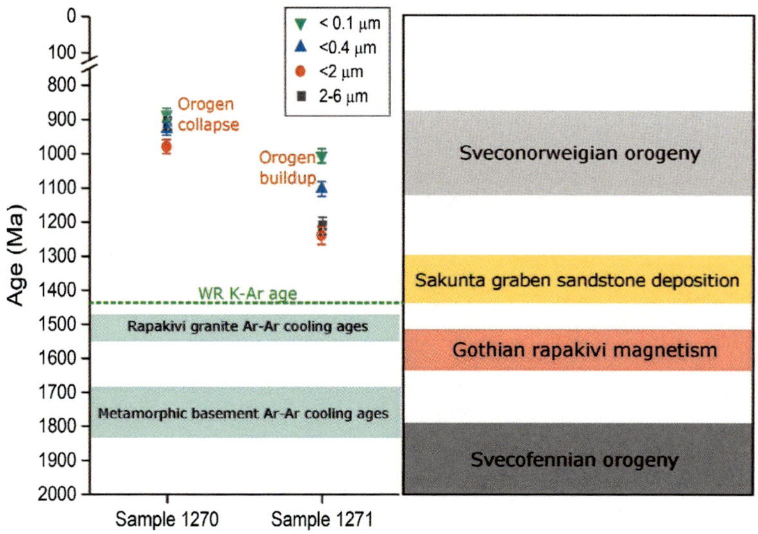

FIG. 7.6 Mesoproterozoic fault dating study, after Viola et al. (2013). Summary of K-Ar age results and main tectonic events that affected the study area. Ages of the finest grain size are interpreted as being the closest to the age of the actual faulting episodes. The <0.1 μm age of sample 1271 dates a phase of faulting interpreted as related to the build-up of the Mesoproterozoic Sveconorwegian orogen in response to overall E-W shortening. The <0.1 μm age of sample 1270 dates reactivation of the same fault during the E-W directed Neoproterozoic extensional collapse of the orogen.

7.2 Dating methods based on thermal reset

7.2.1 Outline of the method

Dating methods are sensitive to temperature, and the temperature at which a mineral-method pair begins to record its age is termed the "closure temperature" (Fig. 7.7). The closure temperature concept was first introduced by Dodson (1973), in which an analytical solution was given by assuming that rock's temperature is proportional to the inverse of time for a relevant part of a rock's thermal history. Around the closure temperature, there is a temperature range in which the clock goes slower than at the ambient temperature because of partial thermal resetting. This is called the "partial retention zone (PRZ)" or "partial annealing zone (PAZ)" (Fig. 7.7). Closure temperatures of various dating methods for the normal geological setting (slow temperature changes) are shown in Table 7.1. Actual temperatures, which need to reset the clock by earthquakes (very fast heating and cooling) should be higher. See, e.g., Tagami (2012) for more details.

Timing and thermal influences of faulting can be inferred by dating methods that have low closure temperatures. The dating is carried out on minerals from rocks in and around the fault core of interest. Thermochronological data place thermo-tectonic constraints on.

(1) the regional thermal history of host rocks, on both footwall and hanging-wall, that may reveal differential amounts of uplift and/or denudation,
(2) signatures of shear heating of fault motion that may help to date ancient faulting episodes and to estimate the paleo-shear stress involved, and/or

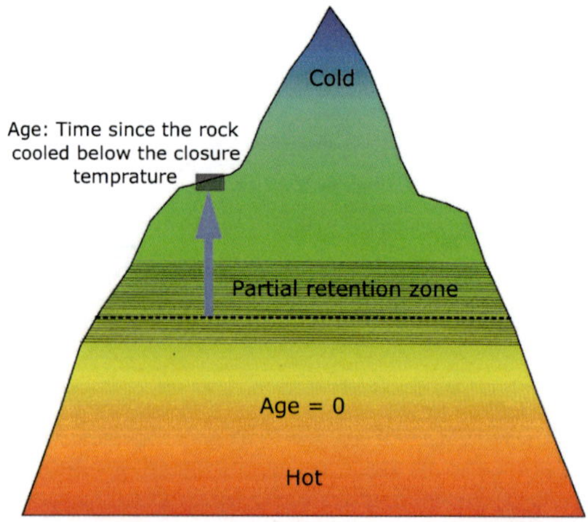

FIG. 7.7 Concept of closure temperature and partial retention zone under the palaeo-geothermal regime in the uplifted and denudated mountains.

TABLE 7.1 Closure temperatures of thermochronometers used for dating fault rocks, after Grün et al. (1999), Reiners et al. (2005), Herman et al. (2010), and King et al. (2016).

Method	Mineral	Closure temperature (°C)
Fission Track	zircon	330–350
	apatite	90–120
(U-Th)/He	zircon	160–200
	apatite	50–115
Electron spin resonance	quartz	50–80
Luminescence	feldspar	40–90
	quartz	30–35

(3) migration of hot fluids in and around the fault zone, which may reflect the faulting history.

Here, we focus on the thermochronological analysis of localized thermal signatures related mainly to (2).

7.2.2 Fission track dating

Fission track (FT) is a linear damage trail produced in non-conductive solid material, caused by the nuclear fission reaction of heavy nuclides, such as ^{238}U, ^{235}U, ^{232}Th, etc. In nature, FTs are formed and accumulated through time effectively by spontaneous fission of ^{238}U in U-bearing minerals, e.g., apatite, zircon and titanite, and in volcanic glasses. A FT age is calculated from quantities of both ^{238}U and spontaneous FTs in a unit volume of the mineral to be dated. The quantity of ^{238}U is measured from the number of FTs that are generated by induced nuclear fission of ^{235}U, as a result of bombardment by thermal neutrons in a nuclear reactor. In recent years, an alternative approach has been introduced to directly measure the number of ^{238}U atoms using Laser Ablation – Inductively Coupled Plasma – Mass Spectrometry (LA-ICP-MS) (e.g., Hasebe et al., 2004; Vermeesch, 2017).

An unparalleled advantage of the FT method is that the FT length information allows quantitative reconstruction of rock's thermal history using inverse modelling techniques. The initial lengths of individual FTs are the same in a specific mineral, and each FT shows a systematic reduction in length (and eventually disappears) as the host mineral is thermally annealed (Fig. 7.8). Because FTs are generated and accumulated at a certain rate through geological time, the distribution of FT lengths reflects a time-temperature pathway of the host mineral within the PAZ (e.g., Green et al., 1989). In quantitative

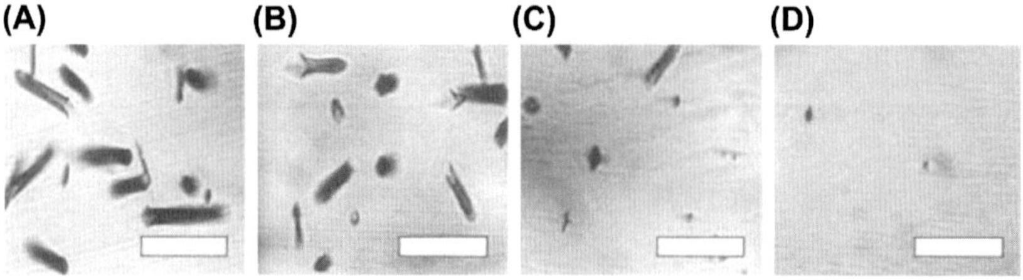

FIG. 7.8 Photographs of spontaneous fission tracks in the Nisatai Dacite zircon. (A) Unannealed, annealed at (B) 650 °C, (C) 700 °C, and (D) 750 °C for 1 h (Tagami, 2005). Scale bar represents 10 μm.

reconstruction of thermal history, time-temperature pathways are randomly produced by numerical approaches, such as the Monte Carlo method, and expected track-length distributions are calculated from individual pathways using the FT thermal annealing kinetics. Then, only those pathways that satisfy the goodness of fitting criterion between the observed and expected distributions are adopted and used to calculate a weighted mean curve as a plausible thermal history. Several modeling packages have been developed and are used widely for thermal history reconstruction, such as QTQt (Gallagher, 2012) and HeFTy (Ketcham, 2005).

A notable disadvantage of the FT method is that the observation and measurement of FTs conventionally, at least for natural rocks, have relied on human skill, and this required time-consuming training. While automatic FT identification and measurement were developed much earlier on for artificial materials and micas with constant background brightness, its application to natural samples with irregular background brightness were not successful for routine analysis until recently. A breakthrough innovation was presented by Gleadow et al. (2009, 2019), who developed an automatic FT analysis system using sophisticated image analysis techniques. The system has been introduced into the leading laboratories, and further refinements in performance are expected. See Tagami & O'Sullivan (2005) and Kohn et al. (2019) for more details of the FT method, and Donelick et al. (2005) and Tagami (2005) for application on apatite and zircon, respectively.

See Tagami (2012) for overviews of fault-zone thermochronology, and Tagami (2019) for more details and recent advancements of FT analyses of fault zones.

7.2.3 (U-Th)/He dating

The (U-Th)/He method utilizes alpha decay products from the radiometric decay series, namely of ^{238}U, ^{235}U and ^{232}Th to ^{206}Pb, ^{207}Pb and ^{208}Pb and from samarium (Sm). A (U-Th)/He age is calculated from the quantities of ^{238}U, ^{235}U, ^{232}Th, ^{137}Sm and ^{4}He in a unit volume of the mineral to be dated. The technique is applicable to U-and Th-rich minerals, such as apatite, zircon and titanite. The mineral grain is first degassed for ^{4}He analysis by appropriate heating, and then chemically dissolved for the determination of the amount of the alpha-emitting elements.

A remarkable advantage of the (U-Th)/He method is its thermal sensitivity due to the high diffusivity of He (Table 7.1), since it is the lightest noble gas. Hence, the method is widely used for thermal history reconstruction for temperatures below 200°C. In addition, (U-Th)/He ages generally have a higher age resolution than FT ages, for instance, for latest Tertiary to Quaternary ages due to the higher production rate of the daughter nuclide (e.g., Danišík et al., 2017). Recent application to hematite opens a new way to constrain temporal evolution of fault zones (Evenson et al., 2014; Ault et al., 2015; see below for details).

One of inherent uncertainties of (U-Th)/He ages is derived from the initial motion of ^{4}He nuclides when alpha decays take place. Because ^{4}He particles are emitted randomly and travel ~20 μm from their original localities within the mineral, which leads to a certain amount of loss out of the grain, an appropriate age correction needs to take account of the size and morphology of the host grain and also the distribution of U and Th atoms in and around the grain. Another issue in terms of thermochronology is systematic change in ^{4}He-diffusivity controlled by accumulation of radiation damage in U- and Th-rich minerals (e.g., Guenthner et al., 2013). Thus, care should be taken when interpreting (U-Th)/He data of old grains, where significant radiation damage has accumulated.

See Farley (2002) for more details of the (U-Th)/He method, and Reiners et al. (2017; Chapter 11) for recent advancements.

7.2.4 Trapped charge dating

Trapped charge dating is a generic term for optically- or thermally-stimulated luminescence (TL, OSL) dating (Aitken, 1985, 1988) and electron spin resonance (ESR) dating (Ikeya, 1993). The method detects unpaired electrons and holes that accumulate over time because they are exposed to natural radiation from radioactive nuclides (U, Th and ^{40}K) and cosmic rays (Fig. 7.9). Such electrons and holes are trapped and stabilized at lattice defects and impurities in the crystal lattice. In quartz (SiO_2) for instance, trapped electrons and holes are localized at Al and Ti impurities, which substitute for Si, and oxygen vacancy or interstitial oxygen (Fig. 7.10).

Although luminescence and ESR dating methods have common physical basis, the way of detecting trapped charge is different. Luminescence is the light emission when a trapped electron is stimulated by light or heat and then recombines with a hole. Measurements of luminescence are usually carried out using sand- or silt-sized mineral grains prepared on metal discs. Measurements are conducted by automated readers, which contain a photomultiplier, stimulation light sources (LEDs), a heating plate, and a beta-ray irradiator (Fig. 7.11). Normally, blue LEDs are used to stimulate quartz, and IR LEDs are used for feldspar.

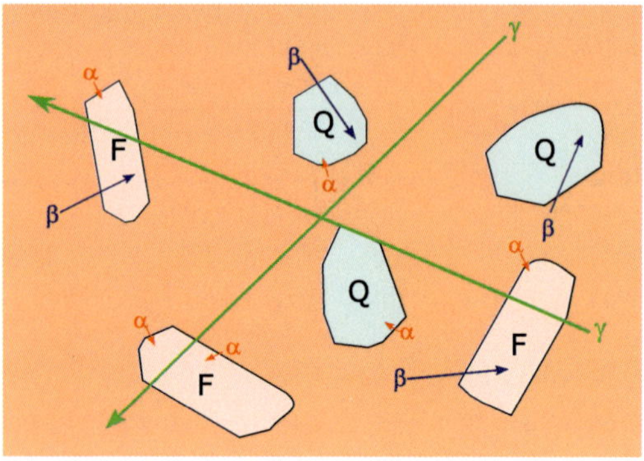

FIG. 7.9 Sources of natural radiations, which produce trapped electrons in minerals.

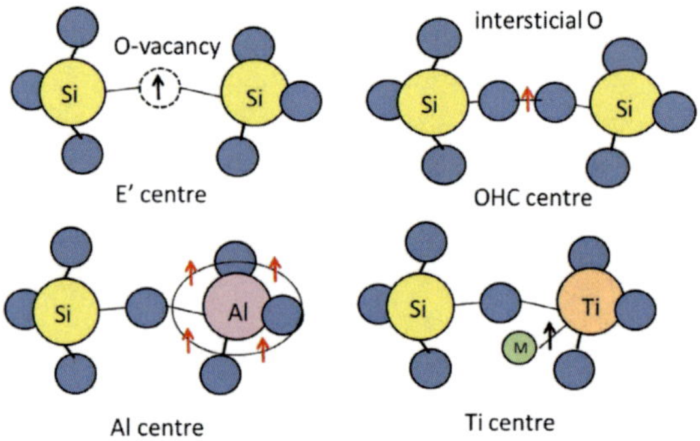

FIG. 7.10 Examples of quartz lattice defects and impurity centres.

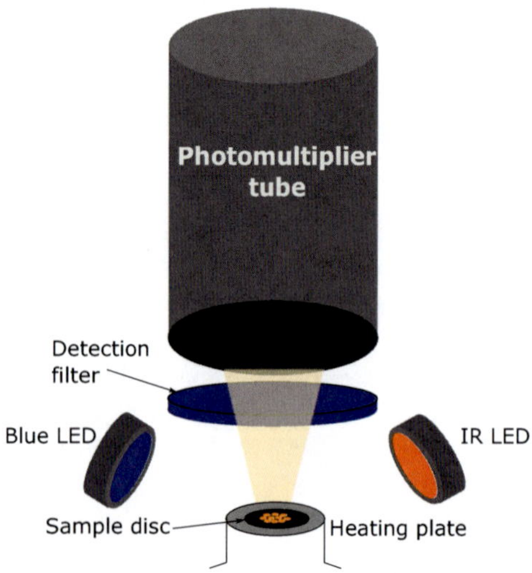

FIG. 7.11 Schematic drawing of luminescence measurement device.

ESR measures the absorption of microwaves. The equipment of ESR consists of electromagnets, a microwave unit and a spectrometer. Samples are filled in quartz glass tubes and placed in a microwave cavity between magnets. When trapped electrons are in the magnetic field, the direction of the electron spin is aligned either parallel or antiparallel to the direction of the magnetic field, because spinning electrons also act as tiny magnets. The two directions of the electron spin have different energy states and electrons absorb microwave energy when the microwave energy is in accordance with the difference of the two energy states. This is electron spin resonance. Measurements of ESR is conducted by scanning the magnetic field, and ESR signals are detected as first derivative of the microwave absorption (Fig. 7.12).

An age, *T*, is derived by dividing the total accumulated energy in the crystal lattice from the natural radiation, called equivalent dose (D_e) with annual radiation dose rate ($\dot{D}$).

$$T = \frac{D_e}{\dot{D}}$$

To calculate the D_e, the natural luminescence or ESR signal is first measured. Then the natural signal intensity is compared with the signal when

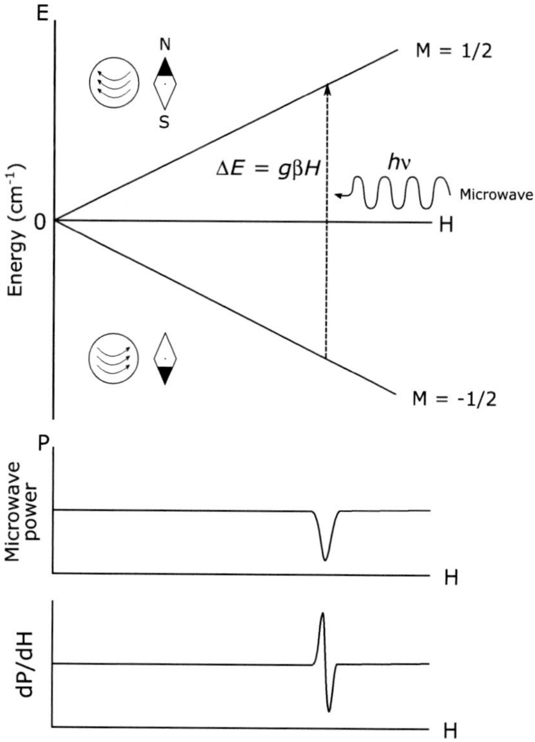

FIG. 7.12 Principle of electron spin resonance.

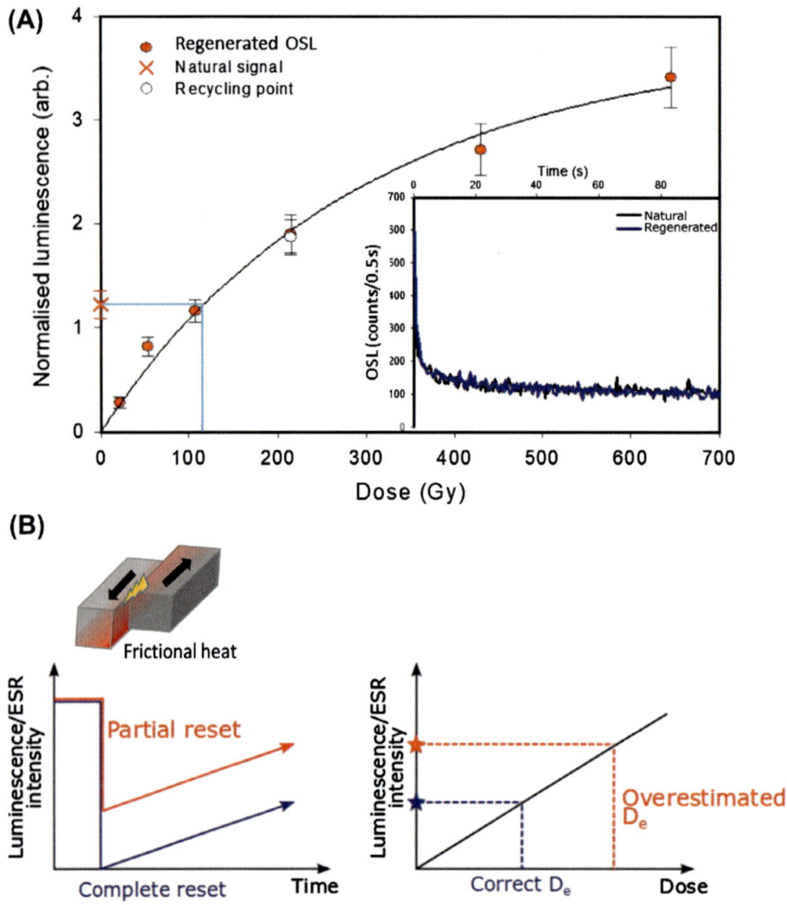

FIG. 7.13 (A) Equivalent dose measurement and example OSL decay curve of quartz. (B) Complete and incomplete resetting and the effect on the D_e estimation.

artificially irradiated in the laboratory (Fig. 7.13A). The annual dose rate can be calculated from U, Th and K concentration in the surrounding material of the sample, e.g., using gamma spectrometry, neutron activation analysis or ICP.

Normally, luminescence and ESR dating deal with quartz and feldspar in sediments, because trapped charges are reset by sunlight exposure during the sediment transport before burial. When the methods are applied to faults, the zeroing event is the fault movement (or earthquake) (Fig. 7.13B). If the frictional heat during the last earthquake was sufficiently high (and long lasting) to reset a signal, the age of the last earthquake can be derived. However if resetting was not complete, the age of the last earthquake is overestimated. This is likely to happen when samples are taken from surface outcrops of faults. However, due to the lower closure temperatures of these methods compared to FT and (U-Th)/He, there is a higher chance to be able to date the last earthquake event using luminescence and ESR.

7.2.5 Case studies

FT thermochronology was applied to apatite separated from rock samples within and adjacent to the San Gabriel Fault Zone, southern California, which was probably active from 13 to 4 Ma (d'Alessio et al., 2003). At the study site, the fault zone consists of a 1-8 cm-wide ultracataclasite that juxtaposes basement gneiss to the north against granodiorite to the south. No significant reduction in apatite FT length and age was found towards the fault plane, and this was interpreted to suggest that either each slip event did not exceed 4 m, or the average apparent coefficient of friction was less than 0.4 (d'Alessio et al., 2003).

After early attempts to date the glassy matrix of pseudotachylytes using Ar/Ar, FT and Rb/Sr methods, FT thermochrology was successfully applied to zircons separated from pseudotachylytes of the Nojima Fault Zone, southwest Japan (Murakami and Tagami, 2004), the Asuke Shear Zone, central Japan (Murakami et al., 2006), the Tsergo Ri Landslide, Nepal (Takagi et al., 2007), and the Median Tectonic Line, southwest Japan (Takagi et al., 2010). Murakami and Tagami (2004) performed zircon FT analysis for a series of rocks collected from a 2−10 mm wide pseudotachylyte layer and adjacent units, and found that both FT age and length show systematic changes towards the pseudotachylyte layer, where the youngest, total reset age was measured at 56 ± 4 (1 s.e.) Ma (Fig. 7.14). By using functions of frictional heat production and conductive transfer and also of zircon FT annealing, the net shear work was estimated as ~50 MPa m for the ancient faulting (Murakami, 2010), which favours a "strong" fault model with higher shear stress (Scholz, 2002).

Another example of constraining the thermo-tectonic evolution of a fault zone was recently reported on conjugate fault zones that were exposed along an artificial pit excavated to reveal the bedrock structure of the nuclear power plant at the Monju site, southwest Japan (Sueoka et al., 2017, Fig. 7.15). The fault zone consists of two individual fault zones, both of which are associated with plastic deformation of biotite grains, suggesting palaeotemperatures >200−300°C. Apatites separated from a fault gouge and adjacent granitic host rocks reveal FT ages of 49.6−17.6 Ma, which were interpreted, along with FT length data, to reflect a secondary heating episode of the fault zone caused by a basaltic intrusion. The spatial distribution of apatite FT data, along with initial cooling history of the granitic body, also allows the estimation of the timing of conjugate faulting as 68-50 Ma and/or ~19 Ma.

(U-Th)/He and FT analyses were conducted on apatites separated from drill cores that transect the Lavanttal Fault System, Eastern Alps (Wölfler et al., 2010). The measured apatite (U-Th)/He and FT ages, along with the apatite FT

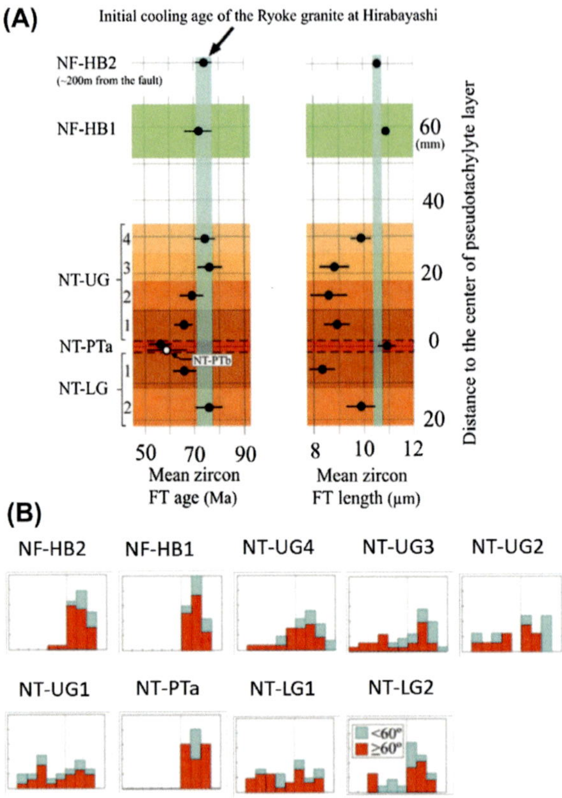

FIG. 7.14 Summary of fission track dating of pseudotachylyte of the Nojima Fault, Japan (Murakami and Tagami, 2004). (A) Measured mean age and mean length and of fission tracks in zircon are plotted against distance from the centre of a pseudotachylyte layer of the Nojima Fault. Error bar represents $\pm$ 1 s.e. (B) FT length distribution of each sample.

lengths, show a reduction towards the core of the fault zone. These results infer that the samples were likely thermally overprinted either by hot fluid flow and/or by frictional heating within the fault zone.

The (U-Th)/He method was also applied to zircon separates from a pseudotachylyte and its tonalite host rock, collected about 100 m north from the Median Tectonic Line, southwest Japan (Yamada et al., 2012). The measured ages of 60.3 ± 2.7 and 55.5 ± 4.5 (1σ) Ma for the pseudotachylyte and its host rock, respectively, suggest that the pseudotachylyte formed at $\sim$60 Ma under the ambient temperature of $\sim$180 °C, i.e., the closure temperature of zircon (U-Th)/He thermochronometry.

(U-Th)/He analyses were also performed on zircon separates from eleven Helminthoid Flysch sandstone samples, collected from a thrust fault zone and surrounding wall rocks in the Ligurian Alps (Maino et al., 2015, Fig. 7.16). X-ray diffraction analysis of clay minerals and fluid inclusion

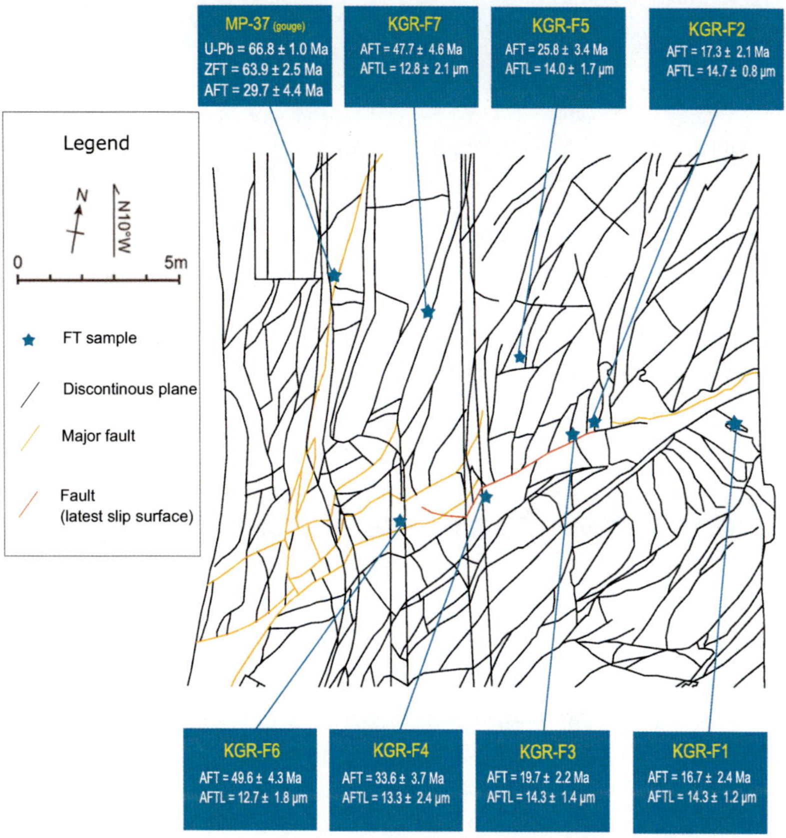

FIG. 7.15 Summary of ZFT and AFT ages from faults near Monju site. *Modified after Sueoka et al. (2017).*

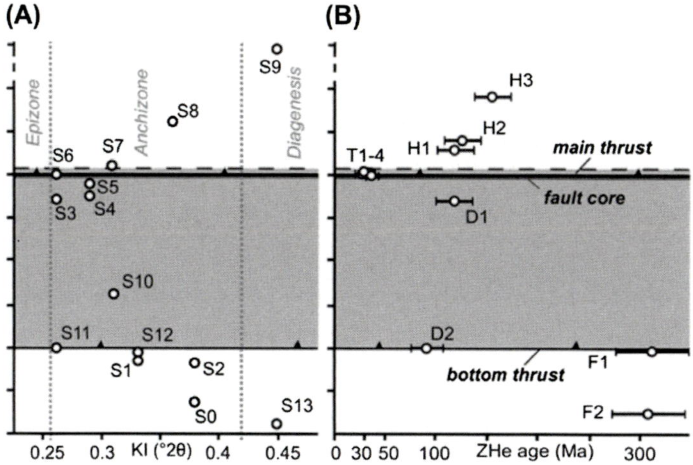

FIG. 7.16 (A) Illite crystallinity and (B) zircon (U-Th)/He ages plotted against distance from the fault core in the Lingurian Alps (Maino et al., 2015). Error bar represents $\pm 2\sigma$.

microthermometry of the vein-filling mineral were also carried out to constrain paleotemperature conditions of the fault zone and wall-rocks. While the wall rocks yield pre-depositional inherited ages of 125.3 ± 15 to 312.3 ± 37 Ma, younger ages of 28.8 ± 3.4 and 33.8 ± 4.0 Ma were obtained from the fault core, which was interpreted to have been reset by thrusting. This implies that the maximum temperature of fault zone exceeded 200°C during thrusting, consistent with the palaeotemperature constraints by illite crystallinity and fluid inclusion data. Thermal modelling of the fault zone suggests that the temperature increase of 50–70°C can be generated by the shear heating during plausible fault motion, with displacement of 10–25 km in 2–10 Myr.

The (U-Th)/He method has recently been applied to hematite that is preserved on minor fault surfaces and in veins of the Wasatch Fault damage zone, Utah (Ault et al., 2015). Hematite (U-Th)/He data were reported for 48 aliquots from 10 samples from 5 fault surfaces, and nine samples yielded intrasample reproducible results of 5.7 ± 0.7 (1σ) to 2.5 ± 0.2 Ma. Along with apatite (U-Th)/He and FT data suggesting that the host rocks cooled through 60 °C by ~4.5 Ma, hematite (U-Th)/He dates of ~4.5 to 2.5 Ma from isolated locations on a single fault surface imply that the dates do not simply record regional cooling. These dates, coupled with observed crystal texture, are better interpreted as recording the timing of localized thermal resetting due to frictional heat at asperities during seismic slip events between ~4.5 and 2.5 Ma.

The Atotsugawa Fault is a right-lateral strike-slip fault and one of the most active faults in central Japan. The Hietsu Earthquake, with an estimated magnitude of 7.0–7.1, was the last activity on the fault that occurred in 1858 (Research Group of Atotsugawa Fault, 1989). Ikeya et al. (1982) reported the first ESR ages of quartz from the Atotsugawa Fault. They observed that the ages from the E' centre of the quartz sample from the fault plane was much

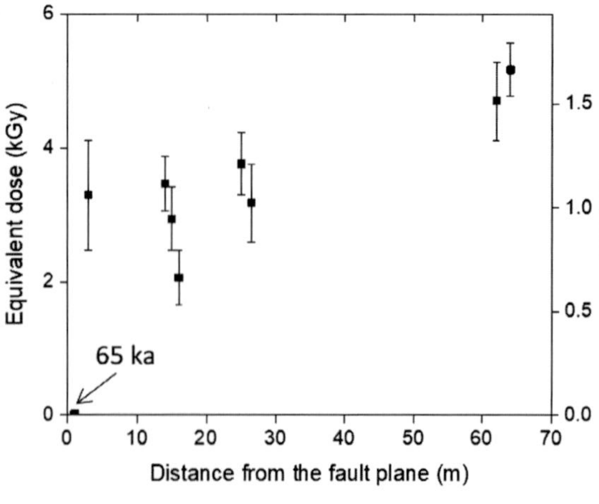

FIG. 7.17 Results of the first study of ESR dating of fault gouges from the Atotsugawa Fault, Japan (after Ikeya et al., 1982).

younger, ~65 ka, than the other parts of the fault core, ~1 Ma (Fig. 7.17), whereas the age of the bedrock at the site could not be measured due to saturation of the signal. Ganzawa et al. (2013) used sand-size quartz from a fault gouge of the Atotsugawa Fault and tested OSL and TL dating. The OSL age of quartz from a gouge sample was consistent with the age of the Hietsu Earthquake (200 ± 200 years), but TL ages were much older (190 and 340 ka).

Lee and Schwarcz (1994, 1996) investigated the San Gabriel Fault zone, which is located parallel to and south of the San Andreas Fault System, southern California. As mentioned above, the San Gabriel Fault is thought to have been active between 13 and 4 Ma, but is suggested to be extinct since then. They measured multiple ESR signals (E' and OHC centres) from quartz of different grain sizes. The ESR ages show a plateau towards smaller grain-size fraction, which possibly indicates a complete resetting at the time of the earthquake (Fig. 7.18). They observed four active periods of faulting over the last 400 ka, at 350, 230, 150 and 4 ka, suggesting that the fault has experienced a number of cycles of activity and inactivity.

After the Great Hanshin Earthquake in 1995, Fukuchi (2001) dated the fault gouge of the Nojima Fault from a deep core at a depth of 388 m. He found that the ESR signal was not completely reset by the 1995 earthquake even from the deep core, and obtained 150 ka age for the gouge. Matsumoto et al. (2001) also used the same deep core and measured quartz grains at intervals of 2 mm across the fault plane. A clear decrease of the signal intensity towards the fault plane was observed for both the Al and E' centres, but similar to Fukuchi (2001), both signal intensities were not at zero for the

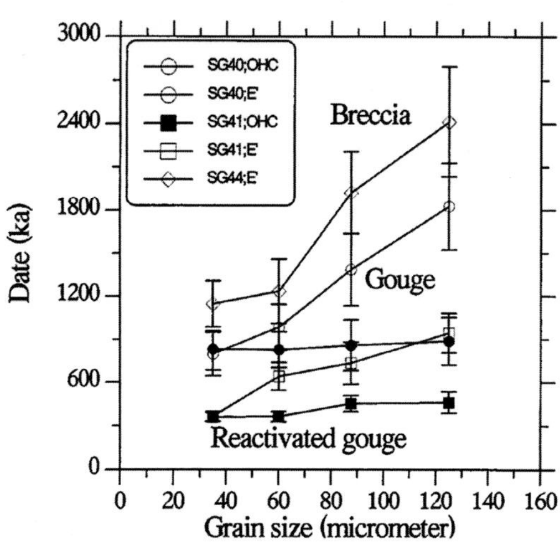

FIG. 7.18 ESR ages quartz with different grain sizes from of the San Gabriel Fault, California (Lee and Schwarcz, 1996).

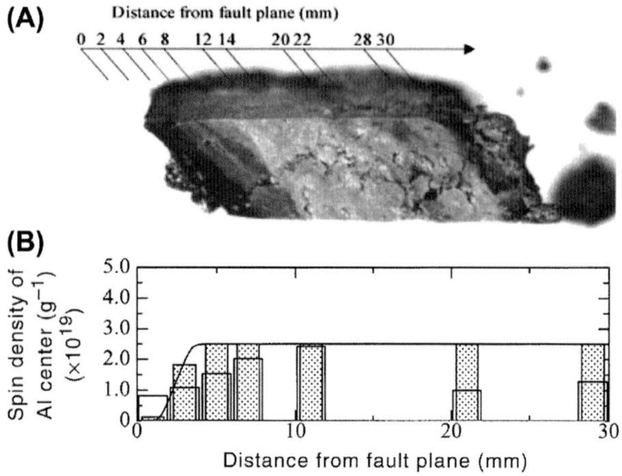

FIG. 7.19 (A) Photo of sample gouge from the Nojima Fault in a deep drilling core and (B) ESR intensity plotted against the distance from the fault plane for the Al centre. *Modified after Matsumoto et al. (2001).*

sample taken 0–2 mm from the fault surface. They further modelled the heat transfer at the time of the earthquake and obtained a broadly consistent intensity of the observed signals (Fig. 7.19).

References

Aitken, M.J., 1985. Thermoluminescence Dating. Academic Press, London.

Aitken, M.J., 1988. An Introduction to Optical Dating: The Dating of Quaternary Sediments by the Use of Photon Stimulated Luminescence. Oxford Scientific Publications.

Ault, A.K., Reiners, P.W., Evans, J.P., Thomson, S.N., 2015. Linking hematite (U-Th)/He dating with the microtextural record of seismicity in the Wasatch fault damage zone, Utah, USA. Geology 43, 771–774.

Clauer, N., Chaudhuri, S., 1995. Clays and Crustal Cycles Isotope Dating and Tracing. Springer-Verlag, Heidelberg-New York, 359 p.

Clauer, N., Lerman, A., 2012. Thermal history analysis of sedimentary basins: an isotopic approach to illitization. In: Harris, N.D., Peters, K. (Eds.), Thermal History Analysis of Sedimentary Basins: Methods and Case Histories, vol. 11. SEPM, Special Publication, pp. 125–146.

Clauer, N., Zwingmann, H., Liewig, N., Wendling, R., 2012. Comparative $^{40}Ar/^{39}Ar$ and K-Ar dating of illite-type clay minerals: a tentative explanation for age identities and differences. Earth Sci. Rev. 115, 76–96.

d'Alessio, M.A., Blythe, A.E., Bürgmann, R., 2003. No frictional heat along the San Gabriel Fault, California: evidence from fission-track thermochronology. Geology 31, 541–544.

Dalrymple, G.B., Lanphere, M.A., 1969. Potassium-Argon Dating. W.H. Freeman, San Francisco, 258 pp.

Danišík, M., Schmitt, A.K., Stockli, D.F., Lovera, O.M., Dunkl, I., Evans, N.J., 2017. Application of combined U-Th-disequilibrium/U-Pb and (U-Th)/He zircon dating to tephrochronology. Quat. Geochronol. 40, 23–32.

Dodson, M.H., 1973. Closure temperature in cooling geochronological and petrological systems. Contrib. Mineral. Petrol. 40, 259–274.

Donelick, R.A., O'Sullivan, P.B., Ketcham, R.A., 2005. Apatite fission-track analysis. Rev. Mineral. Geochem. 58, 49–94.

Evenson, N.S., Reiners, P.W., Spencer, J., Shuster, D.L., 2014. Hematite and Mn oxide (U-Th)/He dates from the Buckskin-Rawhide detachment system, western Arizona: constraining the timing of mineralization and hematite (U-Th)/He systematics. Am. J. Sci. 314, 1373–1435.

Farley, K.A., 2002. (U-Th)/He dating: techniques, calibrations, and applications. Rev. Mineral. Geochem. 47, 819–844.

Faure, G., Mensing, T.M., 2005. Isotopes – Principles and Applications, third ed. J. Wiley and Sons, Inc., Hoboken, New Jersey. 897 pp.

Foland, K.A., Hubacher, F.A., Arehart, G.B., 1992. $^{40}Ar/^{39}Ar$ an encapsulated-vial procedure to overcome the problem of ^{39}Ar recoil loss. Chem. Geol. 102, 269–276.

Fukuchi, T., 2001. Assessment of fault activity by ESR dating of fault gouge; an example of the 500 m core samples drilled into the Nojima earthquake fault in Japan. Quat. Sci. Rev. 20, 1005–1008.

Gallagher, K., 2012. Transdimensional inverse thermal history modeling for quantitative thermochronology. J. Geophys. Res. 117, B02408.

Ganzawa, Y., Takahashi, C., Miura, K., Shimizu, S., 2013. Dating of active fault gouge using optical stimulated luminescence and thermoluminescence. J. Geol. Soc. Jpn. 119, 714–726 (in Japanese with English abstract).

Gleadow, A.J.W., Gleadow, S.J., Belton, D.X., Kohn, B.P., Krochmal, M.S., Brown, R.W., 2009. Coincidence Mapping – A Key Strategy for the Automatic Counting of Fission Tracks in Natural Minerals, 324. Geological Society, London, pp. 25–36. Special Publications.

Gleadow, A.J.W., Kohn, B., Seiler, C., 2019. The future of fission-track thermochronology. In: Fission-Track Thermochronology and its Application to Geology. Springer, pp. 77–92.

Grathoff, G.H., Moore, D.M., 1996. Illite polytype quantification using WILDFIRE—calculated patterns. Clay Clay Miner. 44, 835–842.

Green, P.F., Duddy, I.R., Laslett, G.M., Hegarty, K.A., Gleadow, A.J.W., Lovering, J.F., 1989. Thermal annealing of fission tracks in apatite 4. Quantitative modelling techniques and extension to geological timescales. Chem. Geol. 79, 155–182.

Grün, R., Tani, A., Gurbanov, A., Koshchug, D., Williams, I., Braun, J., 1999. A new method for the estimation of cooling and denudation rates using paramagnetic centers in quartz: A case study on the Eldzhurtinskiy Granite, Caucasus. J. Geophys. Res.: Solid Earth 104 (B8), 17531–17549.

Guenthner, W.R., Reiners, P.W., Ketcham, R.A., Nasdala, L., Giester, G., 2013. Helium diffusion in natural zircon: radiation damage, anisotropy, and the interpretation of zircon (U-Th)/He thermochronology. Am. J. Sci. 313 (3), 145–198. https://doi.org/10.2475/03.2013.01.

Guy, A., Schulmann, K., Clauer, N., Hasalova, P., Seltmann, R., Armstrong, R., Lexa, O., Benedicto, A., 2014. Late Palaeozoic-Mesozoic tectonic evolution of the Trans-Altai and South Gobi Zones in southern Mongolia based on structural and geochronological data. Gondwana Res. 25/1, 309–337. https://doi.org/10.1016/j.gr.2013.03.014.

Haines, S.H., van der Pluijm, B.A., 2008. Clay quantification and Ar/Ar dating of synthetic and natural gouge: application to the Miocene Sierra Mazatan detachment fault, Sonora, Mexico. J. Struct. Geol. 30, 525–538.

Haines, S.H., van der Pluijm, B.A., 2010. Dating the detachment fault system of the Ruby Mountains, Nevada: significance for the kinematics of low angle normal faults. Tectonics 29, 525–538.

Haines, S.H., van der Pluijm, B.A., 2012. Patterns of mineral transformations in clay gouge, with examples from low-angle normal fault rocks in the western USA. J. Struct. Geol. 43, 2–32.

Hasebe, N., Barbarand, J., Jarvis, K., Carter, A., Hurford, A.J., 2004. Apatite fission-track chronometry using laser ablation ICP-MS. Chem. Geol. 207, 135–145.

Herman, F., Rhodes, E.J., Braun, J., Heiniger, L., 2010. Uniform erosion rates and re-lief amplitude during glacial cycles in the Southern Alps of New Zealand, as revealed from OSL-thermochronology. Earth Planet. Sci. Lett. 297, 183–189.

Hobbs, B.H., Means, W.D., Williams, P.F., 1976. An Outline of Structural Geology. John Wiley and Sons, New York, USA, 571 pp.

Holdsworth, R.E., Butler, C.A., Roberts, A.M., 1997. The recognition of reactivation during continental deformation. J. Geol. Soc. London 154, 73–78.

Ikeya, M., 1993. New Applications of Electron Spin Resonance: Dating, Dosimetry and Microscopy. World Scientific.

Ikeya, M., Miki, T., Tanaka, K., 1982. Dating of a fault by electron spin resonance on intrafault materials. Science 215, 1392–1393.

Jackson, M.L., 1979. Soil Chemical Analysis – Advanced Course, second ed. Published by the author, Madison, Wisconsin, USA. 895 p.

Ketcham, R.A., 2005. Forward and inverse modeling of low-temperature thermochronometry data. Rev. Mineral. Geochem. 58, 275–314.

King, G.E., Herman, F., Guralnik, B., 2016. Northward migration of the eastern Himalayan syntaxis revealed by OSL thermochronometry. Science 353, 800–804.

Kohn, B., Ling, C., Gleadow, A., 2019. Fission-track analysis: field collection, sample preparation and data acquisition. In: Fission-Track Thermochronology and its Application to Geology. Springer, pp. 25–48.

Lee, H.K., Schwarcz, H.P., 1994. Criteria for complete zeroing of ESR signals during faulting of the San Gabriel fault zone, southern California. Tectonophysics 235, 317–337.

Lee, H.K., Schwarcz, H.P., 1996. Electron spin resonance plateau dating of periodicity of activity on the San Gabriel fault zone, southern California. Geol. Soc. Am. Bull. 108, 735–746.

Lyons, J.B., Snellenberg, J., 1971. Dating faults. Bull. Geol. Soc. Am. 82, 1749–1752.

Maino, M., Casini, L., Ceriani, A., Decarlis, A., Di Giulio, A., Seno, S., Setti, M., Stuart, F.M., 2015. Dating shallow thrusts with zircon (U-Th)/He thermochronometry— the shear heating connection. Geology 43, 495–498.

Matsumoto, H., Yamanaka, C., Ikeya, M., 2001. ESR analysis of the Nojima Fault gouge, Japan, from the DPRI 500m borehole. Isl. Arc 10, 479–485.

McDougall, I., Harrison, M., 1999. Geochronology and Thermochronology by the 40Ar/39Ar Method, second ed. Oxford University Press, Oxford. 269 pp.

Mancktelow, N., Zwingmann, H., Campani, M., Fügenschuh, B., Mulch, A., 2015. Timing and conditions of brittle faulting on the silltal-brenner fault zone, Eastern Alps (Austria). Swiss J. Geosci. 108 (2–3), 3025–3326. https://doi.org/10.1007/s00015-015-0179-y.

Meunier, A., Velde, B., Zalba, P., 2004. Illite K–Ar dating and crystal growth processes in diagenetic environments: a critical review. Terra Nova 16, 296–304.

Meunier, A., 2005. Clays. Springer, Berlin, 472 pp.

Meunier, A., Velde, B., 2004. Illite: Origins, Evolution and Metamorphism. Springer, Berlin. 286 pp.

Moore, D.A., Reynolds, J.R., 1997. X-ray Diffraction and the Identification and Analysis of Clay Minerals. Oxford University Press, Oxford, 378 pp.

Murakami, M., 2010. Average shear work estimation of Nojima Fault from fission-track analytical data. Earth Monthly 32, 24–29 (in Japanese).

Murakami, M., Kosler, J., Takagi, H., Tagami, T., 2006. Dating pseudotachylyte of the Asuke Shear Zone using zircon fission-track and U–Pb methods. Tectonophysics 424, 99–107.

Murakami, M., Tagami, T., 2004. Dating pseudotachylyte of the Nojima Fault using the zircon fission-track method. Geophys. Res. Lett. 31/12, L12604. https://doi.org/10.1029/2004GL020211.

Pevear, D.R., 1999. Illite and hydrocarbon exploration. Proc. Natl. Acad. Sci. U.S.A. 96, 3440–3446.

Pleuger, J., Mancktelow, N., Zwingmann, H., Manser, M., 2012. K-Ar dating of synkinematic clay gouges from Neoalpine faults of the Central, Western and Eastern Alps. Tectonophysics 550–553, 1–16. https://doi.org/10.1016/j.tecto.2012.05.001.

Reiners, P.W., Carlson, R.W., Renne, P.R., Cooper, K.M., Granger, D.E., McLean, N.M., Schoene, B., 2017. Geochronology and Thermochronology. John Willey & Sons.

Reiners, P.W., Ehlers, T.A., Zeitler, P.K., 2005. Past, present, and future of thermochronology. Rev. Mineral. Geochem. 58, 1–18.

Research Group for the Atotsugawa Fault, 1989. Trenching study of the Atotsugawa Fault at Nokubi, Miyagawa Village, Gifu Prefecture, Central Japan. J. Geogr. 98, 440–463 (in Japanese with English abstract).

Sasseville, C., Tremblay, A., Clauer, N., Liewig, N., 2008. K-Ar time constraints on the evolution of polydeformed fold-thrust belts: the case of the Northern Appalachians (southern Québec). J. Geodyn. 45, 99–119.

Scholz, C.H., 2002. The Mechanics of Earthquakes and Faulting, second ed. Cambridge University Press, Cambridge, UK. 471 pp.

Sherlock, S.C., Kelley, S.P., Zalasiewicz, J., Schofield, D., Evans, J., Merriman, R.J., Kemp, S.J., 2003. Absolute dating of low-temperature deformation: strain-fringe dating by Ar-Ar laser microprobe. Geology 31, 219–222. https://doi.org/10.1130/0091-7613(2003)031<0219:PDOLTD>2.0.CO;2.

Steiger, R.H., Jäger, E., 1977. Subcommission on geochronology: convention on the use of decay constants in geo- and cosmochronology. Earth Planet. Sci. Lett. 36 (3), 359–362.

Sueoka, S., Shimada, K., Ishimaru, T., Niwa, M., Yasue, K., Umeda, K., Danhara, K., Iwano, H., 2017. Fission-track dating of faulting events accommodating plastic deformation of biotites. J. Geophys. Res. Solid Earth 122, 1848–1859.

Surace, I., Clauer, N., Thélin, P., Pfeifer, H.R., 2011. Structural analysis, clay mineralogy and K-Ar dating of fault gouges from Centovalli Line (Central Alps) for reconstruction of their recent activity. Tectonophysics 510, 80–93.

Tagami, T., 2005. Zircon fission-track thermochronology and applications to fault studies. Rev. Mineral. Geochem. 58, 95–122.

Tagami, T., 2012. Thermochronological investigation of fault zones. Tectonophysics 538–540, 67–85.

Tagami, T., 2019. Application of fission-track thermochronology to understand fault zones. In: Fission-Track Thermochronology and its Application to Geology. Springer, pp. 221–233.

Tagami, T., O'Sullivan, P.B., 2005. Fundamentals of fission-track thermochronology. Rev. Mineral. Geochem. 58, 19–47.

Takagi, H., Arita, K., Danhara, T., Iwano, H., 2007. Timing of the Tsergo Ri Landslide, langtang himal, determined by fission track age for pseudotachylite. J. Asian Earth Sci. 29, 466–472.

Takagi, H., Shimada, K., Iwano, H., Danhara, T., 2010. Oldest record of brittle deformation along the Median Tectonic Line: fission-track age for pseudotachylite in the Taki area, Mie Prefecture. J. Geol. Soc. Jpn. 116, 45–50.

van der Pluijm, B.A., Hall, C.M., Vrolijk, P.J., Pevear, D.R., Covey, M.C., 2001. The dating of shallow faults in the Earth's crust. Nature 412, 172–175.

Velde, B., 1965. Experimental determination of muscovite polymorph stabilities. Am. Mineral. 50, 436–449.

Velde, B., 1992. Introduction to Clay Minerals: Chemistry, Origins, Uses, and Environmental Significance. Chapman & Hall, 198 p.

Velde, B., 2004. Green clay minerals. In: Holland, H.D., Turekian, K.K. (Eds.), Treatise on Geochemistry 7. Sediments, Diagnosis, and Sedimentary Rocks, vol. 7. Elsevier, Amsterdam, pp. 309–324.

Vermeesch, P., 2017. Statistics for LA-ICP-MS based fission track dating. Chem. Geol. 456, 19–27.

Viola, G., Zwingmann, H., Mattila, J., Käpyaho, A., 2013. K-Ar illite age constraints on the Proterozoic formation and reactivation history of a brittle fault in Fennoscandia. Terra Nova 25, 236–244.

Viola, G., Scheiber, T., Fredin, O., Zwingmann, H., Margreth, A., Knies, J., 2016. Deconvoluting complex structural histories archived in brittle fault zones. Nat. Commun. 7, 13448. https://doi.org/10.1038/ncomms13448.

Vrolijk, P., van der Pluijm, B.A., 1999. Clay gouge. J. Struct. Geol. 21, 1039–1048.

Vrolijk, P., Pevear, D.R., Covey, M., LaRiviere, A., 2018. Fault gouge dating: history and evolution. Clay Miner. 53, 305–324.

Wölfler, A., Kurz, W., Danišík, M., Rabitsch, R., 2010. Dating of fault zone activity by apatite fission track and apatite (U–Th)/He thermochronometry: a case study from the Lavanttal Fault system (Eastern Alps). Terra Nova 22, 274–282.

Yamada, K., Hanamuro, T., Tagami, T., Shimada, K., Takagi, H., Yamada, R., Umeda, K., 2012. The first (U–Th)/He dating of a pseudotachlylyte collected from the Median Tectonic Line, southwest Japan. Asian J. Earth Sci. 45, 17–23.

Yamasaki, S., Zwingmann, H., Yamada, K., Tagami, T., Umeda, K., 2013. Constraining timing of brittle deformation and faulting in the Toki Granite, Central Japan. Chem. Geol. 351, 168–174.

Zwingmann, H., Mancktelow, N., 2004. Timing of Alpine fault gouges. Earth Planet. Sci. Lett. 223, 415–425.

Zwingmann, H., Mancktelow, N., Antognini, M., Lucchini, R., 2010a. Dating of shallow faults: new constraints from the AlpTransit tunnel site (Switzerland). Geology 38, 487–490.

Zwingmann, H., Yamada, K., Tagami, T., 2010b. Timing of brittle faulting within the Nojima fault zone, Japan. Chem. Geol. 275, 176–185. https://doi.org/10.1016/j.chemgeo.2010.05.006.

Zwingmann, H., Han, R., Ree, J.H., 2011. Cretaceous reactivation of the Deokpori Thrust, Taebaeksan Basin, South Korea, constrained by K-Ar dating of clayey fault gouge. Tectonics 30, TC5015. https://doi.org/10.1029/2010TC002829.

Zwingmann, H., 2014. K-Ar/40Ar/39Ar dating – clays and glauconites. In: Rink, W.J., Thompson, J.W. (Eds.), Encyclopedia of Scientific Dating Methods. Springer, Berlin-Heidelberg, pp. 171–176.

Zwingmann, H., Berger, A., Todd, A., Herweg, M., 2017. Testing high-voltage electrical discharges in disintegrating claystone for isotopic and mineralogical studies: an example using Opalinus claystone. Clay Clay Miner. 65 (5), 342–354. https://doi.org/10.1346/CCMN.2017.064072.

Chapter 8

Fault sealing

Michael Kettermann[a], Luca Smeraglia[b], Christopher K. Morley[c], Christoph von Hagke[d], David C. Tanner[e]

[a]*Department of Geodynamics and Sedimentology, University of Vienna, Vienna, Austria;* [b]*Dipartimento di Scienze della Terra, Sapienza University of Rome, Rome, Italy;* [c]*Department of Geological Sciences, Faculty of Science, Chiang Mai University, Chiang Mai, Thailand;* [d]*Institute of Geology & Palaeontology, RWTH Aachen University, Aachen, Germany;* [e]*Leibnitz Institute for Applied Geophysics (LIAG), Hannover, Germany*

Chapter outline

Understanding Faults. https://doi.org/10.1016/B978-0-12-815985-9.00008-4

8.1 Introduction

Possibly one of the most important aspects of faulting, especially when considering exploitation of a faulted reservoir, for instance, for hydrocarbon production, geothermal energy or CO_2 storage, is whether a fault is permeable to gases and liquids, or not. This subject is termed 'fault seal'. Consideration of fault seal requires fundamental knowledge of the juxtaposition of stratigraphic units due to fault movement over time, as well as the nature of the material in the fault core. This includes the type of material involved, its microstructural, petrophysical and geomechanical properties. The damage zone geometry, with respect to the fault itself and its characteristics, must also be constrained. In part, this information may be gathered by studying surface outcrops and extrapolating findings to greater depth, however with the assumption that the outcrop is representative of other parts of the fault. This may not be the case, particularly as pressure and temperature vary with depth and variations of stratigraphic setup are common in sedimentary systems (i.e., channels). Additional information may be gathered using seismic imaging or borehole data. Borehole data, particularly from cores, provide the best information on the permeability and sealing capacities of a fault. However, commercial wells rarely target fault zones for coring and the data is still only a local 1D snapshot of the fault and damage zone, and therefore may not be representative of the fault's geometrical and petrophysical variability along and across strike. Seismic data provide good constraints on the stratigraphy involved in the faulting process and 3D geometry of the fault itself; however, resolution of conventional seismic is too low (>25 m) to yield a lot of information on fault properties (Chapter 3.2). Recent advances in seismic imaging reach higher resolutions of 5−10 m that greatly improve fault interpretation, but still is not enough to precisely capture structures in the metre to decametre scales that have a large influence on fault seal.

It is important to note that fault seal is not a fixed steady-state value for a fault, but evolves over time during the tectonic history of the fault and often on much shorter timescales when fluid flow properties are governed by the seismic cycle. Therefore, to assess the fault seal, it is important to constrain the complete tectonic history of the fault. It must be analysed when and how the fault moved from its inception to the present-day. This includes diagenesis, metamorphism, cementation and fluid migration on the fault and their changes through time. Additionally, the depth of burial and the stress environment over time must be recovered. All these data and information must be gathered over a range of scales, bridging the nano- to macroscopic scale.

These multiple aspects highlight how complex it is to determine fault seal quantitatively. In this chapter, the state of the art of knowledge on fault sealing processes and evaluation methods is summarized. Starting with general concepts of juxtaposition sealing, we discuss fault seal processes and evaluation methods in both siliciclastic and carbonate rocks. We look into the analogue and numerical experimental approach to fault seal. This is followed by selected case studies that highlight different aspects of fluid flow across and along faults, in the extensional, compressional as well as strike-slip regimes, where field outcrops provide ground truth and applications of the basic concepts.

8.2 How does a fault seal?

As shown in previous chapters, a fault is a discontinuity that shows an offset of the stratigraphic pile. The end-member kinematics of a fault are normal, reverse, or strike slip (Chapter 1). As we will see, kinematics may influence the fault structure and hence permeability. A fault may be subdivided into multiple fault strands or segments that link during progressive deformation (Chapter 6). Each segment or the entire fault has several basic elements, including the fault core, the damage zone, and the process zone (see Chapters 2–6). In each structural element, a range of processes can affect its hydraulic properties. Depending on the deformation structures present, fault zones can baffle, hinder or promote fluid flow either across a fault or parallel to it.

The fault core accumulates the largest strain, and in most rocks, this causes a collapse of pore space and connectivity, either due to reorganisation of grains, grain size reduction (cataclasis), or entrainment of fine-grained material from e.g., clay or shale source beds (Fig. 8.1).

This typically causes a reduction or complete loss of both cross-fault and along-fault permeability. Strong lithologies under low effective stresses (e.g., shallow carbonates, volcanic or igneous rocks) however, can form open tensile fractures (Chapter 2; Mode I fractures) that greatly improve the along-fault permeability, but have little effect on the cross-fault permeability. Cycles of deformation and mineral precipitation can open and reseal those fractures. The damage zone typically shows clusters of either fractures or deformation bands (Chapter 2). Like the fault core, cataclastic grain-size reduction in the deformation band clusters can decrease cross-fault and along-fault permeability while fractures, unless closed by mineral precipitation, promote along-fault flow. Due to the much smaller finite strain within the damage zone, the magnitude of these processes is much smaller here than in the fault core. Essentially the same processes, but on an even smaller scale, can occur in the wider process zone. Assessing cross-fault or along-fault flow is of importance

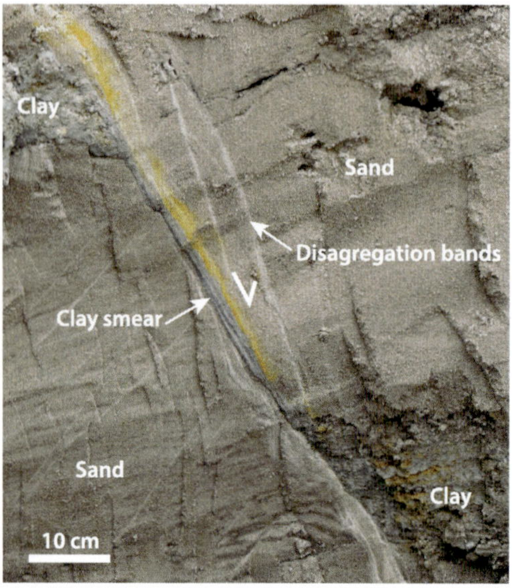

FIG. 8.1 Clay smear in unlithified sand-clay. Open cast lignite mine west of Cologne, Germany.

for different applications or scenarios. Along-fault permeability, for example, is crucial when assessing a faulted top-seal above an anticlinal reservoir (Fig. 8.2A). Cross-fault sealing is important for fault-bound traps (Fig. 8.2B) or compartmentalized reservoirs, where the cost of production wells could be reduced by producing several compartments from a single well (Fig. 8.2C).

As noted before, the sealing capacity of a fault basically depends on the grain-size, or more precisely on the pore throat diameter. Clay minerals make good seals due to their small grain size and sheet-like shape (Chapter 7) that allows for alignment and further reduction of pore throat size. Mineralized veins also often have little to no porosity, and therefore form good seals.

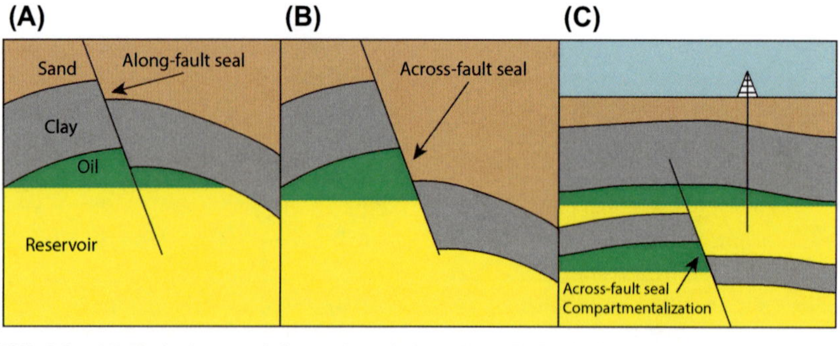

FIG. 8.2 (A) Faulted top-seal shows along-fault sealing. (B) Faulted top-seal shows across-fault sealing. (C) Intra-reservoir fault shows across-fault sealing.

In many applications, two-phase flow problems must be considered. For example, in hydrocarbon exploration, cap-rocks and fault-seals are typically saturated with brine, while the reservoir holds oil or gas. For these two-phase systems, the fine-grained seals represent a capillary- or membrane-seal, as long as the seal remains brine-saturated.

The maximum hydrocarbon column (h) that can be sustained by a capillary seal depends on the capillary entry pressure (Pe), the surface tension of the entering fluid (γ), the wetting angle of the entering fluid (θ), the pore throat diameter (r), gravity (g), and the brine-hydrocarbon density difference ($\Delta\rho$) (Berg, 1975; Ingram et al., 1997; Watts, 1987):

$$h = \frac{2\gamma \cos\theta}{rg\Delta\rho} = \frac{P_e}{g\Delta\rho} \tag{8.1}$$

Capillary entry pressures can also be measured in laboratory tests using mercury injection methods and with recent advances also estimated from scanning electron microscope analysis (Sinn et al., 2017). When the capillary entry pressure is exceeded, a so-called capillary breakthrough occurs, and hydrocarbons can leak through the seal. During this leakage the brine is expelled from the seal and it can become hydrocarbon-wet, preventing further sealing. From this point on, the seal does not provide a membrane seal, but simply acts as a baffle to the flow. For very large hydrocarbon columns, one must additionally consider the risk of hydraulic breakthrough, essentially hydraulic fracturing of the seal, where the buoyancy forces may not exceed the capillary entry pressure, but instead the mechanical strength of the seal (Ingram et al., 1997; Ingram and Urai, 1999).

8.3 General tools for fault seal analysis

In most subsurface applications, little information on mechanics and detailed geometry of faults is available. However, wells, and the increasing quality of 3D seismic reflection data, provide input on the amount of clay in the sequence and the geometry of faults, at least with tens of meters resolution (Chapter 3.1). The very first step in any assessment of faulted top seals or intra-reservoir seals is therefore to build a 3D reservoir and fault model. This is used to investigate the juxtaposition of different lithologies across faults. Assuming a low-permeability clay-rich layer (i.e., a good seal), a seal-seal juxtaposition across a fault implies sealing behaviour across and along fault, while a reservoir-reservoir juxtaposition is potentially not sealing. This, however, becomes more complicated with increasing stratigraphic complexity. If one reservoir bed is juxtaposed to a second reservoir across a fault, one or more shale or clay layers may have travelled past the reservoirs and form a sealing clay- or shale smear. Therefore, the first step in assessing fault seal is always the study of juxtaposition relations, i.e., the existence of juxtaposition seals or leaks. Depending on the available data, this is typically done for 3D data by the means of 2D juxtaposition maps, or in 1D by means of juxtaposition diagrams.

8.3.1 2D juxtaposition and Allan maps

Depending on the available data, i.e., 2D seismic lines or 3D seismic volumes, maybe a well with cuttings, it is possible to evaluate where in the reservoir and trap complex permeable or impermeable layers are juxtaposed. In the simplest case, a juxtaposition of impermeable clay against clay in a faulted sequence will form a sealing fault. Equally, a juxtaposition of sand against sand will allow cross-fault fluid flow. However, more complex sequences of high- and low-permeability lithologies will make the evaluation more challenging.

Imagine, for example, a thick shale layer of 40 m embedded between two sand layers (Fig. 8.3A). Imagine then a normal fault with 20 m displacement cutting through this sequence (Fig. 8.3B). We have now three different juxtapositions: (1) sand-sand at the top and bottom of the sequence, allowing cross fault flow; (2) sand-shale at the top and bottom of the shale layer, inhibiting cross-fault flow, and (3) shale-shale in the middle of the sequence, completely sealing. Now imagine the same sequence, but with a normal fault with 80 m displacement (Fig. 8.3C). We now have only two types of juxtaposition - sand-sand and sand-shale. Again, top and bottom of the sequence show sand-sand juxtaposition, followed by sand- shale juxtaposition. But now, instead of forming a shale-shale juxtaposition, in the middle of the sequence we observe another sand-sand juxtaposition. This juxtaposition could be a potential fluid pathway. However, 40 m of shale travelled past this sand and potentially formed a continuous shale smear, effectively forming a membrane seal between this sand-sand juxtaposition (Fig. 8.3D).

A traditional method to evaluate juxtaposition is the use of fault-plane maps or Allan diagrams (Allan, 1989). As opposed to the previous example, here the reservoir horizons, embedded in sealing horizons, are investigated. With a model of faults and horizons available from seismic interpretation, intersections of top and bottom of reservoir units with the fault surface are plotted on a 2D projection of the fault cutoff-lines from hanging-wall and footwall (Fig. 8.4A). Assigning different colours to hanging-wall and footwall beds allows for easy visualization of potential reservoir-reservoir juxtapositions and hence where potential leaks are located. In the provided example (Fig. 8.4B), one could also estimate the maximum possible hydrocarbon column height that can accumulate in the trap until the spill point is reached.

8.3.2 Juxtaposition diagrams

Almost 10 years after Allan's work, Knipe (1997) developed a juxtaposition diagram that can be used without detailed fault mapping. It shows all possible juxtapositions of a faulted sequence from no displacement up to a throw

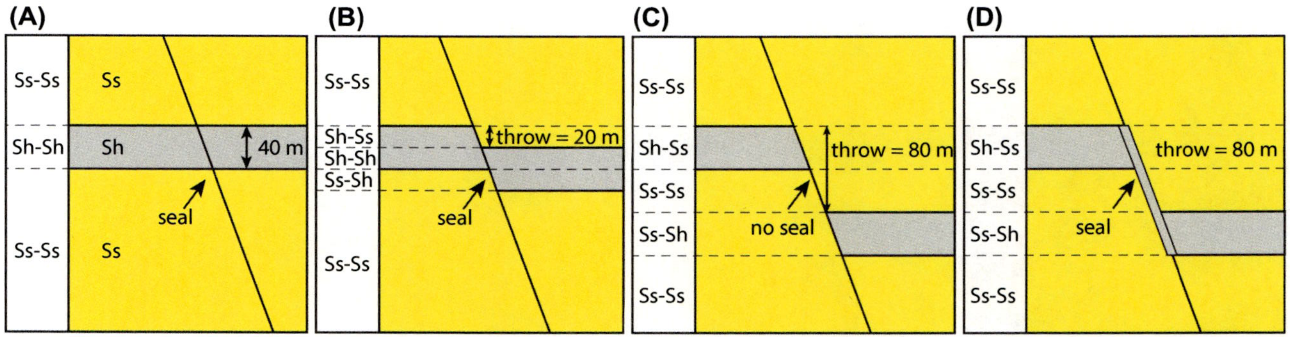

FIG. 8.3 Possible juxtapositions in a sand-shale sequence and inferences for fault seal. (A) Initial situation with a 40 m thick shale embedded in sandstone reservoirs cut by a fault (with initially zero throw). Juctapositions are sandstone (Ss) to sandstone and shale (Sh) to shale. (B) 20 m of throw creates Ss-Ss, Sh-Ss and Sh-Sh juxtapositions. The fault would be sealing, assuming a non-brittle shale. (C) With 80 m throw a Ss-Ss juxtaposition is created between the offset shale source bed. This fault would be potentially not sealing. (D) Same situation as in (C), but with a shale that formed a shale-smear in the fault zone. This fault could be sealing for cross-fault flow.

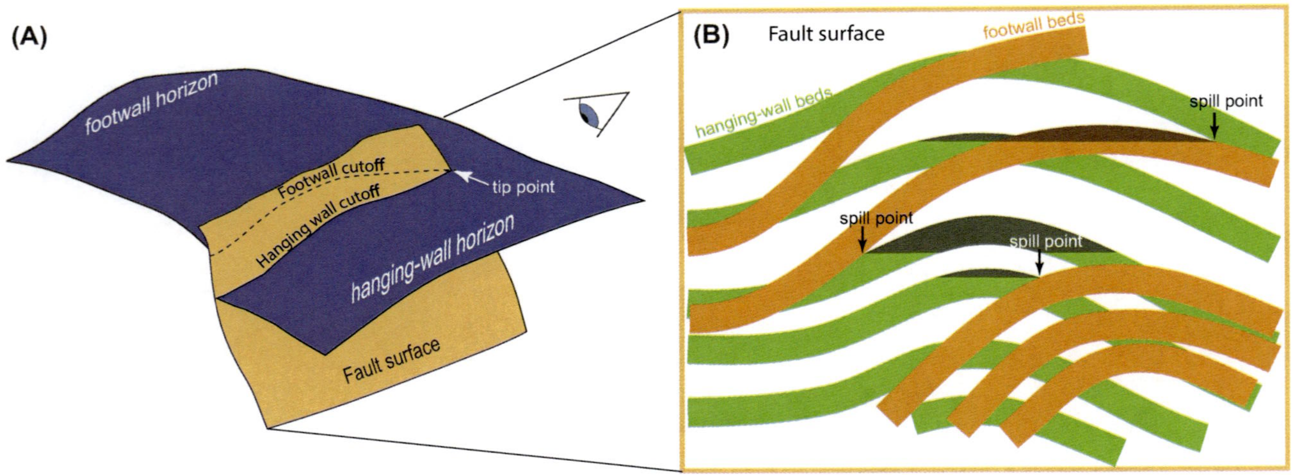

FIG. 8.4 Allan map illustrating where reservoir-reservoir juxtapositions are on the fault plane. This method allows for analysis of traps and possible hydrocarbon column heights in 2D. (A) Blue horizon is cut by yellow fault with smallest throw at the fault tip and largest throw in the center of the fault. Fault cutoff lines are marked for hanging-wall and footwall. These are drawn in the Allan map. (B) Example of an Allan map where hanging-wall and footwall geometry of a number of beds are projected onto the fault plane, allowing to easily identify the juxtaposition of reservoir beds, spill points and possible column heights. *(B) Modified after Allan (1989).*

equalling the height of the sequence (or the maximum expected throw in the study area). This forms a triangular diagram, with the original lithological columns on the left y-axis scaled to its length dimension. Fault throw is plotted on the x-axis, which has the same length and unit as the y-axis. Layer boundaries are marked parallel to the x-axis at the corresponding depths. On the right axis, the lithological column is shifted down either by the maximum throw on the fault or the length of the lithological column to gain a more general view. Lines are then drawn between the horizon tops of the lithological columns on the left and right axis. These lines now intersect with the lines drawn from the original horizontal horizons, outlining several fields, each of which shows a specific juxtaposition at a specific depth and fault throw (Fig. 8.5). This method provides a first-look type of information, in which scenarios can be developed for top- or fault-seal leakage at different fault throws. While these diagrams do not include spatial information, the benefit is an easier implementation of juxtaposition categories for more complex sequences. The user can define for example simple sand-sand juxtapositions, but also define a category for sand-sand juxtapositions that travelled past a shale layer, or different shale content in different layers.

8.4 Fault sealing in siliciclastic rocks

In siliciclastic rocks, three basic observations must be made to assess the sealing capability of a faulted sequence (Yielding et al., 2010). First, as explained before, the juxtaposition of high-permeability reservoir beds against low-permeability units is crucial. Second, the structure and composition of fault core and damage zone can enhance or decrease the sealing capacity of a fault where reservoir-reservoir juxtaposition occurs. Third, the present-day stress state is of importance, as critically-stressed faults might form open

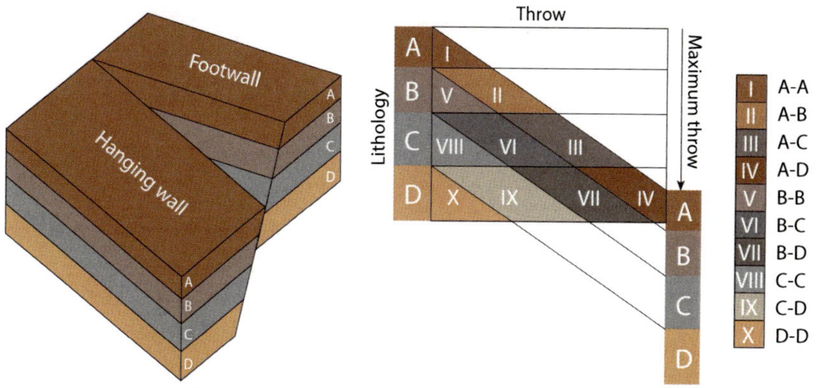

FIG. 8.5 The concept of the triangular juxtaposition diagram. Every possible juxtaposition of a faulted sequence displaced up to a given throw against itself can be evaluated. *After Knipe et al. (1997).*

fractures that cause leakage. In the following sections, we will present important processes in the fault core and damage zone and how their effects on fault seal and reactivation risk can be assessed.

Typical fault rock types found in siliciclastic rocks depend on the available lithologies (mostly clay content) and effective stress at time of faulting (Fig. 8.6; Jolley et al., 2007). With no clay available at low stresses, disaggregation zones form, and with increasing stress sand grains start to break and cataclasites form. Increasing clay content leads to the formation of phyllosilicate framework fault rocks (PFFR) or shaly gouges. High clay content finally leads to the formation of clay smears at low stresses and shale smears or shale gouges at higher stresses. Except for disaggregation zones, all these fault rocks reduce cross-fault and along-fault permeability. However, as explained earlier, the highest sealing capacity results from high clay content in the fault core, produced by clay or shale smearing. In the damage zone, cataclastic deformation bands further reduce the permeability of the fault zone.

8.4.1 Clay smear

The terms clay smear, clay gouge, shale smear or shale gouge are not well defined and often used synonymous for processes that entrain clay or shale into a fault zone, or the clay-rich fault core formed by these processes. Clay smears occur on all scales from millimetre displacements to seismic scale faults (Fig. 8.7). An extensive review and discussion of clay smear processes and applications for fault seal, can be read in Vrolijk et al. (2016). Three basic processes can transport clay from a source bed into the fault rock (Lindsay

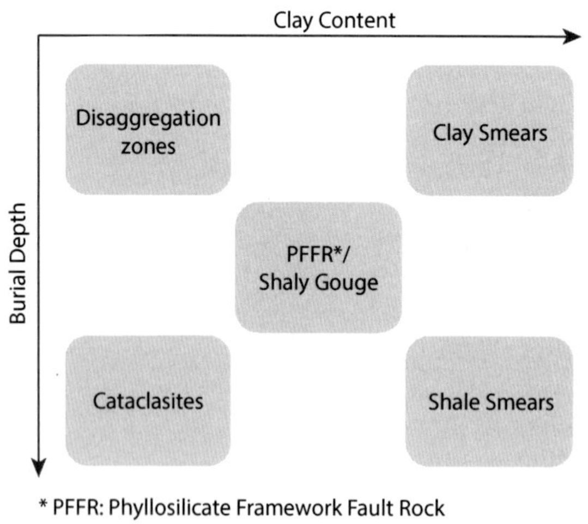

FIG. 8.6 Fault rock types in siliciclastic rocks depend on the clay content and burial depth: fault rock types in siliciclastic rocks depend on the clay content and burial depth. *After Jolley et al. (2007).*

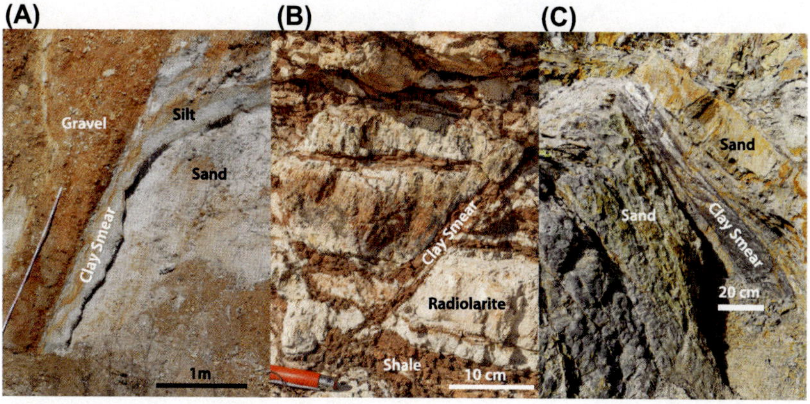

FIG. 8.7 Outcrop examples of clay smears. (A) A silt layer is smeared between gravel in the hanging wall and sand in the footwall. Throw is about 30 m. Outcrop in a gravel-pit southwest of Bonn, Germany. (B) Red shales are smeared between radiolarites in the tectonic mélange of the Batain Coast, east of Sur, Oman. Throw <1 m. (C) 20 cm thick clay smear in Miri/Sarawak in Malaysia (Borneo). Throw >25 m. Detailed discussion on this area in Section 8.7.2.

et al., 1992), depending on clay strength, sand strength and stress state: (1) Injection of ductile clay into dilational jogs, (2) shearing of weak clay, where thickness decreases with distance from the source beds, and (3) abrasion in more lithified rocks, where shale from the source bed is abraded and mixed with sand in the fault rock. For all types of clay/shale smear formation, the chance of discontinuities in the smear increases with displacement. Outcrops commonly show heterogeneous thickness distribution of clay smears (Childs et al., 1997; Doughty, 2003; Kettermann et al., 2016; van der Zee and Urai, 2005), attributed to fault zone complexities, such as relays, horses or Riedel shears. Faults cutting through multiple source layers produce layered clay smears, where the individual smears can be either separated by a 'sand-smear' or amalgamated (e.g., Fig. 8.7B). Increased mixing in multi-layered clay smears increases the thickness and sealing capacity of these faults (Childs et al., 2007; Kettermann et al., 2017, 2016; Schmatz et al., 2010b). These small-scale structures cannot be resolved in seismic data and to date cannot be explicitly accounted for in fault seal prediction.

8.4.2 Deformation bands

Deformation bands, particularly cataclastic deformation bands (Fig. 8.8), are commonly assumed to reduce cross-fault permeability. The magnitude of permeability reduction depends on several factors, such as the burial depth (depth between 1 and 3 km show strongest permeability reduction), the occurrence as single bands or clusters, or the host rock lithology where coarse, well sorted, high-porosity rocks lead to the strongest reduction of permeability

FIG. 8.8 Cataclastic deformation bands in sandstone (Utah/USA).

(Ballas et al., 2015; Fossen and Bale, 2007). While this statement is, in general, true, one must be careful when considering fault-seal rather than permeability. Deformation bands are inherently of small extent and even dense clusters of cataclastic bands will not sustain low-permeability properties over long distances. It is therefore unlikely that deformation bands in the damage zone strongly contribute to the sealing capacity of a fault, but rather act as additional baffle for cross-fault flow in production scenarios (Fossen and Bale, 2007; Medeiros et al., 2010).

8.4.3 Fault seal predicting algorithms

After evaluating simple juxtaposition relationships, the next step is to interrogate the parts of the faults that juxtapose high-permeability units against each other. The questions to ask are: How much clay or shale is entrained into the fault rock? Do we expect a continuous clay smear or shale gouge, forming a membrane seal, or are there potential pathways across the fault zone due to discontinuous distribution of clay in the fault rock? Even without detailed knowledge about the structure of the fault zone on metre-scale or the precise mechanical properties of the involved lithologies, several algorithms are available to estimate fault-seal potential, based on only throw and the amount of available clay or shale in the sequence.

The first algorithm mentioned in literature – the Clay Smear Potential (CSP) - dates back 30 years (Bouvier et al., 1989), although the actual algorithm was first published almost ten years later by Fulljames et al. (1997) and further explained by Lehner and Pilaar (1997). Based on outcrop observations in a shallow lignite mine, Lehner and Pilaar (1997) proposed the idea of a sheared clay veneer that continuously thins from the source clay layer towards the centre between the offset layers. Consequently, their model proposes a higher risk of seal-breach at larger fault throws and particularly at the centre of the fault (Fig. 8.9A). The CSP calculates the thickness of a clay smear at a given distance from the source layer as,

$$\text{CSP} = c\sum\frac{\text{thickness}^2}{\text{distance}} \tag{8.2}$$

where c is a calibration constant that can be used to empirically adjust for variations in stress or mechanical properties.

Yielding et al. (1997) extended the CSP algorithm to the Smear Factor (SF) by implementing two exponents m and n (Fig. 8.9B). The shale bed thickness can now be controlled by the exponent n and the distance from the source bed to the query point by the exponent m. This extension to CSP provides more control for the calibration of the algorithm data to real data.

$$\text{SF} = \frac{\text{thickness}^n}{\text{distance}^m} \tag{8.3}$$

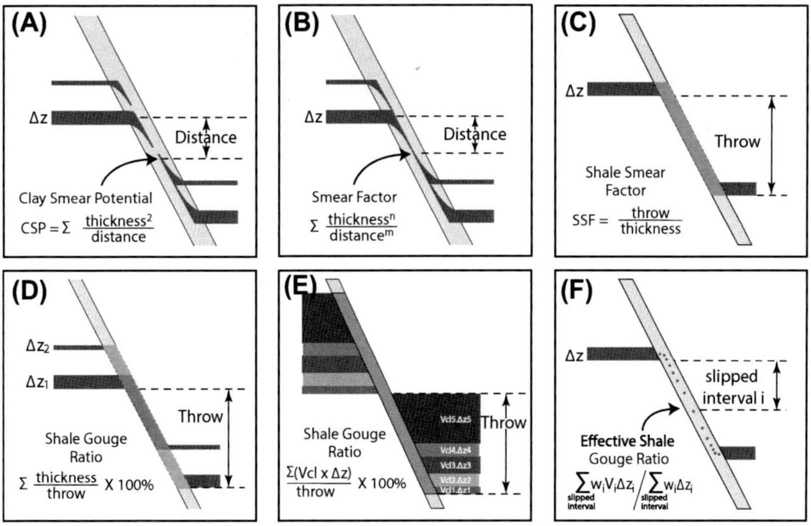

FIG. 8.9 Illustration of fault-seal predicting algorithms. (A) Clay smear potential – CSP. (B) Smear factor. (C) Shale smear factor – SSF. (D) Simple expression of the shale gouge ratio – SGR. (E) Shale gouge ratio, incorporating varying shale content. (F) Effective shale gouge ratio – ESGR. *Redrawn after Yielding et al. (1997).*

Lindsay et al. (1992) introduced the Shale Smear Factor (SSF) to calculate abrasive shale entrainment into the fault rock, a dimensionless parameter that is defined as fault throw divided by source shale thickness (Fig. 8.9C),

$$\mathrm{SSF} = \frac{\text{throw}}{\text{thickness}} \tag{8.4}$$

Based on outcrop data from faulted lithified sand/shale sequences, the authors showed that, for a critical Shale Smear Factor of SSFc ≤ 7, shale smears are likely to be continuous between the offset source layer, but occasionally, even at SSF of 50, no breaching was observed in 2D sections. Given the inherently complex 3D structure of faults, it is not surprising that random 2D sections of a fault do not show the potential locations of a discontinuity in the smear. This has been described by other authors from 3D outcrops and analogue models (Childs et al., 1997; Doughty, 2003; Kettermann et al., 2016; Noorsalehi-Garakani et al., 2013).

In 1997, Yielding and coworkers introduced the Shale Gouge Ratio (SGR), which can be formulated in two ways. The simpler case (Fig. 8.9D) assumes one or more layers of clay or shale being offset by a fault with a given throw. The process that is assumed here is that fine-grained components of the shale beds are homogenously mixed into the fault rock. The SGR is then the ratio of the sum of shale bed thicknesses, divided by the throw, expressed as percentage:

$$\mathrm{SGR} = \frac{\sum(\text{shale bed thickness})}{\text{throw}} \times 100\% \tag{8.5}$$

This equation expresses how much percent of the source clay has slid past any point in the fault and have potentially mixed to the fault gouge. Consequently, larger throws mean smaller amounts of shale in the fault rock and hence a smaller chance of forming a sealing gouge.

The advantage of the SGR algorithm is that it can be easily extended to incorporate, not only multiple shale beds, but also source layers with variable shale content (Fig. 8.9E). This could, for example, be extrapolated from gamma ray logs. The SGR can then be written as:

$$\mathrm{SGR} = \frac{\sum[(\text{Zone thickness}) \times (\text{Zone clay fraction})]}{\text{throw}} \times 100\% \tag{8.6}$$

Yielding et al. (1997) and Yielding (2002) also suggest that SGR equals the volume of shale in the fault rock (V_f).

Knipe et al. (2004) and Freeman et al. (2008) proposed a combination of CSP and SGR, and named it Effective Shale Gouge Ratio (ESGR), essentially a weighted version of the SGR that assumes a non-perfect mixing of shale and host rock. Close to the source beds, the contribution of shale in the fault rock is assumed to be greater (Fig. 8.9F). The ESGR is formulated as

$$\text{ESGR} = \sum_{\substack{\text{slipped}\\ \text{interval}}} w \times V\text{shale} \times \text{thickness} / \sum_{\substack{\text{slipped}\\ \text{interval}}} w \times \text{thickness} \tag{8.7}$$

where w is an empirical weighting factor that can be used to adjust for local field conditions, i.e., brittleness of the source bed.

The Probabilistic Shale Smear Factor (PSSF; Childs et al., 2007) picks up when a critical Shale Smear Factor (SSFc) is exceeded, i.e., when a discontinuity forms in the smear. The basis for the PSSF is the assumption that, after breaching, the length of the shale veneer in the fault core does not increase anymore, and the shale patch is randomly positioned between the hanging-wall and footwall source-bed cutoffs. PSSF then calculates the probability of encountering a hole at a queried point on the fault plane. For multiple shale layers in the sequence, the total chance of encountering a hole in the shale smear is calculated by multiplying the probabilities for the individual shale beds according to

$$\text{PSSF} = \prod 1 - \left(T \times \frac{\text{SSFc} - 1}{\text{throw} - \text{thickness}}\right) \tag{8.8}$$

Çiftçi et al. (2013) found experimentally that increased stress at the time of faulting causes a widening of the fault zone and thereby increased shale content in the fault rock. Furthermore, they added the cutoff angle between the shale bed and the fault in their equation of Scaled SGR (SSGR) that now reads

$$\text{SSGR} = \text{SGR} \times \frac{\sigma_n}{(\sin \alpha \times \sigma_{n-\min})} \tag{8.9}$$

where σ_n is the normal stress, α is the angle between shale bed and fault and $\sigma_{n-\min}$ is the minimum fault normal stress in the data set.

The simplicity of these algorithms, which makes them so widely useable, also implies that they do not account for the vast variety of other processes that influence fault sealing (i.e., cataclasis, deformation bands, brittle fracturing). Consequently, the workflow requires a careful calibration using a local or global calibration dataset to account for the geological setting in the study area. The authors of the original studies propose critical values that separate sealing from non-sealing domains. For the shale smear factor, the critical value is proposed to be SSFc $\geq$7 (Lindsay et al., 1992). For the shale gouge ratio it is SGR $\leq$20% (Yielding et al., 1997). For the clay smear potential, no cut-off value is defined, as this algorithm is intended as probabilistic approach, where higher values of CSP increase the probability of a sealing fault. Compiling cross-fault pressure data from numerous wells worldwide led to a more practical approach to calibrate CSP, SGR and SSF by cross-fault pressure differences (Fig. 8.10; Yielding et al., 1997). These cut-off values further depend on the burial depth. Faults at larger depths can withstand larger cross-fault pressure differences (Yielding, 2002).

8.4.4 Fault permeability from fault seal algorithms

Fault permeability plays an important role in the development of compartmentalized reservoirs. Imagine a good top-seal - e.g., a massive shale anticline - with some faults in the underlying reservoir sand-shale sequence. It is then essential to know whether a production or injection well is required for every compartment in the case of sealing or low permeability faults, or whether multiple compartments can be produced from a single well, in the case of high permeability faults. Two approaches are widely used to estimate fault permeability in sand-shale sequences. Both assume single-phase flow across the fault zone and incorporate shale content in the fault rock V_f, as calculated using SGR.

Manzocchi et al. (1999) base their model on five plug and probe permeability studies on both reservoir and outcrop samples from fault cores, relating volumetric shale fraction (V_f) and permeability (k_f). In their empirical equation (Eq. 8.10), they derive V_f from SGR. Additionally, they integrate a factor for displacement (D) that is used to account for permeability reductions due to cataclasis.

$$\log k_f = -4\text{SGR} - \frac{1}{4}\log(D)(1 - \text{SGR})^5 \tag{8.10}$$

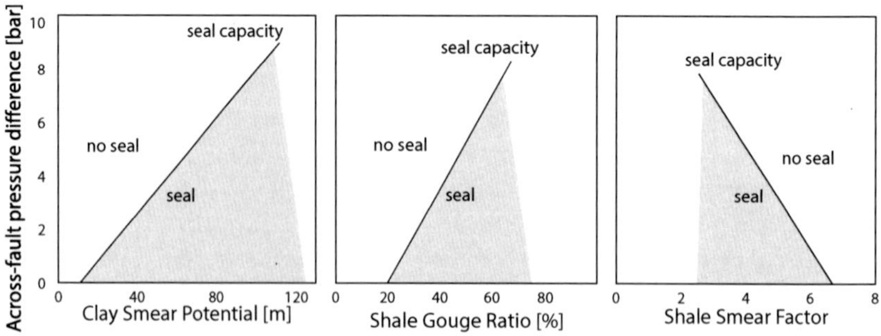

FIG. 8.10 Across-fault pressure differences (*bar*) sustained by faults in subsurface studies plotted against predicting algorithms help to calibrate cut-off values. Across-fault pressure differences sustained by faults in subsurface studies plotted against predicting algorithms help to calibrate cut-off values. *Modified after Yielding et al. (1997).*

Sperrevik et al. (2002) show that k_f further depends on the maximum burial depth z_{max} and the depth of faulting z_f, *empirically incorporating the effects of cataclasis, compaction and diagenesis/cementation*. Using a set of V_f versus permeability data derived from measurements on faults in cores, they propose the following empirical relationship:

$$k_f = a_1 \exp\left\{-\left[a_2 V_f + a_3 z_{max} + (a_4 z_f - a_5)(1 - V_f)^7\right]\right\} \quad (8.11)$$

With the following constants: $a_1 = 80{,}000$; $a_2 = 19.4$; $a_3 = 0.00403$; $a_4 = 0.0055$; $a_5 = 12.5$. The same dataset also yields an empirical relationship of fault rock permeability and capillary threshold pressure (P_f)

$$P_f = 31.838 \times k_f^{-0.3848} \quad (8.12)$$

In reservoir simulations, faults are not modeled in high detail but as contacts between model cells. Naturally, the contacts do not have a thickness, which would be required to physically correctly calculate flux across a fault. Fault properties, such as permeability, therefore, have to be included in the simulation as multipliers at the contacts of cells that are separated by a fault. These multipliers allow the modification of properties, such as flux from one cell to another. These are called Fault Transmissibility Multipliers (FTM or T; Manzocchi et al., 1999). The transmissibility multiplier (T) between two model cells i and j is defined as

$$T = \left[1 + t_f \frac{\left(\frac{2}{k_f} - \frac{1}{k_i} - \frac{1}{k_j}\right)}{\left(\frac{L_i}{k_i} + \frac{L_j}{k_j}\right)}\right]^{-1} \quad (8.13)$$

where T: transmissibility multiplier; t_f: thickness of the fault; k_f: permeability of the fault; k_i & k_j: permeabilities of the host rock in cells i and j; L_i & L_j: length of cells i and j.

Fault rock or fault zone thickness, as it is used as an input parameter in Eq. (8.13) (FTM), is roughly a function of lithology and displacement of the fault, and can vary strongly (Torabi and Berg, 2011). However, in general, it can be assumed that fault cores in clay-rich beds are narrower than in sandstones. Based on this assumption, Sperrevik et al. (2002) formulated an empirical equation to calculate fault thickness (F) for this purpose

$$F = T\left(0.06V_f^2 - 0.12V_f + 0.0659\right) \quad (8.14)$$

where T: local throw; V_f: shale volume in the fault rock, as derived from SGR.

8.4.5 Clay injection and mechanical clay injection potential (MCIP)

Under high stresses, ductile clays have been described to be injected into steep dipping parts of the fault. Two cases of clay injection can be distinguished: (1) Injection of ductile clays from the source beds into the fault between hanging-wall and footwall cutoffs, resulting in an abnormally high volume of clay in the fault core (Fig. 8.11A, van der Zee et al., 2003); (2) injection of ductile clays in dilatant fractures above the footwall cutoff or beneath the hanging-wall cutoff (Fig. 8.11B, Smeraglia et al., 2016). Both cases require the formation of a so-called squeezing block, a contractional dip relay caused by the shallower fault dip in the clay bed (Faerseth, 2006; van der Zee et al., 2003).

Case (1) is of special interest when considering hydrocarbon fault seals, as traditional algorithms, such as the previously noted CSP or SGR, will underestimate the sealing capacity of such a fault.

van der Zee et al. (2003) developed a term called Mechanical Clay Injection Potential (MCIP) to predict the tendency of a ductile clay to be injected into a dilational jog in a fault, based on the clay's cohesion (C), friction angle (φ), and the effective maximum principal stress $\left(\sigma_1'\right)$:

$$\text{MCIP} = \frac{\sigma_1'(1 - \sin\phi)}{(2\cos\phi)C} \tag{8.15}$$

The criterion is valid where the sand in the column is stronger than the clay. If this is true, MCIP calculates the tendency for clay injection, where MCIP $>$ 1 indicates that clay will be ductile enough to be injected under the given stress conditions, and MCIP $<$ 1 suggests that injection is not possible.

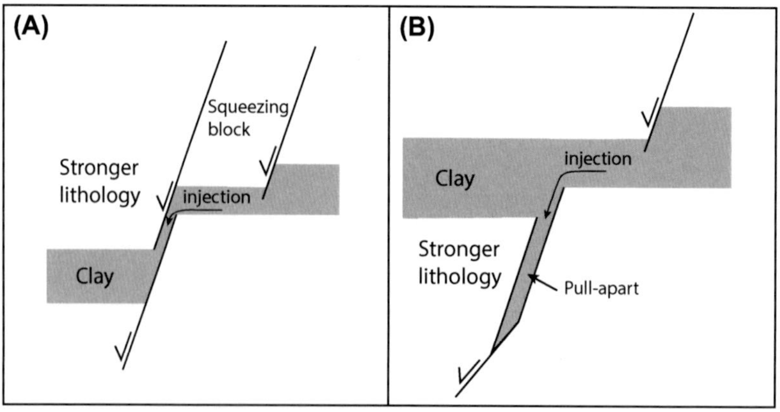

FIG. 8.11 (A) Lateral clay injection into a fault with clay smear. (B) Lateral clay injection into a dilational jog under the source layer.

Case (2), where clay is injected in dilational jogs above or below the source-bed cutoffs, assumes particularly strong beds around the clay layers, e.g., carbonates (Fig. 8.12). This can dramatically increase the portion of the fault covered by clay. As clay behaves as lubricant in fault zones, this sort of clay injection can influence the bulk strength of a fault much more than traditional considerations predict (Smeraglia et al., 2016).

8.4.6 Assessing fault reactivation and seal breach risk

Although juxtaposition seal and membrane seal analysis may indicate that the probability of fault seal is high, there remains a risk that post-hydrocarbon charge, reactivation of a fault, or fault segmentation has occurred and resulted in seal-breach and leakage. Fault reactivation risk for a prospect can be assessed in several ways: (1) looking for indicators of leakage on seismic reflection data, (2) use stress analysis to assess the risk of a fault leaking under appropriate stress fields for the area in question, and (3) assess the presence and thickness of palaeo-hydrocarbon columns from well data.

Most basins will have undergone changes in stress regime since their formation. These changes, including the modern stress state, need to be determined before any quantitative analysis of fault reactivation can be made, and highlight the important role that geomechanics plays in assessing fault

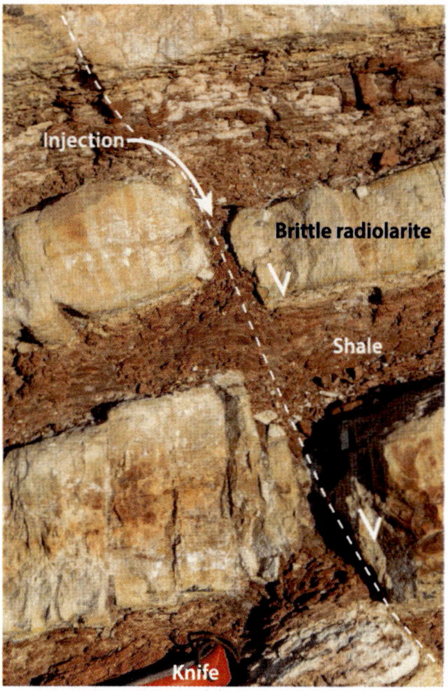

FIG. 8.12 Injection of red shales into dilatant fractures in brittle radiolarites, eastern Oman.

reactivation risk (e.g., King et al., 2012). One of the key examples of fault reactivation affecting hydrocarbon traps, is the NW shelf of Australia, where normal faults, that were formed during the Permian and Mesozoic rifting events, were reactivated during the Miocene-Pliocene by collision of the Australian Plate with Sundaland, in the vicinity of Timor, Indonesia (O'Brien et al., 1996; O'Brien and Woods, 1995). The result was a significant decrease in the initial reserve estimates for the region, coupled with the understanding that significant potential remained in the area (e.g., Vulcan Sub-basin), but that success required detailed understanding of lateral seal risk (e.g., Lisk et al., 1997, 1998). The leakage of hydrocarbons from traps in Mesozoic reservoirs has been determined by the presence of breached, or underfilled traps, coupled with evidence of previously larger hydrocarbon columns determined from fluid inclusion data and the presence of residual hydrocarbon zones (e.g., Lisk et al., 1998, 1997). Bacterial oxidation of migrating hydrocarbons led to the generation of locally pervasive carbonate cements in Eocene sandstones overlying leaks in Jurassic reservoirs (O'Brien et al., 1996; O'Brien and Woods, 1995). These diagenetic effects are manifested as anomalous amplitudes on seismic data, called hydrocarbon-related diagenetic zones (HRDZ's; O'Brien and Woods, 1995), and enable identification of potentially breached or underfilled traps.

Faults zones are typically weaker than the host rock, consequently, depending upon the strength contrast with the host rock, and orientation with respect to the principal stresses, pre-existing fault zones can be reactivated in preference to the creation of a new, ideally-oriented faults (e.g., Jaeger et al., 1976). By estimating the paleo-stress field orientation and magnitudes and pore pressure, and fault rock strength, together with fault orientation, the relative probability of fault reactivation can be determined. The propensity for faults to reactivate can be measured in different ways: e.g., (1) the ratio of shear stress to normal stress acting on a fault surface with frictional characteristics, called slip tendency (Morris et al., 1996), (2) the normal stress operating on the fault plane, normalized to differential stress, (dilation tendency; Ferrill et al., 1999), (3) the increase in pore pressure necessary to cause fault reactivation under cohesionless frictional failure (Wiprut and Zoback, 2000, 2002) and, (4) the increase in pore fluid pressure necessary to cause fault reactivation using fault-rock failure envelopes (fault analysis seal technology method (FAST), (5) the increase in pore pressure necessary to cause fault reactivation under cohesionless frictional failure (Wiprut and Zoback, 2002, 2000) and (6) the increase in pore fluid pressure necessary to cause fault reactivation using fault-rock failure envelopes, or FAST (fault analysis seal technology method, Mildren et al., 2005).

For a cohesionless fault, slip will occur when the resolved shear stress (τ), is equal to, or greater than the frictional resistance of sliding (F), which is proportional to the normal stress (σ_n) that acts perpendicular to the fault surface (Morris et al., 1996). Fault slip is also controlled by the coefficient of static friction (μ) and the cohesive strength of the fault zone. A cohesionless fault will slip, when the shear stress exceeds the frictional resistance:

$$F \leq \tau = \mu \sigma_n \tag{8.16}$$

For this condition, Morris et al. (1996) defined Slip Tendency:

$$(T_s) = \frac{\tau}{\sigma n} = \mu \tag{8.17}$$

For the FAST method the failure criteria are expressed as a function of pore fluid pressure (P_{fp}) (Mildren et al., 2005; Sibson, 1996), where C is the cohesion and T the tensile strength:

Failure mode	Criterion	Condition
Tensile (hydraulic)	$P_{fp} = \sigma_n + T$	$(\sigma_1 - \sigma_3) < 4T$
Hybrid tensile and shear	$P_{fp} = \sigma_n + (4T^2 - \tau^2)/4T$	$4T < (\sigma_1 - \sigma_3) < 6T$
Shear	$P_{fp} = \sigma_n + (C - \tau)/\mu$	$(\sigma_1 - \sigma_3) < 6T$

Fault reactivation risk is typically first assessed on 3-D Mohr diagrams, where the difference in effective normal stress values between a fault position on the diagram and the failure envelope (Chapter 2) is used to determine the risk of reactivation (e.g., Mildren et al., 2005; Morris et al., 1996). The relation of the fault orientation to variation in T_s or ΔP_{fp} can be displayed on polar plots, where the values for the stress magnitude difference for fault reactivation for all orientations are contoured, and then overlain by the plots of the poles to the faults under analysis. The third key display is a contour plot of the risk of reactivation on either faults in map view, or on 3D fault surfaces, where the effects of different fault strike or dip values on reactivation potential can be readily visualized and the locations of potentially-leaky faults or fault segments identified.

Faults can reactivate many times, and in some cases can cause loss of entire hydrocarbon columns. But in other areas, particularly where there are multiple stacked thin reservoirs, fault reactivation may cause some leakage, and remigration into higher reservoirs but not complete loss (e.g., Barmer Basin, India, Naidu et al., 2017). The result often is a stack of underfilled hydrocarbon-bearing reservoirs, with varying levels of thermal maturity that unpredictably alternate with water-bearing reservoirs.

8.4.7 Analogue and numerical experiments of fault clay smear

As often in geology, processes that cause the formation of a sealing fault have hardly ever been observed in nature, and outcrops are often limited to small two-dimensional sections at roadcuts, construction sites, or mines.

Observations from outcrops are crucial, but the risk of overlooking structural elements, such as holes in clay smears, or relays and horses, are high. Laboratory and numerical simulations can greatly improve our understanding of fault zones and corresponding fault sealing behaviour. In experiments, we can observe processes that cause the formation of clay smears. We can investigate three-dimensional structures and we can combine these with direct measurements of cross-fault permeabilities that evolve with increasing displacement. However, a particular model is no universal super-tool. Every type of model is limited by its boundary conditions (e.g., size, stresses, degrees of freedom, materials used, boundary effects, etc.), resolution, and method of observation. Consequently, not every modelling technique has the potential to study the complete range of processes that occur in faults in the brittle crust. For the range of questions that one might ask about the development of fault permeability, a wide range of methods must be deployed to answer smaller, specific subsets of questions.

We must first differentiate laboratory and numerical models. Laboratory models include ring-shear experiments, direct shear experiments, triaxial experiments and sandbox experiments. Numerical simulations, to date, comprise mainly discrete (or distinct) element method (DEM) and the finite element method (FEM) (cf. Chapter 4) or combinations of these methods.

8.4.7.1 Ring shear experiments

Ring shear devices were the first experiments to study clay smears (Mandl et al., 1977; Weber et al., 1978). They allow producing very large shear strains and displacement under in-situ stress conditions, while measuring compaction, shear stress and cross-fault flow. Ring-shear devices are hollow rings, horizontally split in half. The lower half can be rotated to induce shearing, and the top plate can be loaded to apply normal stress on the developing fault. However, ring shear devices impose kinematic constraints because of the pre-described shear plane. (Fig. 8.13).

Generally, the sample container is filled with sand and clay sequences, loaded from the top to achieve a desired stress state and then deformed by rotating the lower ring. A number of studies investigated clay smear formation using a variety of natural and artificial clays and sands and at effective stress

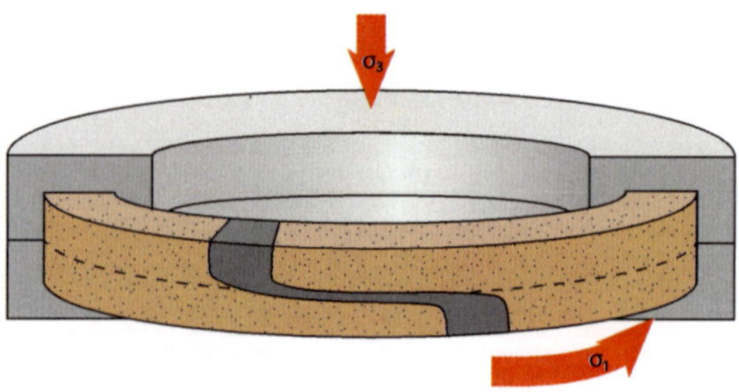

FIG. 8.13 Ring shear device allows for high displacements and strain.

conditions representing depths between 50 and 1500 m (Clausen et al., 2003; Clausen and Gabrielsen, 2002; Cuisiat and Skurtveit, 2010; Sperrevik et al., 2000). First observations by Weber et al. (1978) show that faulted alternating clay-sand sequences form multi-layered clay smears that are often continuous along the shear plane. The same authors first noted a mixing of sand and clay in the shear zone and showed a reduction in cross-fault permeability in parts where a clay smear formed.

At low stresses corresponding to 50 m depth, observations of excavated clay smear surfaces formed from weak clays often show continuous clay smears, with stronger sand-clay mixing and occasional discontinuities at higher displacements and greater distance from the source clay. Stronger clays typically form discontinuous angular clay fragments floating in the fault zone. Multiple clay layers form a layered clay smear, which hosts sand wedges that connect the host-sand lithologies (Sperrevik et al., 2000).

Further experiments, under increasing stress conditions, led to the development of three typical structures: (I) at low stresses, no clay smear forms, angular fragments of clay are dragged into the fault zone, (II) at intermediate stresses, a clay-sand mixture or patches of clay form in the fault zone; (III) at high normal stresses, a partially or fully continuous clay smear forms (Clausen et al., 2003; Clausen and Gabrielsen, 2002). Experiments at high stresses, corresponding to 1500 m overburden, found that cataclasis contributes as much to the reduction of permeability as clay smearing (Cuisiat and Skurtveit, 2010).

8.4.7.2 Direct shear experiments

Direct shear cells are conceptionally similar to ring-shear devices. They consist of a hollow container, mostly cuboid, split in half, parallel to the shear direction, predefining the fault plane (Fig. 8.14). They allow loading normal to the shear plane and measurements of forces that drive the deformation, as well

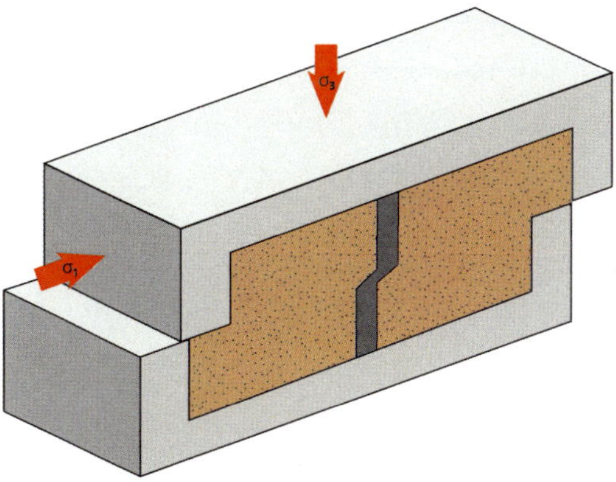

FIG. 8.14 A direct shear device.

as cross-fault fluid flow. Some direct shear devices allow direct observation of structures through a glass barrier (Karakouzian and Hudyma, 2002), showing the development of fault lenses and deformation bands. Other experiments, with multiple different clay layers, showed that the deformation bands are much thinner in clay layers than in sand layers (Urai et al., 2003).

A more sophisticated direct shear apparatus that increases the degree of freedom along the pre-defined shear plane, was developed by Giger et al. (2011). This device includes a 1 cm wide notch around the shear plane that is sealed by high viscosity oils, allowing for a near free development of the fault. Besides cross-fault flow and stress-strain measurements, the sandstone-siltstone samples were studied by CT-scans after deformation. This provides 3D information on the developed clay smear (Çiftçi et al., 2013; Giger et al., 2013). Using this method on a number of samples of varying strength and at varying angle of the source layer to the shear plane, showed an increasing continuity of the clay smear at larger stresses or weaker clays. It led the authors to the formulation of the Scaled Shale Gouge Ratio (SSGR), explained earlier in this chapter. The freedom that the experimental setup provides further allowed the formation of Riedel shears that locally control the thickness of the clay smear.

Briefly straying away from clastics and analogue materials, while staying with the method, direct shear experiments on real travertine rocks show that the permeability of carbonate faults decreases strongly with increasing effective stress, as pore-space collapses and cataclasis causes grain-size reduction (Delle Piane et al., 2016).

8.4.7.3 Triaxial experiments

Triaxial experiments of clay smearing allow the measurement of cross-fault fluid flow and the determination of frictional properties of the developing clay-rich fault gouge. The general setup consists of two cylinders with pre-cut slip planes and a clay or shale sample sandwiched between them (Fig. 8.15).

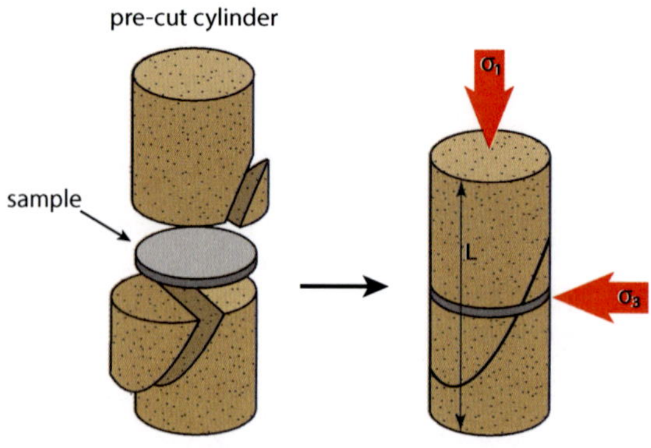

FIG. 8.15 Triaxial test setup for clay smear experiments using a pre-cut cylinder. Modified after Vrolijk et al. (2016).

The sample is then packed in a latex liner and assembled in a triaxial machine. This allows applying minimum principal stress equally around the sample, while increasing the maximum principal stress from the top with a press. While the pre-cut slip planes enforce a non-realistic fault geometry, these tests allow controlled boundary conditions that are best for systematic parameter studies.

Takahashi (2003) performed experiments on a siltstone sample sandwiched between pre-cut sandstone cylinders and measured cross-fault permeability with increasing displacement. He found three regimes of permeability: (1) a strong permeability reduction in early stages of elastic compaction, when pore-space collapses; (2) a phase of slower permeability decrease during early non-elastic deformation; (3) a phase of continuous permeability increase due to abrasive thinning of the siltstone smear. Cataclasis of the host sandstone increased in areas where the smeared siltstone became discontinuous. The reduced grain-size in the cataclasite contributed to a cross-fault permeability, which is lower than the permeability of the host sand.

Systematic studies on the permeability and frictional properties of gouges consisting of various mixtures of fine-grained quartz and kaolinite show an unsurprising decrease of permeability with increasing clay content. At the same time, increasing clay content causes a decrease of frictional strength of the gouge (Crawford et al., 2008, 2002).

More recent studies show a general trend to lower permeabilities of clay gouges at higher effective stresses (Ikari et al., 2009), and that the mineralogy of the phyllosilicates in the gouge has a strong influence on both permeability and strength (Behnsen and Faulkner, 2012; Ikari et al., 2009; Meng et al., 2016).

8.4.7.4 Sandbox experiments

Sandbox experiments of clay smear provide the largest degree of freedom for fault development, and hence provide an overview where clay smear develops above a predefined basement fault. Glass walls allow to observe the developing fault in 2D. Due to the free or only slightly-confined surface conditions, only weak, unlithified materials can be used in these models (Fig. 8.16).

Consequently, the first published experiments on clay smear in sandbox experiments enlightened the process of lateral clay smear injection of ductile clays, leading to the development of the Mechanical Clay Injection Potential algorithm (MCIP), discussed earlier in this chapter (van der Zee et al., 2003).

Close observations of amalgamated clay smears formed from multiple source layers show that mixing of host sand and clay smear (Fig. 8.17) causes an increase of total clay smear volume and large amounts of sand can be accommodated in a forming smear (Schmatz et al., 2010b). These studies also showed a brittle fracturing of stronger clay layers, with subsequent abrasion and reworking of the clay fragments in the shear zone (Schmatz et al.,

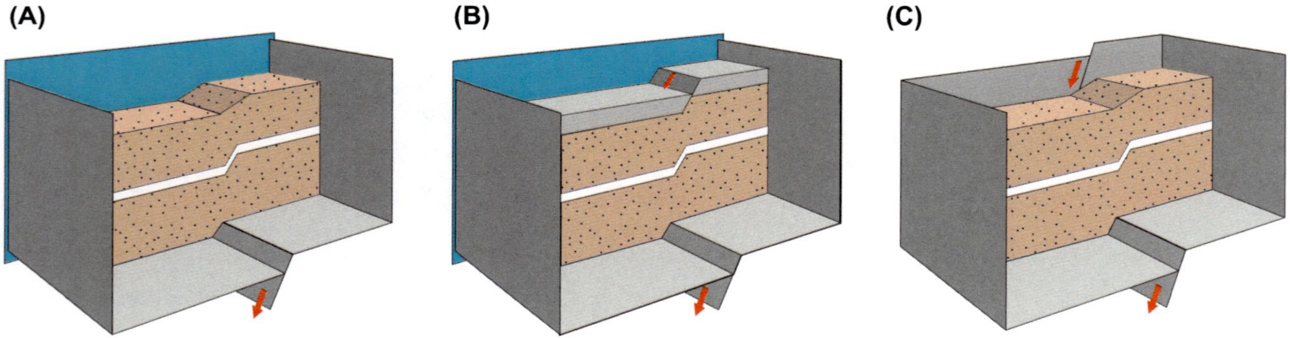

FIG. 8.16 Sandbox setups used to model clay smears. (A) Free surface setup with glass sidewall (blue) used by Schmatz et al. (2010a) and Noorsalehi-Garakani et al. (2013). (B) Top plates (grey) on the surface help to localize the fault more sharply (Schmatz et al., 2010b). (C) Extending the basement fault around the sidewalls allows better sealing of the clay layer and cross-fault fluid flow measurements (Kettermann et al., 2017). Modified after Vrolijk et al. (2016).

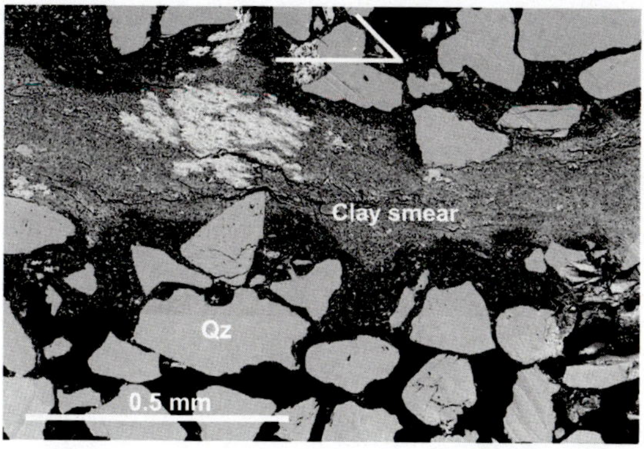

FIG. 8.17 Scanning electron microscope image of an experimentally produced clay smear shows mixing of sand grains and the clay smear. Sand grains add volume to the total clay smear when they are entrained into the clay. Modified after Kettermann et al. (2017).

2010b,a). Particle Imaging Velocimetry (PIV, Adam et al., 2005) further allowed the analysis of detailed shear-zone structure. This method illustrates that the fault forms from overlapping fault segments that grow from the basement fault upwards and the surface downwards. The mechanical contrast between sand and clay further enhances the complexity of the fault zone.

Excavating the clay smear after the experiment ended (Fig. 8.18) provides the opportunity to analyze the clay smear composition. Noorsalehi-Garakani et al. (2013) showed that there are three types of fault gouge: (1) intact thick clay smear; (2) thin clay smear composed of highly mixed sand and clay, and (3) discontinuous clay smear. The general clay smear structure is defined by numerous breached relays or horses that cause displacement partitioning.

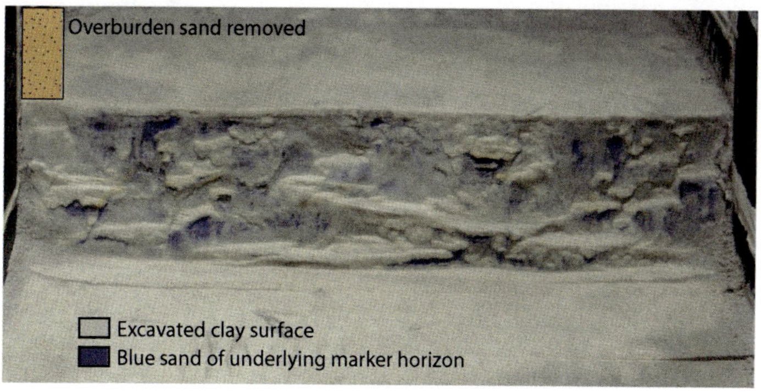

FIG. 8.18 Excavated clay smear surface formed in a sandbox experiment. The normally consolidated kaolinite clay behaves slightly brittle above a 63 degrees dipping basement fault. Blue colors indicate *holes* in the clay smear where underlying blue marker horizons is visible or very thin clay smear with strong mixing. Experiment performed by M. Kettermann at RWTH Aachen University.

Fault segments above and below a breached relay have a smaller throw than the bulk throw of the fault. These small-scale structural effects are of great importance for fault seal prediction, but are not captured by fault seal predicting algorithms, as described earlier.

Using combinations of classic sandbox experiments with cross-fault fluid flow measurements of kaolinite failing in hybrid extension/shear mode, Kettermann et al. (2017) show how brittle failure and reworking of clay in to the fault zone affect the fluid flow. After an early tensile failure of the clay layer causes a sudden increase of fault permeability, a phase of slower increasing permeability follows, while relays and breached relays form. Occasionally, the permeability even decreases at larger throws, as discontinuities in the clay smear are resealed.

8.4.7.5 Numerical simulations

Numerical simulations, while being limited in their representation of natural processes by resolution and implementation of physical laws, do provide reproducibility and control on mechanical properties that analogue models cannot achieve. As a result, carefully interpreted numerical simulations can be beneficial for our understanding of fault structures and processes, and the influence on fault sealing. Two main approaches are used to simulate fault zones: (1) continuum modelling, including advanced Lagrangian finite element method (FEM) codes with remeshing capabilities, and (2) discrete element method (DEM) simulations (Chapter 4).

Continuum models using hypoplasticity (Gudehus and Karcher, 2007) imply a continuous widening of the fault zone with increasing displacement, which leads to a steady clay smear thickness of approximately 15% of the source layer thickness. This fault zone widening could contribute to clay smears in large faults that are thicker than fault-seal algorithms predicted. Models using FEM with remeshing show the development of co-planar shear zones on both sides of a clay smear when the shear crosses a discontinuous clay smear, similar to observations in nature (Kleine Vennekate, 2013). Large strains, as experienced in clay smears, are still challenging even with remeshing, limiting the FEM approach.

The discrete element method approach is not limited by large strains and also allows brittle fracturing. Its limitation lies in the resolution; similar to a sandbox experiment, every single element in the model has to be included in the calculation. High-power computing is required for models with millions of particles, as would be necessary to model the grain size contrast of sand and clay layers. But with the rise of cheap computing power, high-resolution 3D models are technically possible and will probably soon offer new insights. To date, 2D experiments show the correlation of clay friction angle and clay smear thickness, in that smaller friction angles result in larger thicknesses

(Egholm et al., 2008). These models also observe clay injection similar to analogue models and field observations of van der Zee et al. (2003). The effect of clay cohesion on fault evolution and clay smear structure has been systematically studied (Schöpfer et al., 2007a,b, 2006) and extended to varying basement fault dip (Raith, 2012). These studies show an increasing complexity of the fault zone with increasing cohesion of the clay layers. Hybrid failure is observed, as well as reworking in the shear zone, at higher displacements (Raith, 2012). Models with varying particle sizes in the clay layers (TerHeege et al., 2013) reproduce the direct shear experiments of Çiftçi et al. (2013) and illustrate how smaller particle sizes cause a wider intrusion of the clay in the pore space of the sand in the fault zone.

8.5 Fault sealing in carbonates

8.5.1 Introduction

The study of carbonate-hosted faults is crucial both for economic and social issues, as these faults may act as seals and/or conduits for geofluids (i.e., water, hydrocarbons, geothermal fluids) within carbonate reservoirs (61% of the recoverable oil in giant fields; Roehl and Choquette, 1985) and may represent significant seismogenic sources within densely-populated regions (i.e., central Mediterranean area; e.g., Valoroso et al., 2013). For these reasons, over the past 30 years, there has been increasing interest in carbonate-hosted faults and studies have focused on the outcrop to nanoscale fault zone structures, deformation mechanisms, permeability structure, fluid rock interactions, and mixing with other lithologies (i.e., clays) during faulting (for review, see Delle Piane et al., 2017). Accordingly, this section deals with faulting processes and the related outcrop to microscale structures, fluid-rock interactions, and permeability structure of carbonate-hosted faults, (low- to high-porosity carbonates) and of mixed carbonate/clay-hosted faults.

8.5.2 Fault processes in low-porosity carbonates

8.5.2.1 Outcrop observations

Low-porosity carbonates generally consist of massive platform limestones/dolostones, other massive carbonates, and well-bedded pelagic limestones that show different clay content. Surface exposures of faults cutting through such carbonates and exhumed from depths <4 km are mainly characterised by dilatant and brittle processes (i.e., fracturing and cataclasis), whereas ductile processes (i.e., pressure-solution) become important with the increase in clay content (i.e., within pelagic limestones), pressure, and temperature (i.e., with increasing depth). Generally, the fault zone structure shows a gradual transition from a fractured damage zone to a strongly-deformed fault core along the

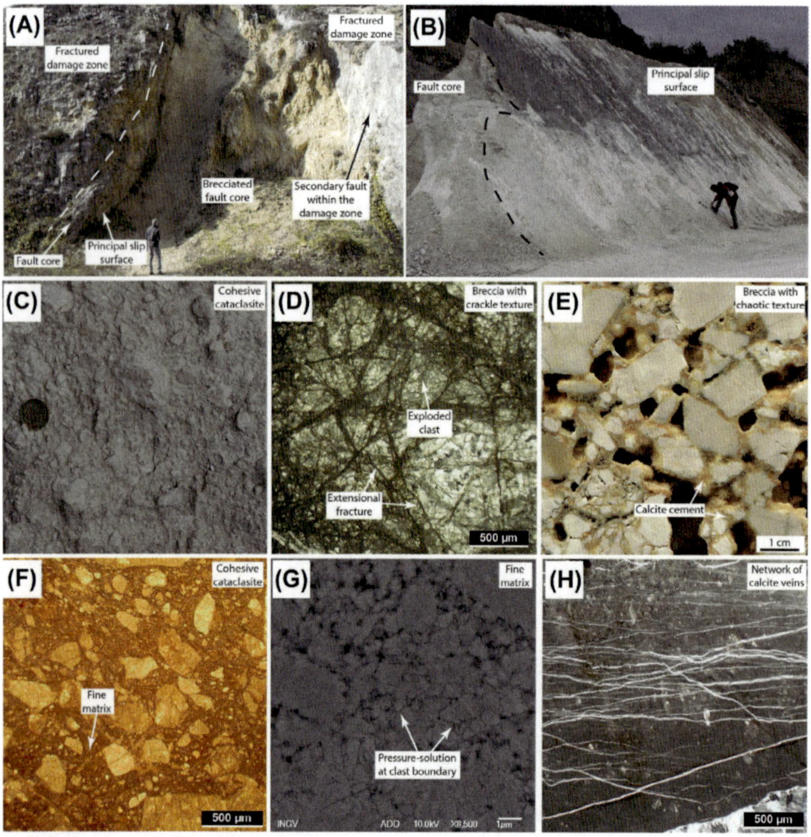

FIG. 8.19 Outcrop images and microstructures of fault cutting through massive and low-porosity limestones. (A) Panoramic view of a normal fault cutting through massive limestones. From left to right, the fault zone is characterized by a fractured damage zone, a cohesive fault core, a principal slip plane, a brecciated and incohesive fault core, and by low displacement faults within the damage zone. (B) Principal slip plane in massive limestones cutting through a cohesive and cataclastic fault core. (C) Detail of a cohesive cataclastic fault core. (D) Crackle breccia texture characterized by radial extensional fractures and "exploded clasts", which show no shear strain. (E) Chaotic breccia texture that show voids between clasts filled by calcite cement. (F) Cataclastic texture at microscale showing large clasts scattered within a fine matrix. (G) Detail of the fine matrix characterized by stylolitic clast boundaries. (H) Network of calcite veins cutting through a cataclasite.

main fault surface (Fig. 8.19A and B; e.g., Agosta and Aydin, 2006; Agosta and Kirschner, 2003; Billi et al., 2003). Low-displacement faults may also occur within the damage zone, indicating complex processes of deformation partitioning (Fig. 8.19A).

Within massive limestones/dolostones, intergranular extensional fracturing is more common in the early stages of deformation, producing host-rock parallelepipeds bounded by a highly connected network of fractures

(Fig. 8.19A). With increasing deformation, and approaching the main fault plane, coarse breccias showing angular clasts occur (Fig. 8.19F). Pre-existing weaknesses and flaws, such as primary bedding, may control the fracture pattern associated with fracturing. In the advanced stages of the cataclastic process, grain rolling, chipping, and further fracturing reduce the grain-size of breccia clasts, producing fault gouges characterized by a few angular/rounded clasts scattered within a fine-grained matrix (Fig. 8.19C; e.g., Billi, 2010). Mature fault zones show an intensely deformed fault core, up to several metres thick, filled by gouges, which are cut by a principal slip zone (Fig. 8.19B). Principal slip zones of high-displacement faults are typically cohesive due to the lithification by compaction and cementation, which generates cataclasites, and are characterized by smooth slip surfaces showing slickenlines, slickenfibres, and grooves (Fig. 8.19B; e.g., Fondriest et al., 2013). Recent studies have shown peculiar features of fault zones in dolostones, which are characterized by the lack of well-defined fault cores and evident shear strain, the preservation of primary sedimentary features (i.e., bedding), and the presence of highly-segmented networks of low-displacement faults (i.e., displacement up to a few tens of metres) faults surrounded by domains of cataclastic rocks. These fault zones are generally interpreted as the result of in-situ shattering-pulverization (e.g., Fondriest et al., 2015).

Fault zones in carbonates with variable clay contents, are characterized by multiple cores, diffuse core/damage zone boundaries, pervasive pressure solution, veining and the occurrence of foliated tectonites (Fig. 8.20A–C). The increased clay content in the host-rock activates ductile deformation processes during faulting, such as pressure-solution. Pressure-solution generates a network of stylolites by limestones and marls dissolution and insoluble materials (i.e., phyllosilicates) concentration within dissolution seams. During deformation and with increasing strain, the fault rocks often become foliated due to the alignment of phyllosilicates, which generate a pervasive scaly fabric (Fig. 8.20A–C). Foliated fault rocks characteristically develop S-C structures by the combination of pressure-solution and frictional sliding along S and C planes (Fig. 8.20A–C). Faults cutting through a heterogeneous sedimentary succession, characterized by carbonates and marly units, show complex fault zones, characterized by the interaction of brittle and ductile processes. By comparing clay-rich and clay-poor fault rocks, it can be inferred that pressure-solution and frictional sliding are the dominant deformation mechanisms in clay-rich zones, characterized by distributed shear planes bounding a continuous and pervasive foliation, whereas cataclasis and slip localization along principal slip zone are favoured in pure massive limestones (e.g., Bullock et al., 2014; Tesei et al., 2013).

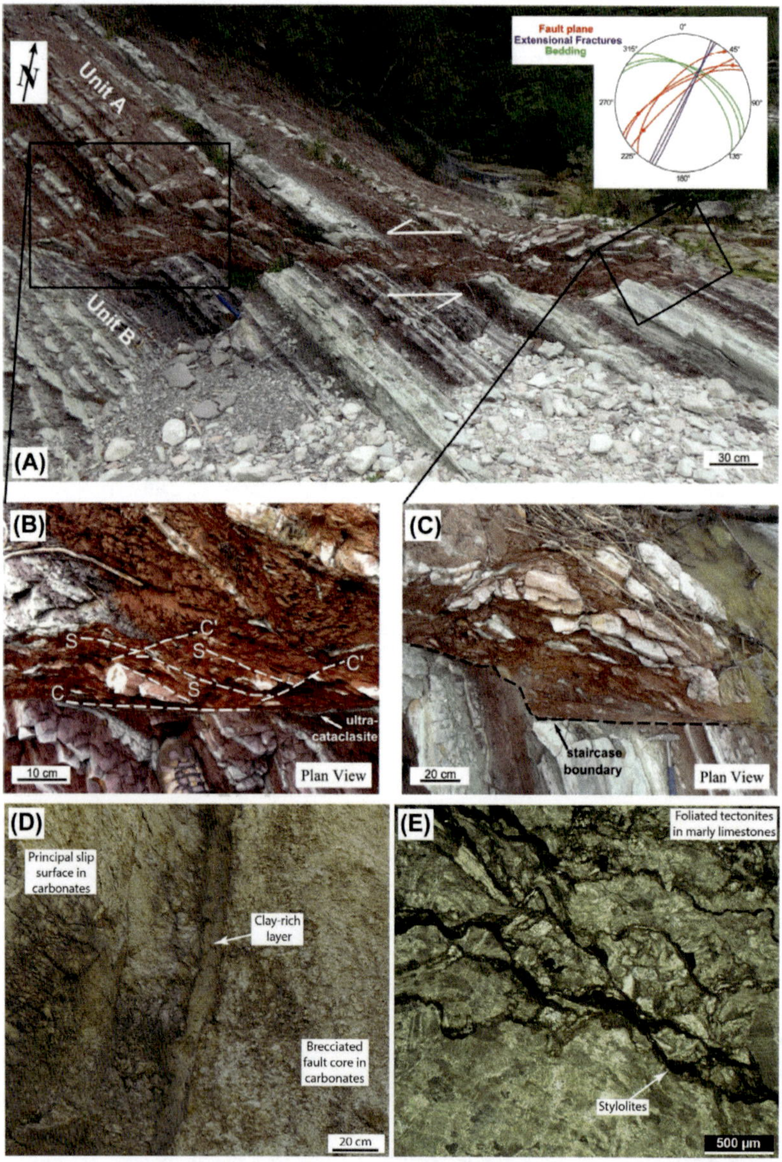

FIG. 8.20 Outcrop images and microstructures of fault cutting through clay-rich limestones and across heterogeneous stratigraphy. (A) Fault with a large (20 m) displacement bringing into contact marly interbeds (unit A) with limestones interbeds (unit B) and showing a staircase trace. The fault surface shows an overall straight trajectory, but its boundaries still maintain the original staircase shape, resulting in a variable thickness of the fault core. (B and C) Detail of the fault core, mostly consisting of reddish marl from the marly unit A, characterized by SCC′ fabric. Note that the fault core preserves the initial staircase geometry (C). (D) Clay-rich injection localized between the principal slip plane and the damage zone, both in carbonate rocks. (E) Microstructure of S-C tectonites in marly limestones, showing a well-developed network of stylolites.

8.5.2.2 Microstructures

A wide range of different microstructures has been observed within faults that cut through pure carbonates and/or carbonates with increasing clay content, deformed at low temperature conditions (<200 °C, e.g., Ferraro et al., 2019). Breccias are characterized by crackle to chaotic textures, depending on how well the clasts fit together (Fig. 8.19D and E). Fractures and voids within breccias are often filled by calcite cements (Fig. 8.19D).

Fault gouges and cataclasites are characterized by the occurrence of few angular/rounded survivor clasts, scattered within a fine-grained matrix (Fig. 8.20F; e.g., Billi, 2010). At microscale, the matrix consist both of very fragmented calcite clasts and of newly-precipitated calcite cement. Pressure solution-processes are limited within fault gouges and are often observed occurring at clasts boundaries (Fig. 8.20G; e.g., Smeraglia et al., 2017). Cataclastic rocks may be characterized by the occurrence of cm to mm-thick layers, showing fluid-like geometries, syn-tectonic calcite vein networks, and reworked clasts of ultracataclasites and veins, indicating multiple stages of deformation and fluid circulation (e.g., Smith et al., 2011). Microstructural characteristics of in-situ shattered fault zones in dolostones include the occurrence of "exploded clasts" with radial fracture patterns, pervasive extensional fractures lacking a dominant orientation down to the micrometre scale and the sharp truncation of relatively large clasts (Fig. 8.20D; e.g., Fondriest et al., 2015).

Deformation intensity has been studied using the particle size distribution analysis of cataclastic rocks from strike-slip, extensional, and thrust faults. Particle size follows a power law distribution, with fractal dimension increasing from 2 in the less deformed parts of the fault zones to 3 in the intensely comminuted gouges and catacalsite within the fault core zones. The increase of fractal dimension is interpreted to reflect a change in grain-size reduction mechanism from fracturing to slip-enhanced clast abrasion/chipping due to clast rotation with increasing fault displacement and slip along the main fault (e.g., Billi, 2005; Storti et al., 2003).

The fault core of carbonate faults with increasing clay content show strongly penetrative scaly foliation, shear bands, shear-related asymmetric folds, and extensive veining, with the original fabric obliterated by pervasive pressure-solution and stylolite generation in foliated tectonites (Fig. 8.20B and C). At microscale, stylolites generate an interconnected network that bound undeformed lenses of the host-rocks (Fig. 8.20E). Stylolites show teeth-shaped morphologies and the infilling material consist of insoluble material from the host-rock, such as quartz, feldspars, clays, and authigenic materials, such as goethite and apatite (Fig. 8.20E; e.g., Viti et al., 2014). With increasing deformation, pressure solution seams become thicker, more pervasive, and aligned parallel to the sense of tectonic transport.

8.5.3 Faulting processes in high-porosity carbonates

8.5.3.1 Outcrop observations

In porous limestones, a different style of deformation is observed during faulting, as strain is generally accommodated through non-dilatant deformation, such as compaction. At outcrop scale, these fault zones are characterized by deformation bands, which consist of zones of reduced porosity, permeability, and grain size (Fig. 8.21A and B). Deformation bands in high porosity carbonates can be of different types based on the dominant deformation mechanisms: shear bands, compaction bands, dilation bands, or hybrids of these two types (see review by Fossen et al., 2007).

Deformation bands occur in high porosity carbonates (20%-30%) and in chalks, a widespread type of highly porous, fine-grained limestones (Fig. 8.21C–E). The thickness of deformation bands varies between 2 and 10 μm (in chalks) and 5 cm for zones of deformation bands in high-porosity grainstones (Tondi et al., 2006) and increases with fault displacement. Deformation bands, with a component of shear movement, typically have high angles to bedding in carbonate rocks, whereas compaction bands in carbonates have been described up to now only parallel to bedding. Stylolites are often associated with deformation bands.

8.5.3.2 Microstructures

Microstructural observations show the occurrence of grain compaction and pore collapse (Fig. 8.21C–E). Pressure-solutions occur at grain boundary and generated discrete stylolites within already compacted bands. Small cataclastic shear zones localize along previously compacted zones and stylolites. In particular, non-lithified porous carbonates develop deformation bands, mainly by grain translation, rotation, and compaction, with no evidence of cataclasis. Deformation bands show strikingly different textures with respect to the porous host rock. Porosity, pore sizes and pore connectivity are dramatically reduced within deformation band, as observed though optical and electron microscopy, X-ray micro-tomography, and mercury injection porosimetry (Fig. 8.21C–E).

8.5.4 Carbonate faults cutting through heterogeneous stratigraphy

If different carbonate lithologies are interbedded within a multilayered sedimentary sequence (i.e., alternance of shale, marls, and limestones) then the fault trajectory and fault rock types change accordingly, resulting in significant heterogeneity in fault types, distribution, and geometries. This, in turn, affects fault propagation and evolution (e.g., Ferrill and Morris, 2008). The effect of mechanical layering leads to refracted fault profiles that show flat-ramp-flat trajectories, with steep to vertical dips through massive limestones, and moderate dips through shale-rich and marly-rich interbeds (Fig. 8.20A–C;

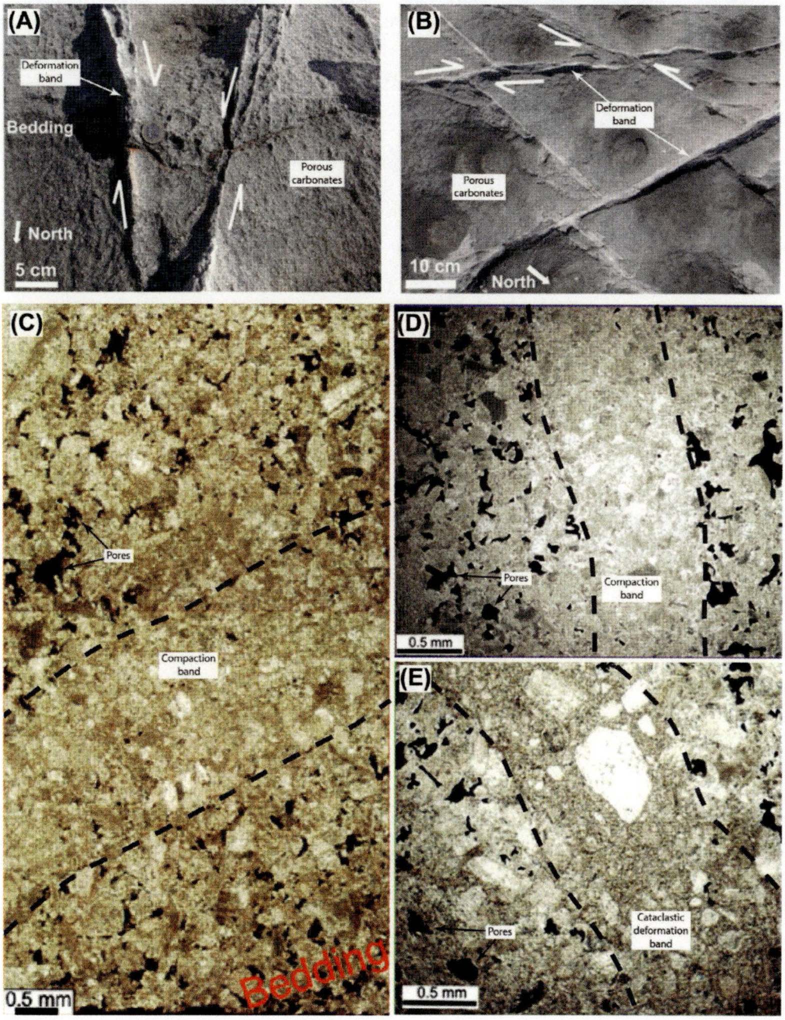

FIG. 8.21 Outcrop images and microstructures of a fault cutting through high-porosity limestones. (A) Shear deformation bands displacing a sedimentary marker. (B) Mutually crosscutting relationships of two conjugate sets of shear deformation bands. (C and D) Compaction band with negligible amounts of cataclasis. (E) Shear band characterized by continuous tabular zone of cataclasis.

e.g., Giorgetti et al., 2016). Fault slip occurs on the moderately-steep segments and concurrent dilation occurs on the steeper and mechanically more-competent segments, which are subsequently cemented with calcite (Ferrill et al., 2017). Dilational zones within competent carbonates can be also filled by shale, which are injected from shaly interbeds located within the sedimentary succession (e.g., Smeraglia et al., 2016). In this context, clay-rich injection localizes between the principal slip plane and the damage zone in carbonate-hosted faults (Fig. 8.20D).

During increasing displacement, the staircase geometry of faults cutting though a multilayer sequence (i.e., alternance of shale, marls, and limestones) is disrupted and the fault trace become smooth. Different protoliths smear and mix, generating mixed fault rocks, which can deform by both brittle and ductile processes, depending on the relative amount of carbonates and shales (Fig. 8.20A–C). An increase in shale content leads to the development of foliated fault rocks characterized by pervasive pressure-solution, whereas a dominance of carbonates leads to the development of cataclastic rocks. However, cataclastic rocks that interact with shale-rich lithologies show the occurrence, at microscale, of phyllosilicates within the fine-grained matrix (e.g., Smeraglia et al., 2017). Such phyllosilicates may enhance pressure-solution processes and compaction of cataclasites.

8.5.5 Normal, thrust, and strike-slip fault architectures in carbonates

Although main deformation mechanisms of carbonate faults are the same, regardless the fault kinematics (i.e., normal, thrust, and strike-slip kinematics), fault zone architecture can varies in different tectonic settings. For instances, for carbonate faults developed at 2–3 km depths, but in different tectonic settings, the fault zone architecture changes accordingly. In particular, normal faults in carbonates are characterized by parallel fault strands showing multiple and broad brecciated damage zones, cataclastic fault cores, and principal slip zones, as highlighted by both field and geophysical observations (e.g., Agosta and Aydin, 2006; Valoroso et al., 2013). Fault strands are often linked by relay ramps (e.g., Peacock, 2002). On the contrary, thrust faults often show a single fault core or a thick shear zone, respectively, if deformation affect pure carbonates or carbonates with increasing clay contents (e.g., Ghisetti et al., 2001; Tesei et al., 2013). This is probably because for the same lithostatic pressure (i.e., depth), normal faults deform under low differential stress (i.e., horizontal minimum principal stress, σ_3, lower than maximum principal stress, σ_1, related to lithostatic pressure) and low confinement, leading to broad fault zones with multiple fault strands. On the contrary, thrust deform under higher differential stress (i.e., horizontal maximum principal stress, σ_1, higher than the minimum principal stress, σ_3, related to lithostatic pressure) and higher confinement, leading to localize fault development. For strike-slip faults, differential stress will be intermediate, so fault zone architecture show both different fault strands with multiple-brecciaed damage zones/fault cores and/or localized single faults (e.g., Billi, 2010; Fondriest et al., 2015).

8.5.6 Fault permeability, fluid circulation, and seal in carbonate hydrocarbon reservoirs

As previously described for siliciclastic rocks, the standard workflow for assessing the ability of normal faults to seal hydrocarbon is based on assessing the juxtaposition of low permeability lithologies with the reservoir rock across the fault using well-established methods (i.e., clay smear potential, smear factor, shale gouge ratio). If shales are available within the carbonate sequence to form a shale smear, or for juxtaposition seal, then the same workflow applies. However, there are oil fields with carbonate-carbonate juxtapositions indicating that seals have developed over either geological or production time spans (e.g., Solum and Huisman, 2017). As the 61% of the recoverable oil in giant fields is hosted in carbonate reservoirs (Roehl and Choquette, 1985), assessing the permeability structure of carbonate faults is crucial for a better oil recovery and reservoir management.

Carbonate fault seals are more complicated than clastics fault seals due to their chemical reactivity, the importance of diagenesis, and their ability to readily fracture and seal by mineral precipitation. The detailed architecture of carbonate fault rocks tends to be highly heterogeneous, depending upon host rock texture (grain size, sorting), porosity, and uniaxial compressive strength (Michie, 2015; Michie and Haines, 2016). Cataclasites tend to develop in coarse-grained, and heterogeneous-grain dominated carbonates, whereas homogenous micritic limestones tend towards more dispersed deformation, with large fracture networks and breccias (e.g., Michie, 2015). More homogeneous carbonates will generate more homogenous fault rocks, whereas significant heterogeneity in the fault rocks will create variable permeability along the fault (e.g., Michie, 2015).

However, field and laboratory studies of faults in low-porosity limestones show complex fault-controlled fluid circulations. In particular, seismically-active faults may act as fluids conduits, allowing fluids pumping and redistribution across different crust levels, regardless of fault kinematics (i.e., thrust, normal, or strike-slip faults). Paleofluid circulation within carbonate faults is evidenced by the occurrence of fault-related mineralization, such as calcite veins and neo-precipitated cements within breccias and cataclasites (Fig. 8.19E and H). In particular, geochemical studies of fault-related mineralization show mineral precipitation temperatures both higher or lower than those of the local host rocks suggesting, respectively, deep-seated, hot fluids pumping or the downward infiltration of cold meteoric-derived fluids along fault cores (Agosta et al., 2007; Smeraglia et al., 2018). In addition, if faults cut through a multilayered sequence characterized by marly/shaly formations, fluid circulation along fault segments cutting through low-permeable shales and marls may be characterized by a closed system with the fluid's chemical composition buffered by the local host rock, which act as an impermeable barrier for infiltration of external fluids (Lacroix et al., 2014).

Overall, microstructural studies on fault-related calcite veins showed that fluid circulation in carbonate faults is heterogeneous, changes during the spatio-temporal evolution of the fault zone, and is episodic, and characterized by multiple fluid flow events (Fig. 8.19H; Nuriel et al., 2011, 2012; Smeraglia et al., 2016).

After fracture cementation and voids closure when faults stop being seismic, the fault zone may act as a barrier for across-fault fluid flow. In particular, fault gouges, cataclasites, and foliated fault rocks show strong decrease in fault permeability, which is 2–3 order of magnitude lower with respect to that of the fractured damage zone (Agosta et al., 2007; Trippetta et al., 2017). Permeability reduction within cataclastic rocks is due to the fine-grained matrix and pore sealing due to cementation by neoformed crystals, whereas in foliated carbonate fault rocks, pressure-solution processes have high impact on permeability reduction due to the effect of phyllosilicate concentration within stylolites, which create low-permeability barrier to fluid flow.

Deformation bands in the damage zones of normal faults in high porosity carbonates, display highly heterogeneous distribution and thicknesses, together with highly variable related permeabilities, which reflect varying degrees of cementation, cataclasis and diagenesis (Kaminskaite et al., 2019). Their evolution is controlled by the stress condition, and texture and composition of the host rock. Within deformation bands, porosity, pore size and pore connectivity are dramatically reduced, as observed by optical and electron microscopy, X-ray micro tomography, and mercury injection porosimetry (Rotevatn et al., 2016) and permeability can be 1–4 orders of magnitude lower than in the host rock (Antonellini et al., 2014).

Given the complexity associated with carbonate fault zones, some simplifying principles are needed for hydrocarbon exploration and development. Cooke et al. (2018) suggest that displacements in the order of 50–200 m are needed to develop a continuous, low-permeability cataclastic zone in their study area in Malta, but in lithologies that favour more distributed damage zones, and multiple slip surfaces, higher displacements may be required. Although the focus in this section is on normal faults, it should be noted that some hydrocarbon traps in carbonate reservoirs in Abu Dhabi are sealed by strike-slip faults, which are further complicated by fault reactivation (e.g., Noufal and Obaid, 2017).

8.6 Evaporites and fault seals

While typical shale seals have permeabilities $\approx 10^{-1}$ to 10^{-5} mD, evaporate permeability is too low to be measured by standard instruments, halite is in the order of 10^{-7} to 10^{-9} mD, while massive anhydrites are $\approx 10^{-5}$ mD (Beauheim and Roberts, 2002). Halite has the potential to hold back fluids to near lithostatic pressures, and exhibits ductile behaviour all the way to the surface, hence it is capable of flow, and of annealing fractures, should they occur (see review in Warren, 2016). Consequently, evaporites are the best lithology that can contribute to fault seal, although they are a far less common lithology in sedimentary basins than shale. In salt basins, evaporites can be present along normal, thrust and strike-slip faults, all of which may be present in the same structure, given the complex structural geometries that develop in areas of salt tectonics. Evaporites commonly migrate from their primary depositional horizon, laterally or towards the surface in a wide variety of differently-shaped bodies (diapirs, salt walls, along-faults) and coalesce at higher levels as salt canopies, and thrusts (see reviews in Hudec and Jackson, 2007; Warren, 2016). As evaporites move through the section they can temporarily grow in a particular place, then thin, leaving behind secondary salt welds. Welds that have no lateral shear offset on them are called salt welds, while those that show some lateral offset are fault welds (Rowan et al., 1999). In terms of fault seal, there is a very wide range of scales, from primary or secondary salt layers that have evolved into low-angled regional detachments, to low and high-angled smaller fault systems restricted to individual structures. Regional salt detachments restrict the vertical migration of hydrocarbons, and welds are important areas where hydrocarbon migration can occur across a salt layer from the underlying source beds into overlying reservoirs and traps (e.g., Gulf of Mexico, Campos Basin, offshore Angola, Wagner and Jackson (2011)). Conversely, in areas like the Gulf of Mexico, offshore Congo and Angola, large fault welds can form lateral seals to hydrocarbon accumulations, where evaporates can be present in substantial thicknesses (e.g., 50 m thickness of halite), but can also thin down to residual veneers less than a metre thick, composed of anhydrite (residual deposit after halite; Jackson et al., 2017; Rowan, 2004; Wagner and Jackson, 2011). Clay gouge and smear can also accompany evaporates in fault welds, and contribute to their sealing potential (Rowan, 2004).

8.7 Case studies of fault seal

8.7.1 The Molasse Basin in Germany and the Rhenish Massif

The foreland basin of the European Alps, the Molasse basin, is one of the best studied foreland basins in the world. It is the result of alpine convergence and consequently the down bending of the lower European Plate (Pfiffner, 1986). The plate is downflexed and a wedge-shaped foreland basin has developed. At the same time, normal faults have formed in the pre-basin stratigraphy (Fig. 8.22). During ongoing convergence, the proximal sections of the basin were incorporated into the orogeny, leading to formation of a fold-thrust belt, the so-called Subalpine Molasse. In the western part of the basin, a more external fold-thrust belt developed associated with mechanically-weak salt layers, the Jura Mountains (Fig. 8.22; Laubscher, 1977). At the eastern

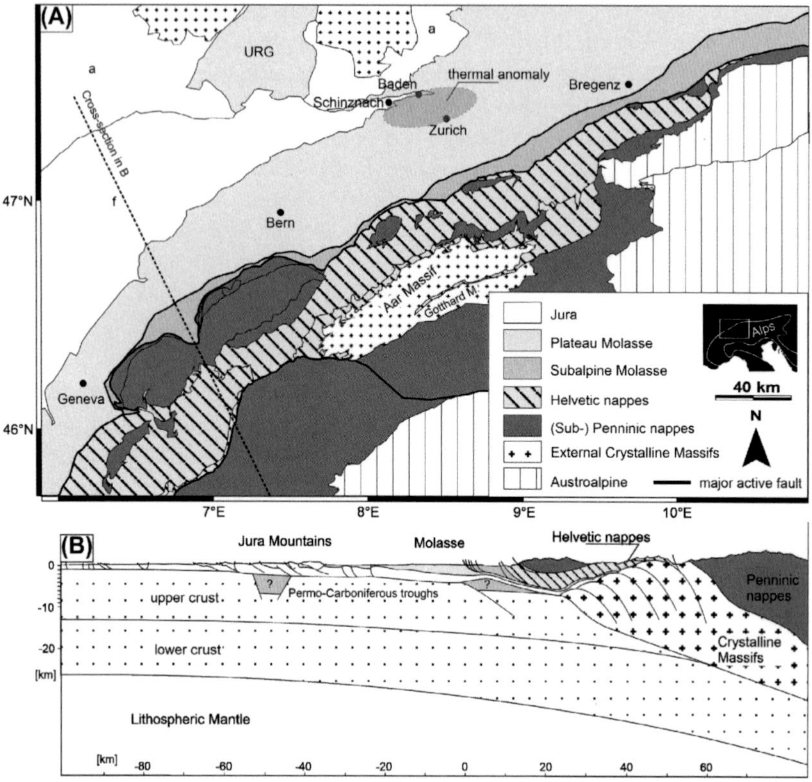

FIG. 8.22 Geological overview of the central Molasse Basin and the external Jura fold-thrust belt. At the eastern boundary of the Jura Mountains, strain is transferred in a complex zone towards the Subalpine Molasse. Structures include normal faults and strike slip faults in the Mesozoic carbonates interacting with younger thrust faults in the Tertiary sequence. This results in fluid flow and consequently thermal anomalies in the basin. Better understanding of the fault geometries at depth, as well as their sealing capacities, helps understand the presence or absence of thermal springs. *Modified after von Hagke et al. (2014).*

boundary of the Jura Mountains, the strain accommodation of alpine convergence links the external Jura Mountains with the Subalpine Molasse in a broad zone, possibly including strike-slip movements (Laubscher, 1992; Mosar, 1999; von Hagke et al., 2014). In the past, both the Jura Mountains and the Subalpine Molasse have been coevally active, and maybe both are connected to the same basal detachment of the Alps (von Hagke et al., 2012, 2014). Furthermore, the region has been affected by recent thick-skinned tectonics and compressive reactivation of Permo-Carboniferous extensional structures (Giamboni et al., 2004; Madritsch et al., 2008; Malz et al., 2015). This complex structural history results in a broad fault network and connection of shallow thrust faults to deeper levels. Wherever these fault systems are permeable, fluids may preferentially percolate and heat flow anomalies may develop. Therefore, the Molasse basin and in particular the eastern termination of the Jura Mountains, is a prime target for geothermal energy exploration. A large geothermal anomaly is reported in the area (Nagra, 2002), associated with several hot springs. These springs and the associated hydrothermal system can be considered an outcrop analogue for fault-related hydrothermal systems in foreland basins.

Thermal springs require deep fluid-flow paths to bring high-temperature fluids to the surface and to overcome lateral heat loss during upward transport. Therefore, a fault damage zone must exist that is permeable enough to allow to bring hot fluids to the surface fast enough. At the same time, the undeformed host rock needs to be tight enough to slow down fluid flow and allow heat conduction into the surrounding areas. This relationship has been modelled for the thermal springs in the vicinity of the geothermal anomalies, the Baden and Schinznach hot springs (Luijendijk, 2019). This study uses discharge and temperature data measured at the spring and assumes a porosity of the rock matrix of 15%, based on standard values. It can be shown that the width of the damage zone, as well as the depth of the permeable section, have a large influence on the modelled spring temperatures. The wider the damage zone and the shallower the fault, the lower the spring temperature (Luijendijk, 2019). This is a straightforward result; however, as the modelled temperatures need to match observations, predictions of fault permeability and depth are possible. In this context, it is important to note that the model outcome also depends on the fault geometry at depth. This is a major concern for all geothermal studies, since without good constraints on the subsurface geometry, different analytical solutions for fault permeability are possible. For this area, but also other fold-thrust belt systems, it can be shown that, even with good seismic data, many different interpretations of the subsurface are possible (von Hagke and Malz, 2018; Wellmann et al., 2010). This highlights the fact that to assess fault permeability, good knowledge of the regional geology is important.

Another example of thermal springs associated with a fold-thrust belt are the hot springs in the Rhenish Massif. As opposed to the structures in the alpine foreland, this fold-thrust belt is the deformation front of the Variscan Orogeny (Carboniferous; Oncken et al., 1999). In the Rhenish Massif, different provinces hosting thermal springs can be distinguished, based on location and water chemistry. The geothermal province around the city of Aachen, Germany is considered one of the most interesting hydrogeological phenomena in Europe (Michel et al., 1998). A series of thermal springs line up along two SE-dipping thrust faults that belong to the Aachen Nappe (von Winterfeld, 1994). The Aachen thrust fault, at the base of the Aachen Nappe, forms the exposed deformation front of the western part of the Rhenish Massif and continues towards the west, where it is known as Midi-Fault in the Belgian Ardennes. The hanging-wall of the thrust consists of a sequence of upper Devonian Limestones and lower Carboniferous siliciclastic sediments. The thermal springs are hosted in the carbonates, which have a thickness of up to 200 m (von Winterfeld, 1994). In the area of the thermal springs, the Aachen thrust has a displacement of 17 km (Hollmann and Von Winterfeld, 1999). For faults with such high displacement, a fault thickness of >100 m is expected (Shipton et al., 2006). The thermal springs in Aachen are all located within few tens of meters distance from the thrust trace, over an along-strike distance of approximately 2 km (Pommerening, 1993). All thermal springs are found to the east of the Laurensberg Fault, a poorly-defined, Cenozoic, NW-SE-striking normal fault, running approximately perpendicular to the Variscan deformation front, associated with rifting of the Lower Rhine Embayment (Tertiary or younger; Geluk et al., 1995). We speculate the eastern termination of the spring series is defined by another normal fault, named hereafter the Lousberg Fault (Fig. 8.23), and the thermal springs are located in a tectonic block bound by two Variscan thrusts and two Cenozoic normal faults. Evidence for the Lousberg Fault is provided by topographic analysis and geochemical data of the spring waters, which are used to define the recharge area towards the NW, SE and SW of the springs (Fig. 8.23; Pommerening, 1993). On its western side,

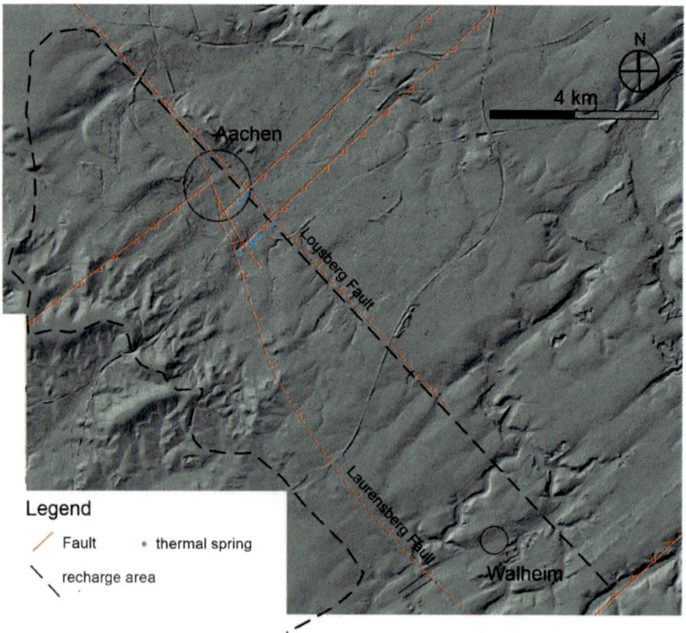

FIG. 8.23 Structural map of the thermal springs in the foreland fold-thrust belt of the Variscan Orogen, Aachen, Germany. The thermal springs are aligned along two faults of the ancient orogen, and only occur in a fault section between two younger (Tertiary) normal faults, namely the Lousberg Fault and the Laurensberg Fault. The recharge area partly includes regions to the northeast of the thrust faults. The springs are associated with the high permeability in the damage zone of the thrust faults.

the recharge area is controlled by the present-day topography. Towards the east, it is bound to the Lousberg Fault. To the south, the fluids infiltrate the upper Devonian limestones that also host the springs (Breddin, 1960; Pommerening, 1993).

Geochemical analyses show that the thermal spring water contains signatures, not only of the highly-permeable carbonate sequences outcropping in the recharge area, but also shows trace elements from the Devonian units, which have a lower permeability (Herbst, 1954; Pommerening, 1993). The water has a residence time of several 1000 years (Pommerening, 1993). This is interesting considering the permeabilities of the faults involved in the system. The Variscan thrust faults guide thermal water flux through the carbonates of the hanging-wall, implying the thrust are impermeable on the timescale of several thousand years. Similarly, the Lousberg Fault and the Laurensberg Fault act as a hydrological barrier, leading to the small area in which thermal springs can occur. Both faults must be less permeable than the Devonian and Carboniferous units. As the thermal springs are located very close to the fault, the damage zone is more permeable than the undamaged limestones. Provided the width of the damage zone is >100 m, the permeability of the damage zone decreases at greater distances from the faults, as expected. However, if the outline of the recharge area is correct, at least the sections of the faults close to the surface must be permeable enough to allow for water flux from the NW. Indeed geochemical data suggests that the thermal spring water is mixing with surface waters (Pommerening, 1993).

These two examples show that, to understand fluid flux through fault systems, it is necessary to study the individual faults at a much smaller scale. This is often challenging, as outcrops of faults, where a systematic study of meso- and microstructural analysis is possible, are rare. The Khao Khwang fold-thrust belt in Thailand, for instance, offers well-exposed sections of a shale detachment between carbonate sequences.

This system offers insights into the macroscopic and microscopic structures present in mixed rheology systems. In this system, many aspects of deformation have been investigated on the macroscopic (Hansberry et al., 2014, 2015; Morley et al., 2018), as well as on the microscopic scale (Morley et al., 2017). As described above, shales within carbonates localize strain and the permeability within the shale layers is significantly lower than in the surrounding carbonates. This can also be observed in outcrop. The system shows a classic evolution of structural style towards the fault, as has also been described for similar systems in the Appalachians (Morley et al., 2017). Above the thrust, the carbonates are brecciated, typically in a zone of several meters thickness. Often bedding surfaces are still recognizable, but crosscut by anastomosing fracture networks, the intensity of which increases towards the fault. Veining and pressure solution is abundant, indicating the fault is affected by both, rapid (seismic) fracturing and slow creep processes. In thin section, intense twinning of calcites can be observed. This zone of brecciation transitions into a zone of limestone cataclasite, sometimes mylonite. This zone may reach thicknesses ranging from few centimetres to typically less than a metre. Veining is intense, and the veins commonly show multiple stages of reactivation and often folding and shearing. Veins often truncate clasts. Clasts within the cataclasite are angular to rounded, and are of variable composition, including carbonate host rock fragments, shale clasts and abundant vein material. Sometimes a transition from brecciated calcite veins to carbonate mylonite can be observed, indicating the mylonite formed at least partly from vein carbonates. Below the limestone cataclasite, the shale shows intense cataclasis and veining. Veins are sheared, and often (sub-)parallel to the fault contact, sometimes perpendicular. With increasing distance from the carbonates, veins within the shales are less abundant. Strain localizes in narrow principal displacement zones, characterized by polished shear surfaces, S-CC′ fabrics and veins that are sub-parallel to foliation. Like close to the limestone-shale contact, veins approximately perpendicular to foliation occur, and a clear cross-cutting relationship between the two vein sets is not possible to determine, indicating they formed coevally. In zones of lesser strain, veins orientations are more variable. These structures stand out morphologically, compared to the high strain ones, since they are less weathered. In the footwall of the shale thrust, mechanical mixing between shales and host rock occurs. Fragments of carbonates are incorporated into the shale sequence. Below this cataclastic zone, a sharp fault surface forms the boundary to the underlying limestones. This surface is polished and features extensive slickensides. Veins are abundant.

The overall structure shows the importance of fluids within the deformation zone. Veins form predominantly at the lithological boundaries, and are later reworked during ongoing shortening. The low permeabilities of the shales resulted in pore fluid pressures exceeding rock strength. The fault may thus act as temporary fluid valve, while being largely impermeable in between faulting events. Fig. 8.24 summarizes the structures observed in the system.

8.7.2 Inboard area of the Baram Delta Province, NW Borneo

The Baram Delta Province of Sarawak, Malaysia, and Brunei, NW Borneo (Fig. 8.25), offers insights into fault seals in growth faults, associated with prograding deltaic deposits of Middle Miocene-Pliocene age, and for deltaic, gravity-driven growth faults, in general. Overpressured shales are important for establishing the weak basal layer to the prograding Middle Miocene-Recent deltaic sequence that acts as the basal detachment for the listric growth faults. One significant, anomalous aspect with respect to deltaic provinces, like the Gulf of Mexico, Nile and Niger deltas, is that the major fields onshore and on the shelf are associated with growth faults that have subsequently been inverted (e.g., Fig. 8.25C and D), giving rise to major anticlinal fields. These fields have multiple-stacked pays that are strongly

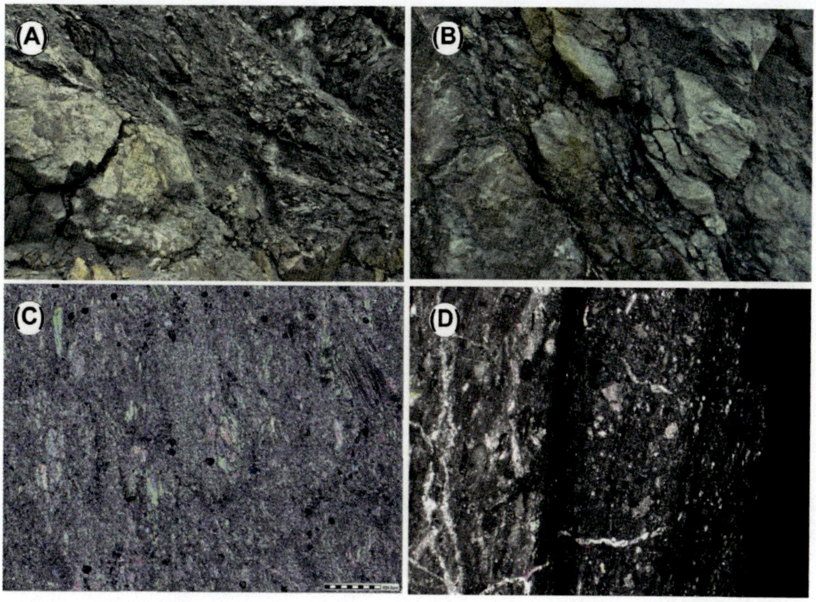

FIG. 8.24 Macroscopic and microscopic structures of a shale thrust between carbonates, as can be expected for the Alpine and the Variscan examples described above. (A) lower boundary between shales and carbonates. The fault surface is very sharp and overlain by intensely deformed shales including carbonate fragments. Base of the image approximately 2 m wide. (B) Detail of the basal thrust, highlighting the sharp boundary between shales and carbonates. The shales contain multiple carbonate clasts. (C) Calcite mylonite from the hanging-wall contact. The material at least in part consists of vein carbonate (large clasts). (D) Hanging-wall thrust in cross section. Image approximately 2 cm wide. Note the veins parallel to main foliation and fewer veins perpendicular to foliation in the more shaly parts. On the left side, calcite dominates the fabric, whereas on the right shales accommodate deformation.

compartmentalised, primarily by normal faults, together with some thrust faults and inverted normal faults (Sandal, 1996). Two important fields lie onshore, Seria in Brunei, and Miri in Sarawak (Fig. 8.25). After Shell conducted the classic exploration strategy of the day (identifying an anticline near oil seeps), they drilled a discovery well (Miri-1) in 1910. When, in 1972 the field was closed, it had produced a total of 80 million barrels of oil (Sorkhabi, 2010).

Seria field, Brunei, was discovered onshore in 1929, and by 1991 it had produced a billion barrels. The field description in Sandal (1996) illustrates the impact of faulting in the fields of the Baram Delta as follows. Over 70 stacked hydrocarbon-bearing reservoirs are dissected by faulting into over 500 different fault and dip-closed accumulations. Possibly the field is still being charged. Continuing structural uplift, and reservoir depletion-induced pressure changes that have caused some partially sealing faults to break down and permit cross-flow between reservoirs, which affects determining individual

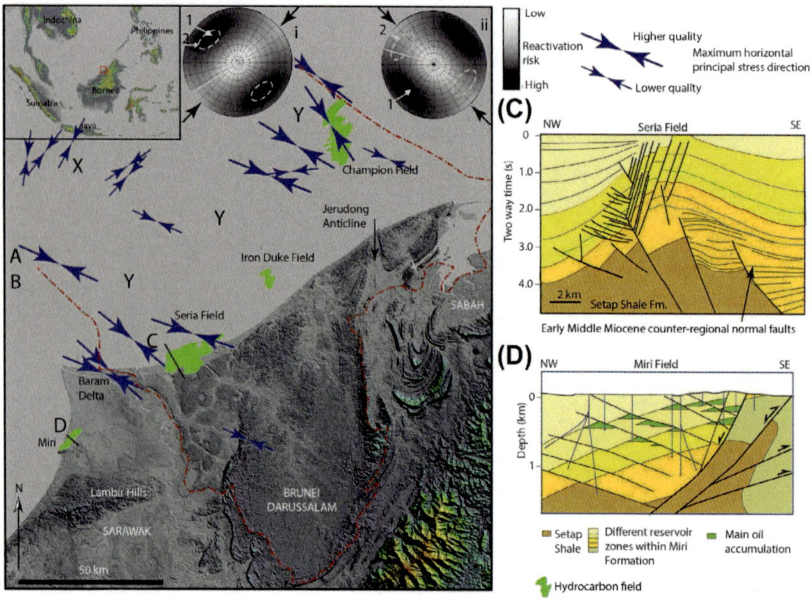

FIG. 8.25 Location map for the Baram Delta Province, NW Borneo. (A) Regional topographic map. (B) Topographic map of Brunei and adjacent areas of Malaysia (Sarawak, Sabah), NW Borneo. (C) Cross-section across the Seria Field, based on a seismic line in Sandal (1996), (D) Cross-section across the Miri Field. *Blue arrows* = modern maximum horizontal stress orientations (King et al., 2009). Stereonets show fault reactivation risk for area X (i), and Y (ii). 1 indicates the area of maximum risk for fault reactivation (dark grey), 2 = area where poles to fault planes plot. In stereonet i, the NE-SW oriented maximum horizontal stress means that 2 plots in the same area as 1, indicating a high risk of fault reactivation. Conversely for stereonet ii (area Y) 1 away from 2, indicating low risk of fault reactivation. *Modified from Bait and Banda (1994).*

reservoir volume from production history. The varying intensity of faulting also affects production: larger fault blocks have higher water drive, and greater recovery (as high as 50%), where faulting is more intense, aquifer influx is small, and gravity drainage and solution gas drive comprise the main recovery mechanism for light oil. The reduced effects of water drive is one of the factors affecting recovery of heavy, biodegraded oil, where recovery can be as low as 10%.

Inversion has not only created unusual antiformal traps, the stress field associated with inversion also affects fault stability and reactivation risk. Fig. 8.25 shows how the modern maximum horizontal stress direction changes from being sub-perpendicular to the shore line onland, and in the near-shore region (area Y, Fig. 8.25), while further offshore, the maximum horizontal stress rotates to subparallel to the coast (area X, Fig. 8.25). The effect of this stress rotation is that faults in area X are much more prone to reactivation than those in area Y (Tingay et al., 2009).

The Miri area is of considerable interest, not only because it provides a good subsurface example of a series of Miocene sandstone reservoirs that are compartmentalised by normal faults (Fig. 8.25C), but because the same fault systems are exposed nearby in outcrop, where the characteristics of the fault zones in poorly-lithified clastics can be studied in detail. Although the outcrops are considered 'world class' in terms of showing detailed fault zone geometries, they have been created by building or road constructing, and are subject ultimately to either weathering, degradation and growth of vegetation, or covering by man-made materials, and are temporary. The classic Airport Road outcrops (Burhannudinnur and Morley, 1997; van der Zee and Urai, 2005, Fig. 8.26) are now degraded, but new outcrops nearby are spectacular replacements (Sosio De Rosa et al., 2018). In places, the faulting was in sequences so poorly lithified that sands as well as shales were smeared out (Fig. 8.27). The fault zones are characterised by broad damage zones of closely-spaced deformation bands that extend up to 10 m into the hangingwalls and footwalls of relatively low displacement (<10 m) faults, together with narrow core, high displacement regions composed of clay smear, clay injection, cataclasis, and sandstone lenses (Fig. 8.26; Burhannudinnur and Morley, 1997; Sosio De Rosa et al., 2018; van der Zee and Urai, 2005). These authors have noted the importance of downward-linking geometries in the fault zones in introducing variations in fault core width and composition, as well as lateral linkage. Deformation bands in the damage zones (Fig. 8.26B–D) variably reduced reservoir porosity from about 26% to between 15% and 23%, indicating that small faults could significantly affect porosity of reservoir sandstones, and consequently have a negative effect on reserve estimates (Burhannudinnur and Morley, 1997). Permeability within the deformation bands is an order of magnitude lower than in adjacent undeformed sandstones (Sorkhabi and Hasegawa, 2005), indicating how both large and small faults need to be considered regarding their impact on reservoir permeability. Detailed analysis of fault zone characteristics shows that where undetected multiple fault strands are present, the reliability of juxtaposition diagrams decreases significantly (van der Zee and Urai, 2005) and patches of thick and thin clay-rich fault-core rocks are highly variable, over short distances of around 3 m (Sosio De Rosa et al., 2018).

The core of the Jerudong Anticline in Brunei (Fig. 8.25) is another area of outcrops in which Miocene normal faults are well exposed (Fig. 8.28). There is a very common occurrence of conjugate faults in this area, which provide good examples of the cross-cutting relationships developed at the zones where faults of opposite dip intersect (Fig. 8.28). Fault damage zones becomes broader and more complex in such areas. The fractal nature of these zones is also apparent, with conjugate faults visible at the seismic scale, all the way down to faults with just a few millimetres of offset (Figs 8.25B and 8.28). Excavations of hills for housing projects have revealed shale intrusion features (Morley, 2003a,b; Morley et al., 1998), which have been very rarely described from outcrop.

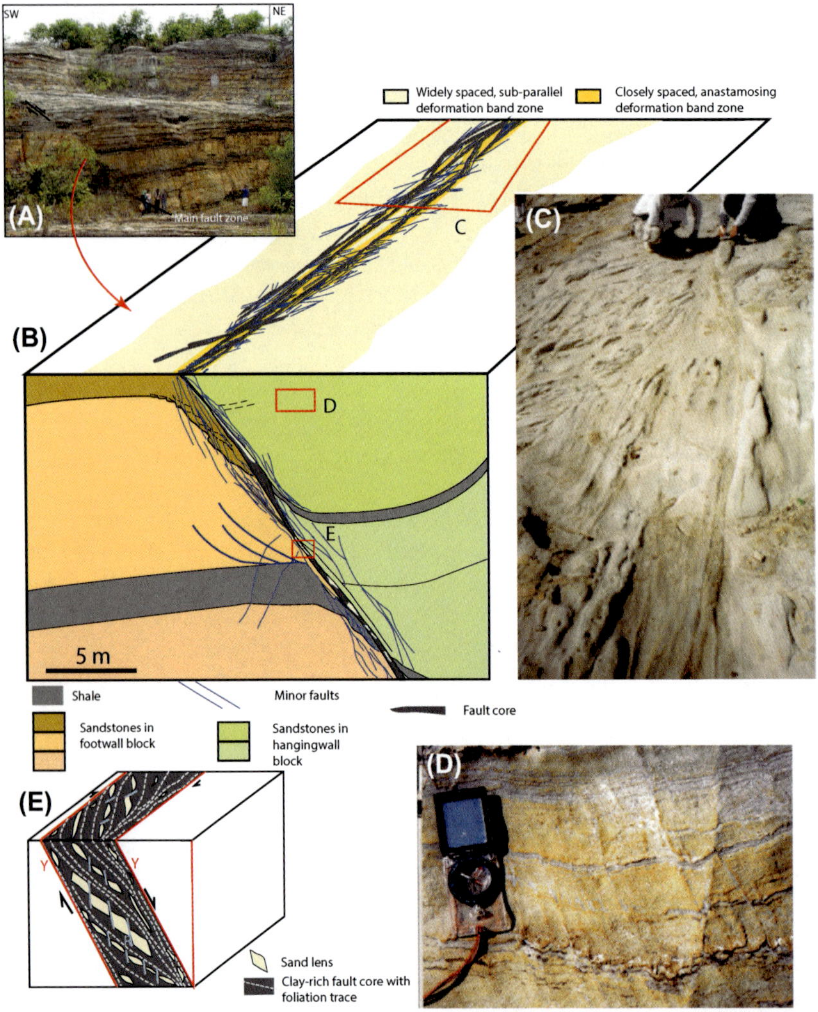

FIG. 8.26 Typical structural geometries associated with well-exposed normal fault zones in the Miri area. (A) Main fault zone of the Airport Road outcrop, c. 2002. (B) Sketch of typical normal fault zone structure in the Miri area, combining observations and outcrop sketches from Burhannudinnur and Morley (1997) and Sosio De Rosa et al. (2018). In map view, the fault zone comprises a number of mappable zones, including the high displacement fault core (shale smear, sand lenses, cataclastic zones), closely-spaced anastomosing deformation bands, and more widely spaced, sub-parallel deformation bands. The fault core can comprise one or more branching, splaying and rejoining zones. Low-displacement (cm-10's cm) fault zones, with a variety of orientations, are present adjacent to the fault core, and lie at a range of angles to the trends of deformation bands. (C) Map view photograph of the damage zone of the main fault, Airport road (A), showing anastomosing deformation bands that weather out are more resistant ridges (lower permeability) than the adjacent host rock. (D) Conjugate deformation bands in cross-section view, in the outer fault damage zone. (F) Detail of deformation in the fault core, showing an example where the clay rich fault core is foliated, and contains numerous sandstone lenses. The fault core walls can be interpreted as Y shears, and the internal faults as R-shears. *Redrawn from Sosio De Rosa et al. (2018).*

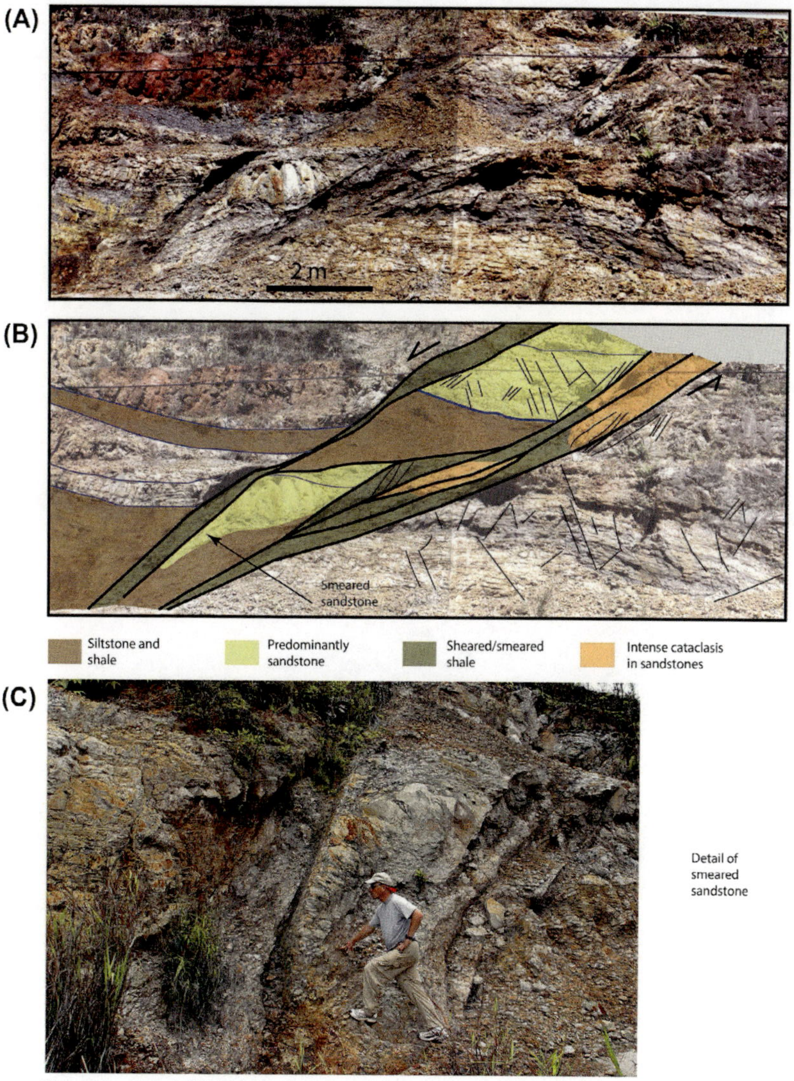

Miocene growth fault, SW of Miri, Lambir Hills. Fault formed in weakly lithified section, where even the sandstones can undergo smearing.

FIG. 8.27 Example of a large normal fault zone, (probably Middle Miocene age), in Middle Miocene clastics, Lambir Hills, Sarawak. (A) Uninterpreted photo. (B) Interpreted photo, showing the broad, multi-strand nature of the fault zones, which can be 2–4 m wide. (C) Detail of the fault zone, where a sandstone lens is smeared (with some internal deformation by deformation bands) by over 1 m into the fault zone.

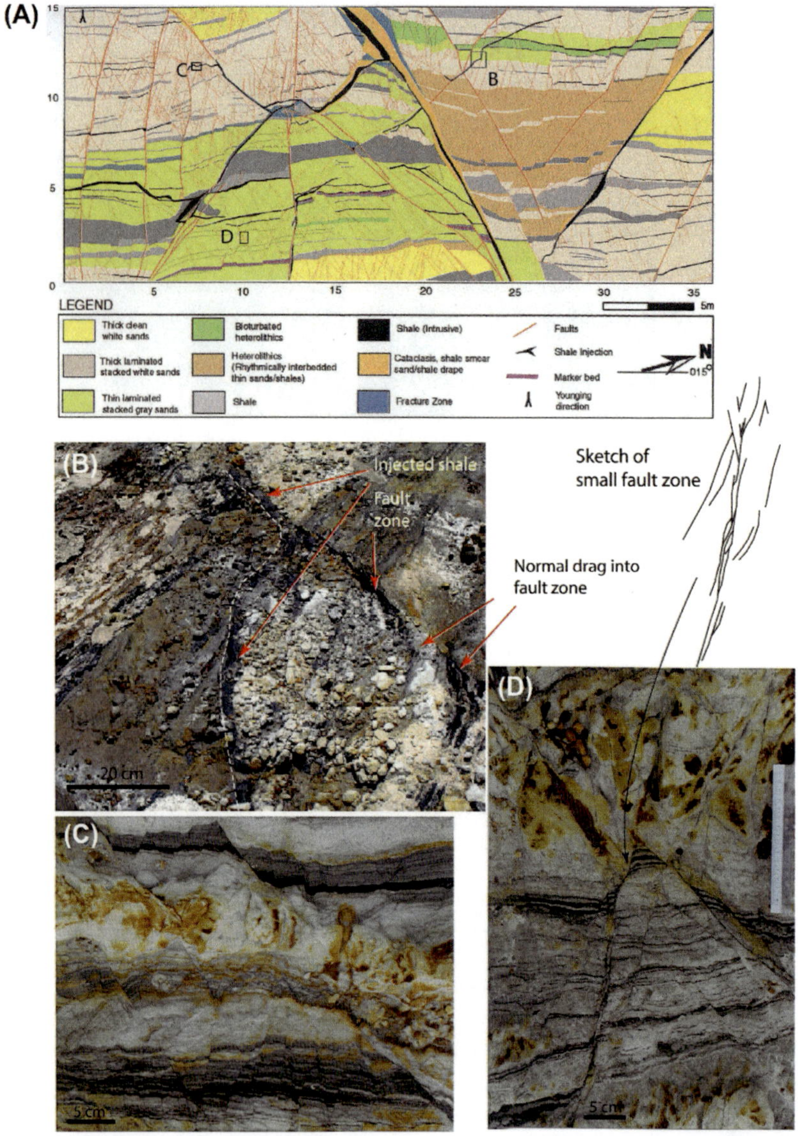

FIG. 8.28 Conjugate faults and shale injection features, Belait Formation (Late Miocene), Jerudong Anticline, Brunei. (A) Map (15 m × 35 m) of conjugate faults, and conjugate deformation bands, made from a rock pavement exposure in sub-vertically dipping beds, showing: (1) the frequent occurrence of injected shales affecting the larger faults in the outcrop, (2) the complex nature of intersection areas of conjugate faults, (3) the intense nature of the deformation down to the scale of small centimetric displacement faults and deformation bands. (B) Injected shale crossing the hanging-wall of a fault and partially running along the fault (see A for location). (C) Small-scale normal faults, illustrating the pervasive nature of deformation in the outcrop, (see A for location). (D) Complex, small-scale fault zone showing offsets of intersecting conjugate faults. Sketch of small fault zone shows how, at a very small scale, the fault zones show complex branches, splays, and multiple strands, similar to much larger faults.

These natural hydraulic fractures that formed by the injection of overpressured mud-rich fluids, are commonly independent of normal faults, but in places are injected along the normal faults (Figs 8.28 and 8.29). In one example, a dyke intruded along a fault turned parallel to bedding, and formed a small laccolith in the footwall (Fig. 8.30). Such intrusions can add sealing shales to a fault zone that are unpredictable when applying juxtaposition and membrane seal techniques. Additionally, shale intrusions, particularly sills and laccoliths, can introduce shale into a sequence after faulting has occurred. However, because in wells intruded shales cannot be distinguished from deposited shales, they will all be counted together in any fault seal algorithm, resulting in over-estimation of sealing potential.

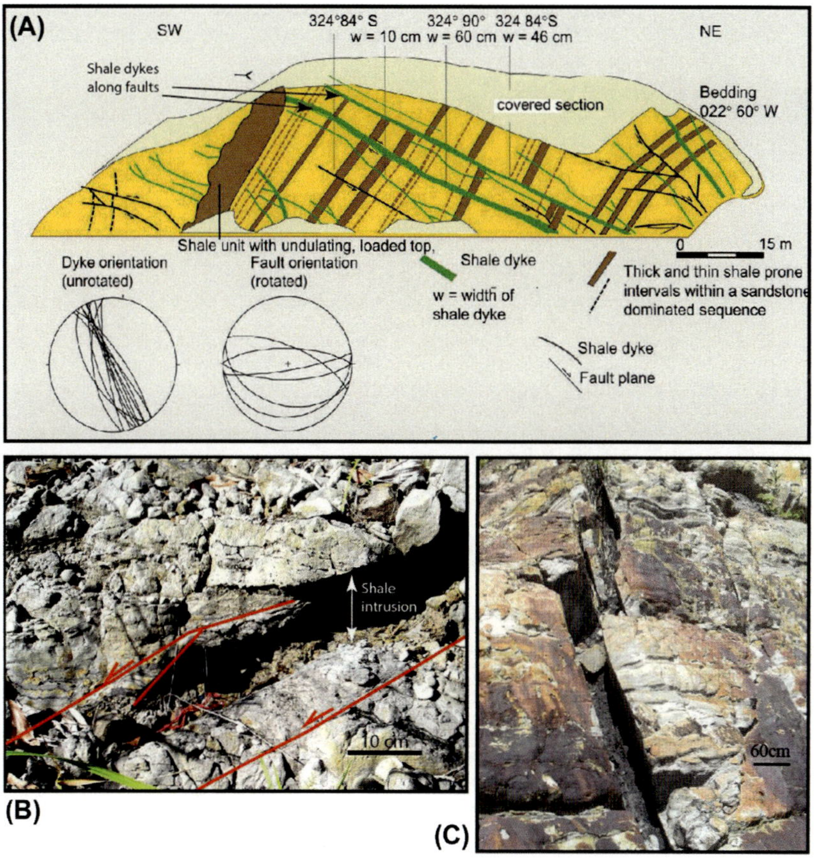

FIG. 8.29 Examples of injection of overpressured shales in the Belait Formation, Jerudong Anticline. (A) Sketch of outcrop in housing estate showing dykes injected along faults (also see B), and dykes injected into unfaulted host-rock (also see C). *Modified from Morley et al. (1998).*

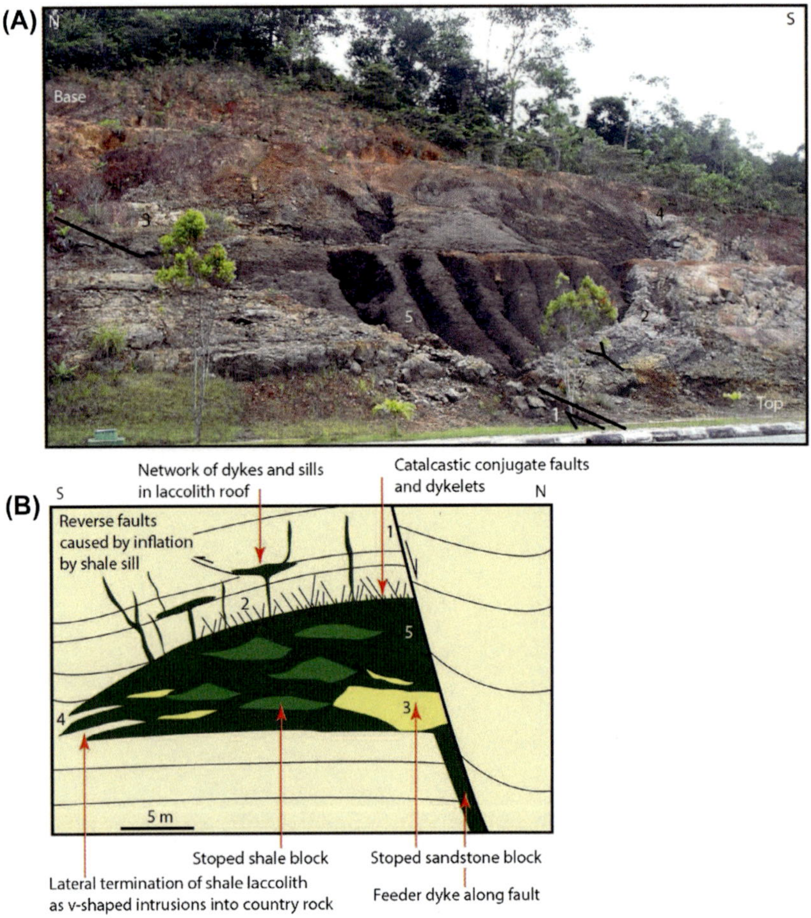

FIG. 8.30 Example of injection of a feeder dyke along a fault, and development of a laccolith adjacent to a fault, Belait Formation, Jerudong Anticline. (A) Photograph of the outcrop where the large shale mass is a laccolith, not a stratigraphic unit. The angle of the exposure is awkward, with the beds dipping steeply, and younging towards the bottom right, and gently dipping slope effectively exposing a cross-section view. (B) Sketch of the outcrop restored to a typical cross-section view. Numbers on the photo (A) and cross-section (B) are in approximately the same location. *Modified after Morley (2003a).*

Overpressured fluids are a drilling hazard, and make exploitation of hydrocarbons in overpressured zones more expensive and complicated. In the Baram Delta, the overpressuring appear to be largely the result of the disequilibrium compaction mechanism (Tingay et al., 2007), and probably also by the hard-to-quantify mechanism of hydrocarbon generation. However, many occurrences of overpressure above the deep, shale-prone sequences are vertically-transferred, inflationary-type overpressuring, whose location is much harder to predict, since they are not associated with the large porosity anomalies that characterise disequilibrium compaction. Changes in pore pressure magnitude are coupled with changes in the total minimum horizontal stress, with a coupling ratio of 0.59 (Tingay et al., 2003). Consequently, conventional models where overpressure-induced brittle failure is independent of total stress are not applicable to the Baram Delta Province, and it suggests that as pore pressure increases, formation of tensile fractures is favoured over fault reactivation (Tingay et al., 2003). Within fields, faults, and thin shales can compartmentalize overpressures, for example, in the Iron Duke Field pressure differences of 20,000 kPa (2,900 psi) were measured across a fault with just 60 m throw (Sandal, 1996). Overpressures have migrated along both new vertical tensile fractures, and along faults (examples of both types are preserved in shale intrusions). Evidence for this fluid movement can be detected in the stable oxygen and carbon isotope data from cements in sandstones (Fig. 8.31). Some cements and nodules in the sandstones formed early in the

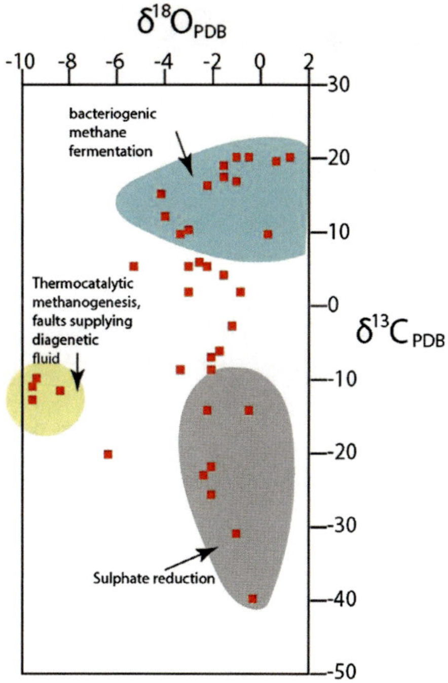

FIG. 8.31 Stable oxygen and carbon isotope analysis of nodules and cements from reservoirs in some fields, offshore Brunei. Nodules formed early in the burial history of the succession plot in a distinctive grouping related to formation in the zone of sulphate reduction, or bocteriogenic methane fermentation, or a combination. Extensive sparry cements found in core, exhibit a different grouping, which is interpreted to be related to thermocatalytic methanogenesis, at greater burial depth. These typically lie at the gas-water contact within reservoir sands, and is thought to indicate ongoing fluid flushing associated with fluid movement along neighbouring faults (J. Warren, pers. Comm., 2019).

burial history, are related to sulphate reduction and bacteriogenic methane formation (low $\delta^{18}O$ values), but other massive sparry-cemented sands cored in the Maharaja Lela and Fairley −Egret fields show higher negative $\delta^{18}O$ values associated with warmer burial conditions. They are explained as a product of thermocatalytic methanogenesis, typically at a gas-water contact, where the reservoir has experienced ongoing fluid flushing that is probably supplied by adjacent faults.

In some highly faulted fields, such as Seria, and Champion, the need to assess the impact of faulting runs all the way through from exploration to secondary recovery, and requires increasingly detailed understanding of the faulting as the project progresses. In the production phase understanding intra-formational seal competence, and fault-compartmentalization risk needs to be assessed, including the impact of sub-seismic faulting, not only for reservoir modelling, but also for optimization of well path planning (Schulz-Rojahn et al., 2004). However, as Figs 8.26, 8.28−8.30 illustrate, fault zones in this region are heterogenous and complex. Heterogeneity is related to many factors, including lithology variations, how fault segments have linked vertically and laterally during propagation, faults of consistent dip direction vs conjugate sets, fluid migration along faults, degree of lithification at the time of deformation, and the wildcard of intrusion by overpressured shale during or following fault movement.

8.7.3 Clay smears in aquifers of the Lower Rhine Embayment

The first systematic observations of clay smearing was made in lignite mines of the Lower Rhine Embayment (LRE) in western Germany (Lehner and Pilaar, 1997; Weber et al., 1978). Faults in the LRE are a perfect natural laboratory, because the sediments in the rift basin are mostly unlithified and many open cast mines, often several hundred metres deep, provide ample opportunity to study freshly-cut faults from surface level down to higher stress regimes at the bottom floor of the mines.

The LRE is part of the larger European Cenozoic Rift System (Michon et al., 2003; Ziegler, 1992). Rifting started in the late Eocene (35 Ma) and continued until the Pliocene (3 Ma). The thick lignite layers, which were extensively mined in the past decades, formed in fluvial-deltaic setting during the Miocene. The normal faults we study today were active between the late Oligocene and Pliocene (Knufinke and Kothen, 1997; Prange, 1958; Quitzow, 1954). They form from faults in the crystalline basement at depths of more than 1000 m (Fig. 8.32) and control the sedimentary structure. Today the LRE still shows tectonic activity and movement on some faults (Kübler, 2012; Winandy et al., 2011). Additionally, the massive groundwater drainage, necessary to keep the deep mines dry, causes heterogeneous subsidence in large areas that further facilitates deformation on the faults (Kübler, 2012).

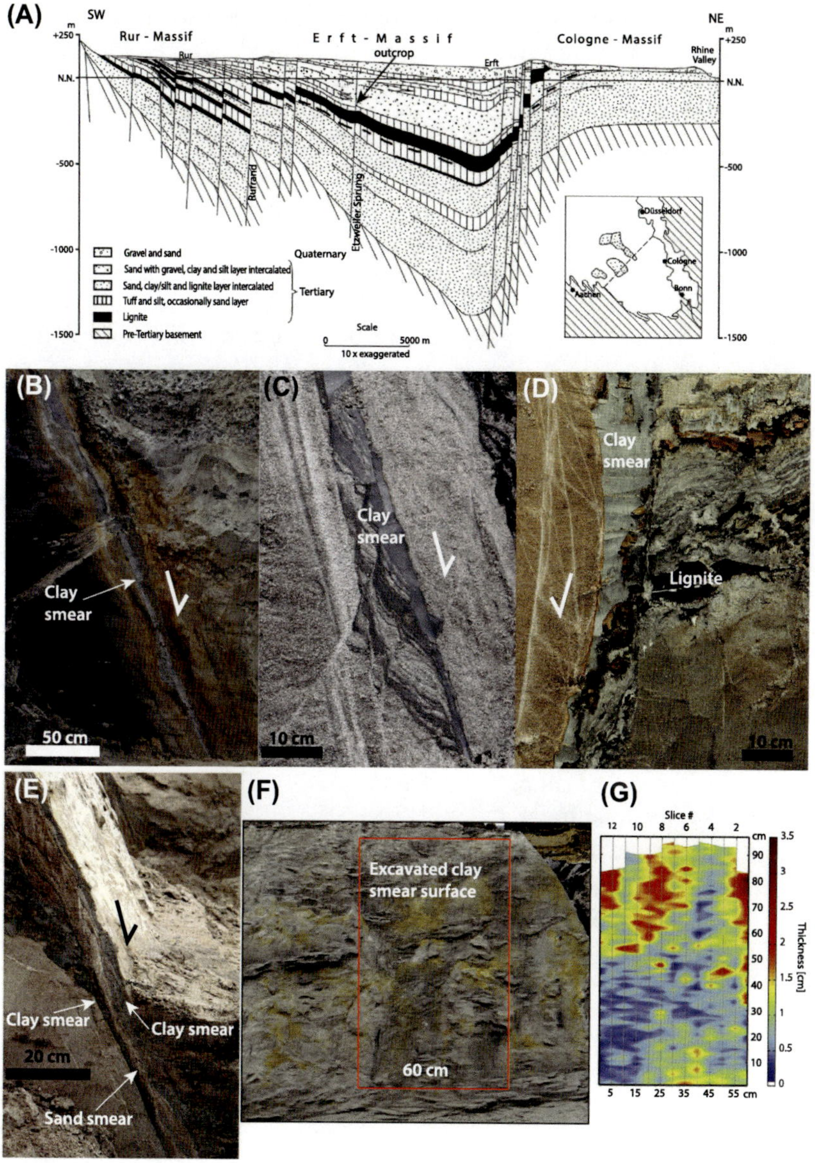

FIG. 8.32 Clay smears in the Lower Rhine Embayment (LRE), western Germany. (A) NE-SW cross-section through the LRE after (Knufinke and Kothen, 1997) location of the outcrops of Lehner and Pilaar (1997) and the following pictures marked by *arrow*. (B) Clay smear in a lignite mine, throw approx. 15 m. Note thickness variations. (C) Distinct footwall deformation of a clay smear. (D) Thick continuous clay smear becomes disrupted by entrained lignite fragments. (E) Sand smear formed between two parallel clay smears. (F) Excavated clay smear surface. Area in *red box* was sequentially sliced, thickness of clay in each slice was measured and constructed to a thickness map (G). *Modified after Kettermann et al. (2016) under CC BY 3.0 License.*

Sediments found in outcrops of the LRE contain ductile clays in layers from millimetres to metres in thickness, thick silt, sand and gravel layers and the previously mentioned lignite. Often clay layers are formed at channel bases as conglomerates of rip-up clasts; centimetre-sized, well-rounded clay fragments in a sand matrix.

Lehner and Pilaar (1997) made their famous observations of clay smear already in 1974 (Bouvier et al., 1989), but published the results more than 20 years later. They described a clay smear formed in a layered fault zone with fault segments overlapping in pull-apart structures. They propose clay injection as likely process to aid in maintaining a continuous clay smear, based on observations of clay being pressed out of freshly cut sections. Finally, they notice a network of secondary faults with small offset that accommodate volume changes during fault and clay smear formation. Lehner and Pilaar's initial observations in the LRE, with the motivation to provide the hydrocarbon industry with a tool to understand the sealing of faults, had immense impact on fault seal research in that they led the authors to the formulation of the first fault seal predicting algorithm – the Clay Smear Potential, or short, CSP (see Section 8.4.3 in this chapter). Their observations and work were the base for an extensive body of research over the past decades (Vrolijk et al., 2016).

More recent studies in the LRE investigated the thickness and continuity of clay smears. Large variations of clay smear thickness are common (Fig. 8.32B and C) both laterally and vertically and in faults of all scales. 2D sections (Navarro, 2002) and 3D data (Kettermann et al., 2016) both show a log-normal distribution of clay smear thickness. This can be attributed to multiple processes: (1) secondary faulting in the hanging-wall (Lehner and Pilaar, 1997); (2) secondary faulting in the footwall (Fig. 8.32C; Kettermann et al., 2016); (3) fractal undulations of the fault (Navarro, 2002) or simply heterogenous source beds.

Another noteworthy observation, so far only made in the LRE outcrops, is caused by the heterogeneous lithology where clay beds are often next to lithologies with larger fragments: Large gravel clasts or fragments of lignite can be transported into the clay smears, but also promote or cause the disruption of an otherwise continuous clay smear (Fig. 8.32D; Bense et al., 2003b; Kettermann et al., 2016). Mixing of clay smear and host rock can occur on this large scale but is also abundant on sand-grain scale. Especially in multilayer sequences (e.g., Fig. 8.32E) sand can be smeared but is commonly also mixed into the clay smear. Layers of rip-up clasts, due to the sand matrix are not a sealing layer in undeformed state, can easily form thick sealing clay smears due to the intense mixing of the sand matrix with sheared clay clasts (Kettermann et al., 2016).

Clay smears in the LRE do not only play a role as analogue for hydrocarbon motivated subsurface studies, but also greatly affect groundwater flow in the high-permeability sand aquifers. The abundant clay in faults causes a compartmentalization of fault blocks. Bense et al. (2003a) noticed variations in ground moisture at crop fields associated with faults. Subsequently, they mapped hydraulic heads across multiple faults in an area of more than 2000 km^2. Groundwater extraction to drain the mines causes an artificial pressure-decline in some compartments, while in some neighbouring fault blocks pressure is maintained to different levels. This is direct evidence of the permeability-reducing effect of faults with clay smear, although quantitative inferences into the sealing capacity or actual permeability of the faults is difficult as the authors note the existence of many relay ramps that affect cross-fault pressure communication (Bense and Van Balen, 2004).

Consequently, assessing fault permeability is also of importance for the operation of large open pit mines. Drainage of groundwater in depths of several hundred metres is expensive and the operators must know where they need to set up pumping stations to keep the mines dry. Occasionally, the open mines are separated from a water-bearing aquifer by a sealing fault. Instances are known where they suddenly started to leak, amongst other issues, due to production-related (decompaction) movement on the faults (Spiller et al., 2004). These incidents are rare and unpredictable but show that neither the problems nor the solutions in these well accessible areas are much different from deep subsurface applications.

References

Adam, J., Urai, J.L., Wieneke, B., Oncken, O., Pfeiffer, K., Kukowski, N., Lohrmann, J., Hoth, S., van der Zee, W., Schmatz, J., 2005. Shear localisation and strain distribution during tectonic faulting—new insights from granular-flow experiments and high-resolution optical image correlation techniques. J. Struct. Geol. 27, 283–301. https://doi.org/10.1016/j.jsg.2004.08.008.

Agosta, F., Aydin, A., 2006. Architecture and deformation mechanism of a basin-bounding normal fault in Mesozoic platform carbonates, central Italy. J. Struct. Geol. 28, 1445–1467. https://doi.org/10.1016/j.jsg.2006.04.006.

Agosta, F., Kirschner, D.L., 2003. Fluid conduits in carbonate-hosted seismogenic normal faults of central Italy. J. Geophys. Res. Solid Earth 108. https://doi.org/10.1029/2002JB002013.

Agosta, F., Prasad, M., Aydin, A., 2007. Physical properties of carbonate fault rocks, fucino basin (Central Italy): implications for fault seal in platform carbonates. Geofluids 7, 19–32. https://doi.org/10.1111/j.1468-8123.2006.00158.x.

Allan, U.S., 1989. Model for hydrocarbon migration and entrapment within faulted structures. Am. Assoc. Pet. Geol. Bull. 803–811.

Antonellini, M., Cilona, A., Tondi, E., Zambrano, M., Agosta, F., 2014. Fluid flow numerical experiments of faulted porous carbonates, Northwest Sicily (Italy). Mar. Pet. Geol. 55, 186–201. Fluid-rock-tectonics interactions in basins and orogens. https://doi.org/10.1016/j.marpetgeo.2013.12.003.

Bait, R., Banda, R.M., 1994. Tertiary basins of North Sarawak-Malaysia: A field excursion in the Tatau-Bintulu and Miri areas. Presented at the International Conference and Exhibition, American Association of Petroleum Geology, Miri, Sarawak.

Ballas, G., Fossen, H., Soliva, R., 2015. Factors controlling permeability of cataclastic deformation bands and faults in porous sandstone reservoirs. J. Struct. Geol. 76, 1−21. https://doi.org/10.1016/j.jsg.2015.03.013.

Beauheim, R.L., Roberts, R.M., 2002. Hydrology and hydraulic properties of a bedded evaporite formation. J. Hydrol. 259, 66−88. https://doi.org/10.1016/S0022-1694(01)00586-8.

Behnsen, J., Faulkner, D.R., 2012. The effect of mineralogy and effective normal stress on frictional strength of sheet silicates. J. Struct. Geol. 42, 49−61. https://doi.org/10.1016/j.jsg.2012.06.015.

Bense, V.F., Van Balen, R., 2004. The effect of fault relay and clay smearing on groundwater flow patterns in the Lower Rhine Embayment. Basin Res. 16, 397−411. https://doi.org/10.1111/j.1365-2117.2004.00238.x.

Bense, V.F., Van Balen, R.T., Vries, J.J.D., De Vries, J.J., 2003a. The impact of faults on the hydrogeological conditions in the Roer Valley Rift System: an overview. Neth. J. Geosci./Geol. Mijnb. 82, 41−54.

Bense, V.F., Van den Berg, E.H., Van Balen, R.T., 2003b. Deformation mechanisms and hydraulic properties of fault zones in unconsolidated sediments; the Roer Valley Rift System, The Netherlands. Hydrogeol. J. 11, 319−332. https://doi.org/10.1007/s10040-003-0262-8.

Berg, R.R., 1975. Capillary pressures in stratigraphic traps. Am. Assoc. Pet. Geol. Bull. 59, 939−956. https://doi.org/10.1306/83D91EF7-16C7-11D7-8645000102C1865D.

Billi, A., Salvini, F., Storti, F., 2003. The damage zone-fault core transition in carbonate rocks: implications for fault growth, structure and permeability. J. Struct. Geol. 25, 1779−1794. https://doi.org/10.1016/S0191-8141(03)00037-3.

Billi, A., 2005. Grain size distribution and thickness of breccia and gouge zones from thin (<1m) strike-slip fault cores in limestone. J. Struct. Geol. 27, 1823−1837. https://doi.org/10.1016/j.jsg.2005.05.013.

Billi, A., 2010. Microtectonics of low-P low-T carbonate fault rocks. J. Struct. Geol. 32, 1392−1402. Faulting and fracturing of carbonate rocks: new insights into deformation mechanisms, petrophysics and fluid flow properties. https://doi.org/10.1016/j.jsg.2009.05.007.

Bouvier, J.D., Kaars-Sijpesteijn, C.H., Kluesner, D.F., Onyejekwe, C.C., van der Pal, R.C., 1989. Three-dimensional seismic interpretation and fault sealing investigations, Nun River Field, Nigeria. Am. Assoc. Pet. Geol. Bull. 73, 1397−1414.

Breddin, H., 1960. Neue Erkenntnisse zur Geologie der Aachener Thermalquellen. Geol. Mitt. 1, 211−238.

Bullock, R.J., De Paola, N., Holdsworth, R.E., Trabucho-Alexandre, J., 2014. Lithological controls on the deformation mechanisms operating within carbonate-hosted faults during the seismic cycle. J. Struct. Geol. 58, 22−42. https://doi.org/10.1016/j.jsg.2013.10.008.

Burhannudinnur, M., Morley, C.K., 1997. Anatomy of growth fault zones in poorly lithified sandstones and shales; implications for reservoir studies and seismic interpretation; Part 1, Outcrop study. Petrol. Geosci. 3, 211−224. https://doi.org/10.1144/petgeo.3.3.211.

Childs, C., Walsh, J.J., Watterson, J., 1997. Complexity in fault zone structure and implications for fault seal prediction. In: Møller-Pedersen, P., Koestler, A.G. (Eds.), Norwegian Petroleum Society Special Publications : Hydrocarbon Seals Importance for Exploration and Production. Elsevier, pp. 61−72. https://doi.org/10.1016/S0928-8937(97)80007-0.

Childs, C., Walsh, J.J., Manzocchi, T., Strand, J., Nicol, A., Tomasso, M., Schöpfer, M.P.J., Aplin, A.C., 2007. Definition of a Fault Permeability Predictor from Outcrop Studies of a Faulted Turbidite Sequence, Taranaki, New Zealand, vol. 292. Geological Society, London, pp. 235−258. Special Publications. https://doi.org/10.1144/SP292.14.

Çiftçi, N.B., Giger, S.B., Clennell, M.B., 2013. Three-dimensional structure of experimentally produced clay smears: implications for fault seal analysis. Am. Assoc. Pet. Geol. Bull. 97, 733–757. https://doi.org/10.1306/10161211192.

Clausen, J.A., Gabrielsen, R.H., 2002. Parameters that control the development of clay smear at low stress states: an experimental study using ring-shear apparatus. J. Struct. Geol. 24, 1569–1586. https://doi.org/10.1016/S0191-8141(01)00157-2.

Clausen, J.A., Gabrielsen, R.H., Johnsen, E., Korstgård, J.A., 2003. Fault architecture and clay smear distribution. Examples from field studies and drained ring-shear experiments. Norw. J. Geol. 83, 131–146.

Cooke, A.P., Fisher, Q.J., Michie, E.A.H., Yielding, G., 2018. Investigating the controls on fault rock distribution in normal faulted shallow burial limestones, Malta, and the implications for fluid flow. J. Struct. Geol. 114, 22–42. https://doi.org/10.1016/j.jsg.2018.05.024.

Crawford, B.R., Myers, R.D., Woronow, A., Faulkner, D.R., Rutter, E.H., 2002. Porosity-permeability relationships in clay-bearing fault gouge. In: SPE/ISRM Rock Mechanics Conference. Society of Petroleum Engineers. https://doi.org/10.2118/78214-MS.

Crawford, B.R., Faulkner, D.R., Rutter, E.H., 2008. Strength, porosity, and permeability development during hydrostatic and shear loading of synthetic quartz-clay fault gouge. J. Geophys. Res. 113, B03207. https://doi.org/10.1029/2006JB004634.

Cuisiat, F., Skurtveit, E., 2010. An experimental investigation of the development and permeability of clay smears along faults in uncemented sediments. J. Struct. Geol. 32, 1850–1863. https://doi.org/10.1016/j.jsg.2009.12.005.

Delle Piane, C., Giwelli, A., Clennell, M.B., Esteban, L., Nogueira Kiewiet, M.C.D., Kiewiet, L., Kager, S., Raimon, J., 2016. Frictional and hydraulic behaviour of carbonate fault gouge during fault reactivation — an experimental study. Tectonophysics 690, 21–34. https://doi.org/10.1016/j.tecto.2016.07.011.

Delle Piane, C., Clennell, M.B., Keller, J.V.A., Giwelli, A., Luzin, V., 2017. Carbonate hosted fault rocks: a review of structural and microstructural characteristic with implications for seismicity in the upper crust. J. Struct. Geol. 103, 17–36. https://doi.org/10.1016/j.jsg.2017.09.003.

Doughty, P.T., 2003. Clay smear seals and fault sealing potential of an exhumed growth fault, Rio Grande rift, New Mexico. Am. Assoc. Pet. Geol. Bull. 87, 427–444. https://doi.org/10.1306/10010201130.

Egholm, D.L., Clausen, O.R., Sandiford, M., Kristensen, M.B., Korstgård, J. a., 2008. The mechanics of clay smearing along faults. Geology 36, 787. https://doi.org/10.1130/G24975A.1.

Faerseth, R.B., 2006. Shale smear along large faults: continuity of smear and the fault seal capacity. J. Geol. Soc. 163, 741–751. https://doi.org/10.1144/0016-76492005-162.

Ferraro, F., Agosta, F., Ukar, E., Grieco, D.S., Cavalcante, F., Belviso, C., Prosser, G., 2019. Structural diagenesis of carbonate fault rocks exhumed from shallow crustal depths: an example from the central-southern Apennines, Italy. J. Struct. Geol. 122, 58–80. https://doi.org/10.1016/j.jsg.2019.02.008.

Ferrill, D.A., Morris, A.P., 2008. Fault zone deformation controlled by carbonate mechanical stratigraphy, Balcones fault system, Texas. Am. Assoc. Pet. Geol. Bull. 92, 359–380. https://doi.org/10.1306/10290707066.

Ferrill, D.A., Winterle, J., Wittmeyer, G., Sims, D., Colton, S., Armstrong, A., Morris, A.P., 1999. Stressed rock strains groundwater at Yucca mountain, Nevada. GSA Today 32.

Ferrill, D.A., Evans, M.A., McGinnis, R.N., Morris, A.P., Smart, K.J., Wigginton, S.S., Gulliver, K.D.H., Lehrmann, D., de Zoeten, E., Sickmann, Z., 2017. Fault zone processes in mechanically layered mudrock and chalk. J. Struct. Geol. 97, 118–143. https://doi.org/10.1016/j.jsg.2017.02.013.

Fondriest, M., Smith, S.A.F., Candela, T., Nielsen, S.B., Mair, K., Di Toro, G., 2013. Mirror-like faults and power dissipation during earthquakes. Geology 41, 1175–1178. https://doi.org/10.1130/G34641.1.

Fondriest, M., Aretusini, S., Di Toro, G., Smith, S.A.F., 2015. Fracturing and rock pulverization along an exhumed seismogenic fault zone in dolostones: the Foiana Fault Zone (Southern Alps, Italy). Tectonophysics 654, 56–74. https://doi.org/10.1016/j.tecto.2015.04.015.

Fossen, H., Bale, A., 2007. Deformation bands and their influence on fluid flow. Am. Assoc. Pet. Geol. Bull. 91, 1685–1700. https://doi.org/10.1306/07300706146.

Fossen, H., Schultz, R.A., Shipton, Z.K., Mair, K., 2007. Deformation bands in sandstone: a review. J. Geol. Soc. 164, 755–769. https://doi.org/10.1144/0016-76492006-036.

Freeman, S.R., Harris, S.D., Knipe, R.J., 2008. Fault Seal Mapping - Incorporating Geometric and Property Uncertainty, vol. 309. Geological Society, London, pp. 5–38. Special Publications. https://doi.org/10.1144/SP309.2.

Fulljames, J.R., Zijerveld, L.J.J., Franssen, R.C.M.W., 1997. Fault seal processes: systematic analysis of fault seals over geological and production time scales. In: Møller-Pedersen, P., Koestler, A.G. (Eds.), Norwegian Petroleum Society Special Publications : Hydrocarbon Seals Importance for Exploration and Production. Elsevier, pp. 51–59. https://doi.org/10.1016/S0928-8937(97)80006-9.

Geluk, M., Duin, E.T., Dusar, M., Rijkers, R., Van den Berg, M., Van Rooijen, P., 1995. Stratigraphy and tectonics of the Roer Valley Graben. Geol. Mijnb. 73, 129.

Ghisetti, F., Kirschner, D.L., Vezzani, L., Agosta, F., 2001. Stable isotope evidence for contrasting paleofluid circulation in thrust faults and normal faults of the central Apennines, Italy. J. Geophys. Res. Solid Earth 106, 8811–8825. https://doi.org/10.1029/2000JB900377.

Giamboni, M., Ustaszewski, K., Schmid, S.M., Schumacher, M.E., Wetzel, A., 2004. Plio-Pleistocene transpressional reactivation of Paleozoic and Paleogene structures in the Rhine-Bresse transform zone (northern Switzerland and eastern France). Int. J. Earth Sci. 93, 207–223. https://doi.org/10.1007/s00531-003-0375-2.

Giger, S.B., Clennell, M.B., Harbers, C., Clark, P., Ricchetti, M., Ter Heege, J.H., Wassing, B.B.T., Orlic, B., 2011. Design, operation and validation of a new fluid-sealed direct shear apparatus capable of monitoring fault-related fluid flow to large displacements. Int. J. Rock Mech. Min. Sci. 48, 1160–1172. https://doi.org/10.1016/j.ijrmms.2011.09.005.

Giger, S.B., Clennell, M.B., Çiftçi, N.B., Harbers, C., Clark, P., Ricchetti, M., 2013. Fault transmissibility in clastic-argillaceous sequences controlled by clay smear evolution. Am. Assoc. Pet. Geol. Bull. 97, 705–731. https://doi.org/10.1306/10161211190.

Giorgetti, C., Collettini, C., Scuderi, M.M., Barchi, M.R., Tesei, T., 2016. Fault geometry and mechanics of marly carbonate multilayers: an integrated field and laboratory study from the Northern Apennines, Italy. J. Struct. Geol. 93, 1–16. https://doi.org/10.1016/j.jsg.2016.10.001.

Gudehus, G., Karcher, C., 2007. Hypoplastic simulation of normal faults without and with clay smears. J. Struct. Geol. 29, 530–540. https://doi.org/10.1016/j.jsg.2006.09.011.

Hansberry, R.L., King, R., Collins, A.S., Morley, C.K., 2014. Complex structure of an upper-level shale detachment zone: Khao Khwang fold and thrust belt, Central Thailand. J. Struct. Geol. 67, 140–153.

Hansberry, R.L., Collins, A.S., King, R.C., Morley, C.K., Giże, A.P., Warren, J., Löhr, S.C., Hall, P., 2015. Syn-deformation temperature and fossil fluid pathways along an exhumed detachment zone, khao khwang fold-thrust belt, Thailand. Tectonophysics 655, 73–87.

Herbst, G., 1954. Untersuchungen an den Burtscheider Thermalquellen in Aachen. Z. Dtsch. Geol. Ges. 183–202.

Hollmann, G., Von Winterfeld, C., 1999. Laterale Strukturvariationen eines Vorlandüberschiebungsgürtels. Z. Dtsch. Geol. Ges. 431–450.

Hudec, M.R., Jackson, M.P.A., 2007. Terra infirma: understanding salt tectonics. Earth Sci. Rev. 82, 1–28. https://doi.org/10.1016/j.earscirev.2007.01.001.

Ikari, M.J., Saffer, D.M., Marone, C., 2009. Frictional and hydrologic properties of clay-rich fault gouge. J. Geophys. Res. 114. https://doi.org/10.1029/2008JB006089.

Ingram, G.M., Urai, J.L., 1999. Top-Seal Leakage Through Faults and Fractures: The Role of Mudrock Properties, vol. 158. Geological Society, London, pp. 125–135. Special Publications. https://doi.org/10.1144/GSL.SP.1999.158.01.10.

Ingram, G.M., Urai, J.L., Naylor, M.A., 1997. Sealing processes and top seal assessment. In: Møller-Pedersen, P., Koestler, A.G. (Eds.), Hydrocarbon Seals: Importance for Exploration and Production. Elsevier, Singapore, pp. 165–174. https://doi.org/10.1016/S0928-8937(97)80014-8.

Jackson, C.A.-L., Bell, R.E., Rotevatn, A., Tvedt, A.B.M., 2017. Techniques to Determine the Kinematics of Synsedimentary Normal Faults and Implications for Fault Growth Models. Geological Society, London. Special Publications SP439.22. https://doi.org/10.1144/SP439.22.

Jaeger, J.C., Cook, N.G.W., Zimmermann, R., 1976. Fundamentals of Rock Mechanics, second ed. Chapman and Hall, London.

Jolley, S.J., Dijk, H., Lamens, J.H., Fisher, Q.J., Manzocchi, T., Eikmans, H., Huang, Y., 2007. Faulting and fault sealing in production simulation models: Brent Province, northern North Sea. Petrol. Geosci. 13, 321–340. https://doi.org/10.1144/1354-079306-733.

Kaminskaite, I., Fisher, Q.J., Michie, E.A.H., 2019. Microstructure and petrophysical properties of deformation bands in high porosity carbonates. J. Struct. Geol. 119, 61–80. https://doi.org/10.1016/j.jsg.2018.12.001.

Karakouzian, M., Hudyma, N., 2002. A new apparatus for analog modeling of clay smears. J. Struct. Geol. 24, 905–912. https://doi.org/10.1016/S0191-8141(01)00106-7.

Kettermann, M., Thronberens, S., Juarez, O., Urai, J.L., Ziegler, M., Asmus, S., Krüger, U., 2016. Mechanisms of clay smear formation in unconsolidated sediments - insights from 3-D observations of excavated normal faults. Solid Earth 7, 789–815. https://doi.org/10.5194/se-7-789-2016.

Kettermann, M., Urai, J.L., Vrolijk, P.J., 2017. Evolution of structure and permeability of normal faults with clay smear: insights from water-saturated sandbox models and numerical simulations. J. Geophys. Res. Solid Earth 122, 29. https://doi.org/10.1002/2016JB013341.

King, R.C., Hillis, R.R., Tingay, M.R.P., Morley, C.K., 2009. Present-day stress and neotectonic provinces of the Baram Delta and deep-water fold–thrust belt. J. Geol. Soc. 166, 197–200. https://doi.org/10.1144/0016-76492008-062R.

King, R., Holford, S., Hillis, R., Tuitt, A., Swierczek, E., Backé, G., Tassone, D., Tingay, M., 2012. Reassessing the in-situ stress regimes of Australia's petroleum basins. APPEA J. 52, 415. https://doi.org/10.1071/AJ11033.

Kleine Vennekate, G., 2013. Numerische Simulationen zur Scherfugenentwicklung in Sand-Ton-Wechselfolgen (Dissertation). RWTH Aachen University.

Knipe, R.J., 1997. Juxtaposition and Seal Diagrams to Help Analyze Fault Seals in Hydrocarbon Reservoirs. AAPG Bulletin 81, 187–195. https://doi.org/10.1306/522B42DF-1727-11D7-8645000102C1865D.

Knipe, R.J., Freeman, S.R., Harris, S.D., Davies, R.K., 2004. Structural uncertainty and scenario modelling for fault seal analysis. In: Proceedings of the AAPG Annual Convention Abstracts, 13. Presented at the AAPG Annual Convention, 18th–21st April, Dallas, Texas, p. A77.

Knufinke, H.U., Kothen, H., 1997. Die Tektonik der Niederrheinischen Bucht vor, während und nach der Hauptflözbildung. Braunkohle Surf. Min. 49, 473–479.

Kübler, S., 2012. Active Tectonics of the Lower Rhine Graben (NW Central Europe) Based on New Paleoseismological Constraints and Implications for Coseismic Rupture Processes in Unconsolidated Gravels (Dissertation). Ludwig-Maximilians-Universität, München, Germany.

Lacroix, B., Travé, A., Buatier, M., Labaume, P., Vennemann, T., Dubois, M., 2014. Syntectonic fluid-flow along thrust faults: example of the South-Pyrenean fold-and-thrust belt. Mar. Pet. Geol. 49, 84–98. https://doi.org/10.1016/j.marpetgeo.2013.09.005.

Laubscher, H., 1977. Fold development in the Jura. Tectonophysics 37, 337–362. https://doi.org/10.1016/0040-1951(77)90056-7.

Laubscher, H., 1992. Jura kinematics and the Molasse basin. Eclogae Geol. Helv. 85, 653–675.

Lehner, F.K., Pilaar, W.F., 1997. The emplacement of clay smear in synsedimentary normal faults: inferences from field observations near Frechen, Germany. In: Møller-Pedersen, P., Koestler, A.G. (Eds.), Norwegian Petroleum Society Special Publications : Hydrocarbon Seals Importance for Exploration and Production. Elsevier, pp. 39–50. https://doi.org/10.1016/S0928-8937(97)80005-7.

Lindsay, N.G., Murphy, F.C., Walsh, J.J., Watterson, J., 1992. Outcrop studies of shale smears on fault surfaces. In: Flint, S.S., Bryant, I.D. (Eds.), The Geological Modelling of Hydrocarbon Reservoirs and Outcrop Analogues. Blackwell Publishing Ltd., Oxford, UK, pp. 113–123. https://doi.org/10.1002/9781444303957.ch6.

Lisk, M., O'Brien, G.W., Brincat, M.P., 1997. Gas displacement: an important control on oil and gas distribution in the Timor Sea? APPEA J. 37, 259–271. https://doi.org/10.1071/aj96016.

Lisk, M., Eadington, P.J., O'Brien, G.W., 1998. Unravelling Complex Filling Histories by Constraining the Timing of Events Which Modify Oil Fields After Initial Charge, vol. 144. Geological Society, London, p. 189. Special Publications. https://doi.org/10.1144/GSL.SP.1998.144.01.15.

Luijendijk, E., 2019. Beo V1.0: numerical model of heat flow and low-temperature thermochronology in hydrothermal systems. Geosci. Model Dev. Discuss. https://doi.org/10.5194/gmd-2018-341.

Madritsch, H., Schmid, S.M., Fabbri, O., 2008. Interactions between thin- and thick-skinned tectonics at the northwestern front of the Jura fold-and-thrust belt (Eastern France). Tectonics 27, TC5005. https://doi.org/10.1029/2008tc002282.

Malz, A., Madritsch, H., Kley, J., 2015. Improving 2D-Seismic interpretation in challenging settings by integration of restoration techniques: a case study from the Jura fold-and-thrust belt (Switzerland). Interpretation 3. SAA37–SAA58. https://doi.org/10.1190/INT-2015-0012.1.

Mandl, G., de Jong, L.N.J., Maltha, A., 1977. Shear zones in granular material. Rock Mechanics 9, 95–144. https://doi.org/10.1007/BF01237876.

Manzocchi, T., Walsh, J.J., Nell, P., Yielding, G., 1999. Fault transmissibility multipliers for flow simulation models. Petrol. Geosci. 5, 53–63. https://doi.org/10.1144/petgeo.5.1.53.

Medeiros, W.E., do Nascimento, A.F., Alves da Silva, F.C., Destro, N., Demétrio, J.G. a., 2010. Evidence of hydraulic connectivity across deformation bands from field pumping tests: two examples from Tucano Basin, NE Brazil. J. Struct. Geol. 32, 1783–1791. https://doi.org/10.1016/j.jsg.2009.08.019.

Meng, L., Fu, X., Lv, Y., Li, X., Cheng, Y., Li, T., Jin, Y., 2016. Risking fault reactivation induced by gas injection into depleted reservoirs based on the heterogeneity of geomechanical properties of fault zones. Pet. Geosci. https://doi.org/10.1144/petgeo2016-031.

Michel, G., Adams, U., Schollmayer, G., 1998. Mineral-und Heilwässervorkommen in Nordrhein-Westfalen und angrenzenden Gebieten. Geologisches Landesamt Nordrhein-Westfalen.

Michie, E.A.H., Haines, T.J., 2016. Variability and heterogeneity of the petrophysical properties of extensional carbonate fault rocks, Malta. Petrol. Geosci. 22, 136−152. https://doi.org/10.1144/petgeo2015-027.

Michie, E.A.H., 2015. Influence of host lithofacies on fault rock variation in carbonate fault zones: a case study from the Island of Malta. J. Struct. Geol. 76, 61−79. https://doi.org/10.1016/j.jsg.2015.04.005.

Michon, L., Van Balen, R.T., Merle, O., Pagnier, H., 2003. The Cenozoic evolution of the Roer Valley Rift System integrated at a European scale. Tectonophysics 367, 101−126. https://doi.org/10.1016/S0040-1951(03)00132-X.

Mildren, S.D., Hillis, R.R., Lyon, P.J., Meyer, J.J., Dewhurst, D.N., Boult, P.J., 2005. FAST: a new technique for geomechanical assessment of the risk of reactivation-related breach of fault seals. In: Boult, P.J., Kaldi, J. (Eds.), Evaluating Fault and Cap Rock Seals, AAPG Hedberg Series. American Association of Petroleum Geologists, Tulsa, Oklahoma.

Morley, C.K., Crevello, P., Ahmad, Z.H., 1998. Shale tectonics and deformation associated with active diapirism: the Jerudong Anticline, Brunei Darussalam. J. Geol. Soc. 155, 475−490. https://doi.org/10.1144/gsjgs.155.3.0475.

Morley, C.K., von Hagke, C., Hansberry, R.L., Collins, A.S., Kanitpanyacharoen, W., King, R., 2017. Review of major shale-dominated detachment and thrust characteristics in the diagenetic zone: Part I, meso- and macro-scopic scale. Earth Sci. Rev. 173, 168−228. https://doi.org/10.1016/j.earscirev.2017.07.019.

Morley, C.K., von Hagke, C., Hansberry, R., Collins, A., Kanitpanyacharoen, W., King, R., 2018. Review of major shale-dominated detachment and thrust characteristics in the diagenetic zone: part II, rock mechanics and microscopic scale. Earth Sci. Rev. 176, 19−50. https://doi.org/10.1016/j.earscirev.2017.09.015.

Morley, C.K., 2003a. Outcrop examples of mudstone intrusions from the Jerudong anticline, Brunei Darussalam and Inferences for Hydrocarbon Reservoirs, vol. 216. Geological Society, London, pp. 381−394. Special Publications.

Morley, C.K., 2003b. Mobile Shale Related Deformation in Large Deltas Developed on Passive and Active Margins, vol. 216. Geological Society, London, pp. 335−357. Special Publications.

Morris, A., Ferrill, D.A., Brent Henderson, D.B., 1996. Slip-tendency analysis and fault reactivation. Geology 24, 275. https://doi.org/10.1130/0091-7613(1996)024<0275:STAAFR>2.3.CO;2.

Mosar, J., 1999. Present-day and future tectonic underplating in the western Swiss Alps: reconciliation of basement/wrench-faulting and decollement folding of the Jura and Molasse basin in the Alpine foreland. Earth Planet. Sci. Lett. 173, 143−155.

Nagra, 2002. Projekt Opalinuston Synthese der geowissenschaftlichen Untersuchungsergebnisse. Nagra Technischer Bericht NTB 02−03, pp. 139−142.

Naidu, B.S., Sunder, V.R., Kothari, V., Mohapatra, P., Burley, S.D., Dolson, J., Farrimond, P., 2017. Hydrocarbon Generation and Migration Modeling in the Barmer Basin of Western Rajasthan, India: Lessons for Exploration in Rift Basins with Late-Stage Inversion, Uplift, and Tilting. AAPG Memoir, pp. 61−94. Petroleum Systems Analysis.

Navarro, M., 2002. Fault Roughness and Fault Complexity: Field Study, Multi-Scale Analysis and Numerical Fault Model (Ph.D. thesis). Rheinische Friedrich-Willhelms-Universität Bonn, Bonn.

Noorsalehi-Garakani, S., Kleine Vennekate, G.J., Vrolijk, P., Urai, J.L., 2013. Clay-smear continuity and normal fault zone geometry - first results from excavated sandbox models. J. Struct. Geol. 57, 58−80. https://doi.org/10.1016/j.jsg.2013.09.008.

Noufal, A., Obaid, K., 2017. Sealing faults: a bamboozling problem in Abu Dhabi fields. In: Presented at the Abu Dhabi International Petroleum Exhibition & Conference. Society of Petroleum Engineers. https://doi.org/10.2118/188775-MS.

Nuriel, P., Rosenbaum, G., Uysal, T.I., Zhao, J., Golding, S.D., Weinberger, R., Karabacak, V., Avni, Y., 2011. Formation of Fault-Related Calcite Precipitates and Their Implications for Dating Fault Activity in the East Anatolian and Dead Sea Fault Zones, vol. 359. Geological Society, London, pp. 229–248. Special Publications. https://doi.org/10.1144/SP359.13.

Nuriel, P., Weinberger, R., Rosenbaum, G., Golding, S.D., Zhao, J., Tonguc Uysal, I., Bar-Matthews, M., Gross, M.R., 2012. Timing and mechanism of late-Pleistocene calcite vein formation across the Dead Sea Fault Zone, northern Israel. J. Struct. Geol. 36, 43–54. https://doi.org/10.1016/j.jsg.2011.12.010.

O'Brien, G.W., Woods, E.P., 1995. Hydrocarbon-related diagenetic zones (HRDZs) in the Vulcan Sub-basin, Timor Sea: recognition and exploration implications. APEA J. 35, 220.

O'Brien, G.W., Duddy, M.L.I., Eadington, P.J., Cadman, S., Fellows, M., 1996. Late tertiary fluid migration in the Timor Sea: a key control on thermal and diagenetic histories? APPEA J. 36, 399–427. https://doi.org/10.1071/aj95022.

Oncken, O., von Winterfeld, C., Dittmar, U., 1999. Accretion of a rifted passive margin: the late Paleozoic Rhenohercynian fold and thrust belt (middle European Variscides). Tectonics 18, 75–91.

Peacock, D.C.P., 2002. Propagation, interaction and linkage in normal fault systems. Earth Sci. Rev. 58, 121–142. https://doi.org/10.1016/S0012-8252(01)00085-X.

Pfiffner, O.A., 1986. Evolution of the north alpine foreland basin in the central Alps. In: Allen, P.A., Homewood, P.W. (Eds.), Foreland Basins, Special Publications. International Association of Sedimentologists, Oxford, pp. 219–228.

Pommerening, J., 1993. Hydrogeologie, Hydrogeochemie und Genese der Aachener Thermalquellen. Lehrstuhl für Ingenieurgeologie und Hydrogeologie der RWTH Aachen.

Prange, W., 1958. Tektonik und Sedimentation in den Deckschichten des Niederrheinischen Hauptbraunkohlenflözes in der Ville. Fortschr. Geol. Rheinld. u. Westf. 2, 651–682.

Quitzow, H.W., 1954. Tektonik und Grundwasserstockwerke im Erftbecken. Geol. Jb. 69, 455–464.

Raith, A., 2012. The Evolution of Fault Zones in Brittle Ductile-Layered Rocks (MSc thesis). RWTH Aachen Univeristy.

Roehl, P.O., Choquette, P.W. (Eds.), 1985. Carbonate Petroleum Reservoirs, Casebooks in Earth Sciences. Springer-Verlag, New York.

Rotevatn, A., Thorsheim, E., Bastesen, E., Fossmark, H.S.S., Torabi, A., Sælen, G., 2016. Sequential growth of deformation bands in carbonate grainstones in the hangingwall of an active growth fault: implications for deformation mechanisms in different tectonic regimes. J. Struct. Geol. 90, 27–47. https://doi.org/10.1016/j.jsg.2016.07.003.

Rowan, M.G., Jackson, M.P.A., Trudgill, B.D., 1999. Salt-related fault families and fault welds in the northern Gulf of Mexico. Am. Assoc. Pet. Geol. Bull. 83, 1454–1484.

Rowan, M.G., 2004. Do salt welds seal? In: Post, P.J., Olson, D.L., Lyons, K.T., Palmes, S.L., Harrison, P.F., Rosen, N.C. (Eds.), Salt Sediment Interactions and Hydrocarbon Prospectivity Concepts, Applications and Case Studies for the 21st Century. SEPM Society for Sedimentary Geology, p. 0. https://doi.org/10.5724/gcs.04.24.0390.

Sandal, S.T., 1996. The Geology and Hydrocarbon Resources of Negara Brunei Darussalam. Bandar Seri Begawan, Muzium Brunei.

Schmatz, J., Holland, M., Giese, S., Zee, W., Urai, J.L., 2010a. Clay smear processes in mechanically layered sequences — results of water-saturated model experiments with free top surface. J. Geol. Soc. India 75, 74–88. https://doi.org/10.1007/s12594-010-0025-9.

Schmatz, J., Vrolijk, P.J.J., Urai, J.L.L., 2010b. Clay smear in normal fault zones – the effect of multilayers and clay cementation in water-saturated model experiments. J. Struct. Geol. 32, 1834–1849. https://doi.org/10.1016/j.jsg.2009.12.006.

Schöpfer, M.P.J., Childs, C., Walsh, J.J., 2006. Localisation of normal faults in multilayer sequences. J. Struct. Geol. 28, 816–833. https://doi.org/10.1016/j.jsg.2006.02.003.

Schöpfer, M.P.J., Childs, C., Walsh, J.J., 2007a. Two-dimensional distinct element modeling of the structure and growth of normal faults in multilayer sequences: 1. Model calibration, boundary conditions, and selected results. J. Geophys. Res. 112, B10401. https://doi.org/10.1029/2006JB004902.

Schöpfer, M.P.J., Childs, C., Walsh, J.J., 2007b. Two-dimensional distinct element modeling of the structure and growth of normal faults in multilayer sequences: 2. Impact of confining pressure and strength contrast on fault zone geometry and growth. J. Geophys. Res. 112, B10404. https://doi.org/10.1029/2006JB004903.

Schulz-Rojahn, J.P., Walshe, P., Suhaili, H.I., 2004. Champion West Field Development, Brunei: A Study in Seal, Compartmentalisation and Fluid Fill Uncertainty. Presented at the SPE Asia Pacific Oil and Gas Conference and Exhibition. Society of Petroleum Engineers. https://doi.org/10.2118/88518-MS.

Shipton, Z.K., Soden, A.M., Kirkpatrick, J.D., Bright, A.M., Lunn, R.J., 2006. How thick is a fault? Fault displacement-thickness scaling revisited. In: Earthquakes: Radiated Energy and the Physics of Faulting, Geophys. Monogr. Ser. AGU, Washington and DC, pp. 193–198.

Sibson, R.H., 1996. Structural permeability of fluid-driven fault-fracture meshes. J. Struct. Geol. 18, 1031–1042. https://doi.org/10.1016/0191-8141(96)00032-6.

Sinn, C.J.A., Klaver, J., Fink, R., Jiang, M., Schmatz, J., Littke, R., Urai, J.L., 2017. Using BIB-SEM imaging for permeability prediction in heterogeneous shales. Geofluids 2017, 1–19. https://doi.org/10.1155/2017/4709064.

Smeraglia, L., Aldega, L., Billi, A., Carminati, E., Doglioni, C., 2016. Phyllosilicate injection along extensional carbonate-hosted faults and implications for co-seismic slip propagation: case studies from the central Apennines, Italy. J. Struct. Geol. 93, 29–50. https://doi.org/10.1016/j.jsg.2016.10.003.

Smeraglia, L., Bettucci, A., Billi, A., Carminati, E., Cavallo, A., Di Toro, G., Natali, M., Passeri, D., Rossi, M., Spagnuolo, E., 2017. Microstructural evidence for seismic and aseismic slips along clay-bearing, carbonate faults. J. Geophys. Res. Solid Earth 122, 3895–3915. https://doi.org/10.1002/2017JB014042.

Smeraglia, L., Bernasconi, S.M., Berra, F., Billi, A., Boschi, C., Caracausi, A., Carminati, E., Castorina, F., Doglioni, C., Italiano, F., Rizzo, A.L., Uysal, I.T., Zhao, J., 2018. Crustal-scale fluid circulation and co-seismic shallow comb-veining along the longest normal fault of the central Apennines, Italy. Earth Planet. Sci. Lett. 498, 152–168. https://doi.org/10.1016/j.epsl.2018.06.013.

Smith, S.A.F., Billi, A., Toro, G.D., Spiess, R., 2011. Principal slip zones in limestone: microstructural characterization and implications for the seismic cycle (Tre Monti Fault, Central Apennines, Italy). Pure Appl. Geophys. 168, 2365–2393. https://doi.org/10.1007/s00024-011-0267-5.

Solum, J.G., Huisman, B. a. H., 2017. Toward the creation of models to predict static and dynamic fault-seal potential in carbonates. Petrol. Geosci. 23, 70–91. https://doi.org/10.1144/petgeo2016-044.

Sorkhabi, R., Hasegawa, S., 2005. Fault zone architecture and permeability distribution in the neogene clastics of northern Sarawak (Miri Airport road outcrop), Malaysia. In: Sorkhabi, R., Tsuji, Y. (Eds.), Faults, Fluid Flow, and Petroleum Traps. American Association of Petroleum Geologists, p. 0.

Sorkhabi, R., 2010. Miri 1910. Presented at the GEOExPro.

Sosio De Rosa, S., Shipton, Z.K., Lunn, R.J., Kremer, Y., Murray, T., 2018. Along-strike fault core thickness variations of a fault in poorly lithified sediments, Miri (Malaysia). J. Struct. Geol. 116, 189–206. https://doi.org/10.1016/j.jsg.2018.08.012.

Sperrevik, S., Faerseth, R.B., Gabrielsen, R.H., 2000. Experiments on clay smear formation along faults. Petrol. Geosci. 6, 113–123. https://doi.org/10.1144/petgeo.6.2.113.

Sperrevik, S., Gillespie, P.A., Fisher, Q.J., Halvorsen, T., Knipe, R.J., 2002. Empirical estimation of fault rock properties. In: Norwegian Petroleum Society Special Publications. Elsevier, pp. 109–125. https://doi.org/10.1016/S0928-8937(02)80010-8.

Spiller, M., Forkel, C., Köngeter, J., 2004. Case study: inflow of groundwater into open-cast mine Hambach, Germany. J. Hydraul. Eng. 130, 608–615. https://doi.org/10.1061/(ASCE)0733-9429(2004)130:7(608).

Storti, F., Billi, A., Salvini, F., 2003. Particle size distributions in natural carbonate fault rocks: insights for non-self-similar cataclasis. Earth Planet. Sci. Lett. 206, 173–186. https://doi.org/10.1016/S0012-821X(02)01077-4.

Takahashi, M., 2003. Permeability change during experimental fault smearing. J. Geophys. Res. 108, 2234. https://doi.org/10.1029/2002JB001984.

TerHeege, J.H., Wassing, B.B.T., Orlic, B., Giger, S.B., Clennell, M.B., 2013. Constraints on the sealing capacity of faults with clay smears from discrete element models validated by laboratory experiments. Rock Mech. Rock Eng. 46, 465–478. https://doi.org/10.1007/s00603-013-0383-x.

Tesei, T., Collettini, C., Viti, C., Barchi, M.R., 2013. Fault architecture and deformation mechanisms in exhumed analogues of seismogenic carbonate-bearing thrusts. J. Struct. Geol. 55, 167–181. https://doi.org/10.1016/j.jsg.2013.07.007.

Tingay, M.R.P., Hillis, R.R., Morley, C.K., Swarbrick, R.E., Okpere, E.C., 2003. Pore Pressure/Stress Coupling in Brunei Darussalam — Implications for Shale Injection, vol. 216. Geological Society, London, pp. 369–379. Special Publications. https://doi.org/10.1144/GSL.SP.2003.216.01.24.

Tingay, M.R., Hillis, R.R., Swarbrick, R.E., Morley, C.K., Damit, A.R., 2007. 'Vertically transferred'overpressures in Brunei: evidence for a new mechanism for the formation of high-magnitude overpressure. Geology 35, 1023–1026.

Tingay, M.R.P., Hillis, R.R., Morley, C.K., King, R.C., Swarbrick, R.E., Damit, A.R., 2009. Present-day stress and neotectonics of Brunei: implications for petroleum exploration and production. Bulletin 93, 75–100. https://doi.org/10.1306/08080808031.

Tondi, E., Antonellini, M., Aydin, A., Marchegiani, L., Cello, G., 2006. The role of deformation bands, stylolites and sheared stylolites in fault development in carbonate grainstones of Majella Mountain, Italy. J. Struct. Geol. 28, 376–391. https://doi.org/10.1016/j.jsg.2005.12.001.

Torabi, A., Berg, S.S., 2011. Scaling of fault attributes: a review. Mar. Pet. Geol. 28, 1444–1460. https://doi.org/10.1016/j.marpetgeo.2011.04.003.

Trippetta, F., Carpenter, B.M., Mollo, S., Scuderi, M.M., Scarlato, P., Collettini, C., 2017. Physical and transport property variations within carbonate-bearing fault zones: insights from the Monte Maggio Fault (Central Italy). Geochem. Geophys. Geosyst. 18, 4027–4042. https://doi.org/10.1002/2017GC007097.

Urai, J.L., Schmitz, R.M., van der Zee, W., Holland, M., Vrolijk, P.J., 2003. Experimental investigation of clay smear processes in a geotechnical direct shear apparatus, faults and top seals: what do we know and where do we go?. In: EAGE, Montpellier, France, 8e11 September 2003, p. 43.

Valoroso, L., Chiaraluce, L., Piccinini, D., Di Stefano, R., Schaff, D., Waldhauser, F., 2013. Radiography of a normal fault system by 64,000 high-precision earthquake locations: the 2009 L'Aquila (central Italy) case study. J. Geophys. Res. Solid Earth 118, 1156–1176. https://doi.org/10.1002/jgrb.50130.

van der Zee, W., Urai, J.L., 2005. Processes of normal fault evolution in a siliciclastic sequence: a case study from Miri, Sarawak, Malaysia. J. Struct. Geol. 27, 2281–2300. https://doi.org/10.1016/j.jsg.2005.07.006.

van der Zee, W., Urai, J.L., Richard, P.D., 2003. Lateral clay injection into normal faults. GeoArabia 8, 501–522.

Viti, C., Collettini, C., Tesei, T., 2014. Pressure solution seams in carbonatic fault rocks: mineralogy, micro/nanostructures and deformation mechanism. Contrib. Mineral. Petrol. 167, 970. https://doi.org/10.1007/s00410-014-0970-1.

von Hagke, C., Malz, A., 2018. Triangle zones—Geometry, kinematics, mechanics, and the need for appreciation of uncertainties. Earth Sci. Rev. 177, 24–42.

von Hagke, C., Cederbom, C.E., Oncken, O., Stöckli, D.F., Rahn, M.K., Schlunegger, F., 2012. Linking the northern Alps with their foreland: the latest exhumation history resolved by low-temperature thermochronology. Tectonics 31, TC5010. https://doi.org/10.1029/2011tc003078.

von Hagke, C., Oncken, O., Ortner, H., Cederbom, C.E., Aichholzer, S., 2014. Late Miocene to present deformation and erosion of the Central Alps — evidence for steady state mountain building from thermokinematic data. Tectonophysics 632, 250–260. https://doi.org/10.1016/j.tecto.2014.06.021.

von Winterfeld, C.-H., 1994. Variszische Deckentektonik und devonische Beckengeometrie der Nordeifel-ein quantitatives Modell: Profilbilanzierung und Strain-Analyse im Linksrheinischen Schiefergebirge. Verlag der Augustinus-Buchh.

Vrolijk, P.J., Urai, J.L., Kettermann, M., 2016. Clay smear: review of mechanisms and applications. J. Struct. Geol. 86, 95–152. https://doi.org/10.1016/j.jsg.2015.09.006.

Wagner, B.H., Jackson, M.P.A., 2011. Viscous flow during salt welding. Tectonophysics 510, 309–326. https://doi.org/10.1016/j.tecto.2011.07.012.

Warren, J.K., 2016. Evaporites: A Geological Compendium, second ed. Springer International Publishing.

Watts, N.L., 1987. Theoretical aspects of cap-rock and fault seals for single- and two-phase hydrocarbon columns. Mar. Pet. Geol. 4, 274–307. https://doi.org/10.1016/0264-8172(87)90008-0.

Weber, K., Mandl, G., Pilaar, W., Lehner, B., Precious, R., 1978. The role of faults in hydrocarbon migration and trapping in Nigerian growth fault structures. In: Proceedings of Offshore Technology Conference. The Offshore Technology Conference. https://doi.org/10.4043/3356-MS.

Wellmann, J.F., Horowitz, F.G., Schill, E., Regenauer-Lieb, K., 2010. Towards incorporating uncertainty of structural data in 3D geological inversion. Tectonophysics 490, 141–151.

Winandy, J., Grützner, C., Reicherter, K., Wiatr, T., Fischer, P., Ibeling, T., 2011. Is the Rurrand Fault (Lower Rhine Graben, Germany) responsible for the 1756 Düren earthquake series?. In: 2nd INQUA-IGCP-567 International Workshop on Active Tectonics, Earthquake Geology, Archeology and Engineering, Corinth, Greece, pp. 286–289.

Wiprut, D., Zoback, M.D., 2000. Fault reactivation and fluid flow along a previously dormant normal fault in the northern North Sea. Geolgy 28, 595–598.

Wiprut, D., Zoback, M., 2002. Fault reactivation, leakage potential, and hydrocarbon column heights in the northern North Sea. In: Koestler, A.G., Hunsdale, R. (Eds.), Hydrocarbon Seal Quantification, NPF Special PublicationS. Elsevier Scie, Amsterdam, pp. 203–219. https://doi.org/10.1016/S0928-8937(02)80016-9.

Yielding, G., Freeman, B., Needham, D.T., 1997. Quantitative fault seal prediction. Am. Assoc. Pet. Geol. Bull. 81, 897–917. https://doi.org/10.1306/522B498D-1727-11D7-8645000102C1865D.

Yielding, G., Bretan, P., Freeman, B., 2010. Fault Seal Calibration: A Brief Review, vol. 347. Geological Society, London, pp. 243–255. Special Publications. https://doi.org/10.1144/SP347.14.

Yielding, G., 2002. Shale gouge ratio — calibration by geohistory. In: Koestler, A.G., Hunsdale, R. (Eds.), Hydrocarbon Seal Quantification, NPF Special Publication. Elsevier Science B.V., Amsterdam, pp. 1–15. https://doi.org/10.1016/S0928-8937(02)80003-0.

Ziegler, P.A., 1992. European Cenozoic rift system. Tectonophysics 208, 91–111. https://doi.org/10.1016/0040-1951(92)90338-7.

Conclusions

David C. Tanner and Christian Brandes

This book gives the state of the art concerning fault analysis in geological and geophysical studies. Based on this, an outlook for future research can be formulated. The best way to describe this is to first answer the grand challenges in fault analysis (Lay et al., 2009; Forsyth et al., 2009; Gudmundsson, 2013; Huntington et al., 2018) that were put forward in Chapter 2.

- *How do fault zones behave from the Earth's surface to the base of the lithosphere?*
  The state of knowledge on faults decreases understandingly with depth. However, new experimental apparatuses can now simulate nearly all crustal stresses and geological strain rates (Chapter 5), so this will improve. Especially, the rate-dependency of deformation that can shift the transition between brittle and ductile fault processes has to be taken into account. At the same time, resolution is improving in geophysical methods, so that clearer images of faults at depth are becoming available.
- *How are new faults initiated and how are they reactivated throughout Earth history and how do fault networks evolve on different spatial and temporal scales?*
  This is the most difficult challenge to answer. Fault growth models are shown in Chapter 6, and fault reactivation is discussed in Chapter 8, but these are processes that are very difficult, if not impossible, to study in nature. Consequently, analogue and numerical models have to be used to generate this knowledge. Especially using 3D numerical models to follow analogue models is now a possibility, given the computing power of today. Sealing is a crucial process during the evolution of a fault. This does not only affect the permeability of a fault, but also shows how variable faults are and how their host material can strongly control their structure and composition.
- *How do faults slip? What causes large seismogenic fault-slip?*
  The physics of these processes have been under investigation for many decades now, but no uniform model has been forthcoming. This is because faults are very different (i.e., in their dimensions, displacements, kinematics and the composition of the rocks involved in faulting). However, some of the questions have now been partially answered by analogue friction experiments. Fault lubrication processes and their effects on rupture behaviour are now better understood. Nevertheless, the experiments still need to bridge time and length-scales.

- *What controls the transition from seismic rupturing to aseismic fault creep?* This crucial question has a strong impact on the hazard potential of a fault. The controlling factors for the change in fault behaviour are far from being understood. Until now, only some general parameters, such as friction and pore-pressure have been identified as having an impact on the fault behaviour. Again, the complexity of fault zones is a limiting factor for the development of a unified model that can explain the transition from seismic rupturing to aseismic fault creep.

An important step ahead in fault research is being made by improving dating techniques. As demonstrated in Chapter 7, groundbreaking results were recently achieved in direct dating of fault movement. This allows quantifying more accurately the timing of fault movement over geological time. This has impact on the analysis of seismic recurrence intervals and thus a major component of a profound hazard analysis.

Many of the limitations in fault analysis result from an insufficient knowledge of the fault zone structure and composition (see Chapter 1, Introduction). This can only be overcome with a combination of detailed outcrop analyses to derive the small-scale structures (e.g., Chapter 8) and improved geophysical methods (Chapter 3) that allow a better visualization of the fault zone and potentially an extrapolation of the small-scale outcrop observation over a larger rock volume in order to create a holistic static fault model. This in turn can serve as input for numerical simulation that should be calibrated by e.g., seismic data from active faults.

Another important step in fault research are the improvements in numerical simulations that were achieved in the last years. This allows to analyse fault behaviour full-scale without the spatial limitations of outcrops or experiments. A major advantage of numerical simulations is their holistic approach, which makes it possible to test different controlling factors for fault behaviour such as the lithology of the fault zone, friction along the principle slip-surface, pore-pressure and temperature. These parameters can easily varied and quickly tested to derive their impact on fault movement. Numerical simulations can be further modified and repeated (e.g., in different codes) without much effort.

We argue that future research should develop an (open access) numerical fault model that is a realistic representation of natural faults. This can be achieved by taking the fault geometry and structure from outcrops and by the implementation of the results derived from laboratory friction experiments, dating of fault gouge and implementation of geometrical fault models. Especially the rate-dependent deformation behaviour and the dynamic friction should be included in the model.

References

Forsyth, D.W., Lay, T., Aster, R.C., Romanowicz, B., 2009. Grand challenges for seismology. EOS 90, 361–362.

Gudmundsson, A., 2013. Great challenges in structural geology and tectonics. Front. Earth Sci. 1 https://doi.org/10.3389/feart.2013.00002.

Huntington, K.W., Klepeis, K.A., With 66 Community Contributors, 2018. Challenges and Opportunities for Research in Tectonics: Understanding Deformation and the Processes that Link Earth Systems, from Geologic Time to Human Time. A Community Vision Document Submitted to the U.S. National Science Foundation. University of Washington, p. 84. https://doi.org/10.6069/H52R3PQ5.

Seismological Grand Challenges in Understanding Earth's Dynamic Systems. In: Lay, T. (Ed.), 2009 Report to the National Science Foundation, IRIS Consortium, p. 76.

Index

Note: 'Page numbers followed by "*f*" indicate figures and "*t*" indicate tables'.

C

D

E

F

G

J

K

L

M

Q

R

S

T

U

V

Made in the USA
Las Vegas, NV
13 October 2025

32263397R00228